Undergraduate Texts in Mathematics

Editors

S. Axler
F. W. Gehring
K. A. Ribet

Springer
New York
Berlin
Heidelberg
Barcelona
Hong Kong
London
Milan
Paris
Singapore
Tokyo

Undergraduate Texts in Mathematics

(continued after index)

Steven Roman

Introduction to Coding and Information Theory

With 50 illustrations

Springer

Steven Roman
Department of Mathematics
California State University at Fullerton
Fullerton, CA 92634
USA

Mathematics Subject Classification (2000): 94-01, 62B10, 94A15, 94B

Library of Congress Cataloging-in-Publication Data
Roman, Steven
    Introduction to coding and information theory / Steven Roman.
       p. cm – (Undergraduate texts in mathematics)
    Includes bibliographical references and index.
    ISBN 0-387-94704-3 (hard : alk. paper)
    1. Coding theory.   2. Information theory.   I. Title.   II. Series.
    QA268.R66   1996
    005.7′2–dc20   96-11738

Printed on acid-free paper.

Production managed by Robert Wexler; manufacturing supervised by Jacqui Ashri.
Photocomposed copy prepared using LaTeX and Springer's utm macro.
Printed and bound by Hamilton Printing Co., Rensselaer, NY.
Printed in the United States of America.

9 8 7 6 5 4 3 2

ISBN 0-387-94704-3 Springer-Verlag New York Berlin Heidelberg SPIN 10773354

To Donna

Preface

Generally speaking, there are three reasons to encode data that is about to be transmitted (through space, for instance) or stored (on a computer disk, for instance). One reason is for *efficiency*. It clearly makes sense to compress data as much as possible in order to save transmission time or storage space. In fact, data compression is very big business in the computer world. The second reason to encode data is for *error detection and/or correction*. The third reason is for *secrecy*, that is, so that unauthorized persons cannot read the data.

While the goals of encoding for efficiency, error correction, and secrecy are distinct, the first two are related, if not by their purpose, then by their simultaneous desirability. To put it bluntly, if you are going to transmit or store some data, then you want to both compress it as much as possible *and* protect it from errors.

Encoding for efficiency falls under the rubric of information theory and encoding for error correction falls under the rubric of coding theory. (Encoding for secrecy is the subject of cryptology.) Indeed, elementary information theory is a beautiful application of discrete probability theory to the problem of encoding for efficiency, and elementary coding theory is a beautiful application of algebra (in our case linear algebra) and combinatorics to the problem of error detection and correction.

The subjects of information theory and coding theory began in 1948 with a famous paper by Claude Shannon, of Bell Labs, entitled *A Mathematical Theory of Communication*.

The purpose of this book is to introduce these two fields to undergraduate students of mathematics and computer science (from motivated sophomores to more experienced seniors). Formal prerequisites are kept to a minimum. We do use elementary probability theory fairly heavily at times, especially in the information theory portion of the book, but we use only the basic notions (conditional probability, the Theorem on Total Probabilities and Bayes' Theorem) as applied to finite sample spaces. (I teach this much probability to my liberal arts students, although I don't expect them to use it quite this fluently.) These topics are reviewed in Chapter 0. We also use the basic algebraic properties of the vector space $\mathbb{Z}_p^n$ of all strings of length n over the finite field $\mathbb{Z}_p$. However, these properties are discussed in the first section of Chapter 5, just before they are needed. (I cannot deny that students would benefit from an introduction to linear algebra prior to studying this book, but it is not mandatory.)

The information theory portion of the book consists of Chapters 2 and 3. The coding theory portion is Chapters 4, 5, and 6. We emphasize that these two portions are independent and either one may be skipped. Thus, a course on error-correcting codes could consist of Chapters 1, 4, 5, and 6 and a minicourse on information theory (using probability but no linear algebra) could consist of Chapters 1–3. More specifically, the chapter descriptions are as follows.

Chapter 0 is devoted to a brief discussion of prerequisites, such as a bit of probability theory. The reader should look over this chapter before proceeding to the main topics, if only to establish the terminology used throughout the book.

Chapter 1 contains a general discussion of codes and some issues related to variable length codes (such as unique decipherability). We also establish Kraft's Theorem, which characterizes the codeword lengths of instantaneous codes. In Chapter 2, we discuss the efficiency of encoding schemes and describe Huffman's method for constructing efficient variable length codes. In Chapter 3, we define and discuss the entropy of an information source and state and prove the main result of noiseless (error free) encoding—The Noiseless Coding Theorem.

Chapter 4 is devoted to a discussion of communications channels, decision rules (ideal observer, maximum likelihood, and nearest neighbor) for decoding in the presence of errors and general remarks about the maximum size of a code that can correct a given number of errors (the so-called *main coding theory problem*).

In Chapter 5, we develop some linear algebra and discuss the general theory of linear codes and their decoding. Finally, in Chapter 6, we consider specific linear and nonlinear codes; to wit, the Hamming codes, the simplex codes, the Golay codes, the (first order) Reed-Muller codes, the ISBN code, single- and double-error-correcting decimal codes, and some codes obtained from Latin squares.

I would like to express my thanks to Tom von Foerster, my editor at Springer; Robert Wexler, production editor; and Fred Bartlett, Springer's TeX specialist. Also thanks to Steve Zicree for his help in proofreading the manuscript.

— Steven Roman

Contents

Introduction

In this book, we discuss two distinct aspects of the problem of transmitting data (called *source data*) from one location to another or, what amounts to the same thing, storing data and then retrieving the data at a later time. Both of these aspects involve encoding of the source data.

In Part I of the book, on information theory, we discuss the issue of encoding for *efficiency*, that is, encoding so that the source data takes up as little space as possible. (This is also known as *data compression*.) Our focus will be on the theoretical aspects of the problem, rather than on the practical aspects. As we will see, encoding for efficiency is best accomplished using *variable-length* encoding schemes, where the most frequently used source symbols are encoded with the shortest codewords. Also, we will assume for this discussion that errors do not occur in the handling of source symbols. Hence, this type of encoding is *noiseless*.

There are various ways in which we might model source data. The model we will adopt is that of a "black box" that emits source symbols from a given finite source alphabet at regular intervals. Each source symbol has a fixed probability of being emitted at any time. We must encode the source data, either a single symbol at a time or in predefined blocks of symbols, as it is being emitted from the black box—the point is that we are not privy to the entire message before encoding.

In such a situation, it is the (finite) probability distribution of the source symbols that is important, and not the actual symbols themselves. As we will see, there is associated to each probability distribution a quantity, known as the *entropy* of the distribution, that is a measure of the total amount of "information" in the source. The goal of efficient encoding is to encode the source data in such a way as to add as little additional information beyond the entropy as possible.

As an example, linguists have done statistical calculations to obtain approximate probability distributions for letters in English text. For instance, the following table gives the probabilities associated with the six most common characters.

Character	Probability
space	0.1859
E	0.1031
T	0.0796
A	0.0642
O	0.0632
I	0.0575

Now, if we were to encode each of the 26 letters, together with the space character, using binary words of a fixed length, then since $2^4 = 16 < 27 < 2^5$, we would need to use codewords of length 5. Hence, each source character would take 5 bits of storage, for instance. On the other hand, the so-called Huffman encoding scheme, which we will study in detail, allows us to store these characters with an average codeword length of only 4.1195 bits per character. This is reasonably efficient encoding, since the entropy of this probability distribution is 4.07991 bits per character.

Apropos of efficient encoding, we will describe and prove a famous theorem of information theory known as *The Noiseless Coding Theorem*. Roughly stated, this theorem says that, by clever encoding, we can arrange it so that the total information in the encoded message is as close to the entropy of the original source message as desired. (Of course, there is a penalty to pay for this efficiency.) To put this in more concrete terms suppose, for example, that the entropy of a given source is 5 bits per source symbol. (A *bit* is a 0 or a 1.) The Noiseless Coding Theorem tells us that, given any $\epsilon > 0$, we can, by clever encoding, encode each source symbol with a binary codeword (string of 0s and 1s) in such a way that the average length of the codewords is $5 + \epsilon$ bits.

In many situations, such as the archival storage of data on computer disk, the entire message is at our disposal, and so the black box model may not be the most appropriate one. In this case, for instance, we may scan the entire message and compile actual frequencies for the source symbols. This removes the uncertainties associated with a probability distribution. In fact, under the previous model, on a particularly unlucky day, the encoded message may actually be (physically) longer than the the source message in its original form! One of the difficulties of this two-pass approach, however, is that we must store additional frequency data along with the message, thus partially defeating the purpose of the encoding.

We could also take a different approach to the black box model and keep a running frequency count of the symbols as they appear, using this frequency to encode the source symbols as they appear. Thus, while the form of the encoding remains the same, the substance of the encoding is constantly changing. This dynamic approach is called *adaptive encoding* and offers distinct practical improvements over static approaches.

Unfortunately however, a further discussion of the variety of available data compression methods would take us too far afield from our more theoretical approach to the Noiseless Coding Theorem. Hence, we will stick to our simple black box—static probability distribution model.

In Part 2 of the book, on coding theory, we will turn to the issue of how to encode source information in such a way as to detect, and even correct, errors in transmission (or storage). This is referred to as *noisy coding*. As we have seen, in encoding for efficiency, the goal is to minimize the average amount of information in each codeword, over the given probability distribution of the source symbols. However, in our concern for accurate encoding, emphasis will shift away from the source and its probability distribution to the issue of how to minimize the amount of additional information (called *redundancy*) that we *must* add to the source (through encoding) in order to detect and/or correct the desired number of errors. Indeed, we will quickly assume (as is customary) that the probability distribution for the source is uniform, in order to get a good handle on how best to make decoding decisions.

Most (but by no means all) of the work done on encoding for error detection/correction involves the use of *fixed-length* encoding schemes, where all codewords have the same length. Thus, in Part 2, we will confine our attention to *fixed-length codes*, or *block codes*.

Let us illustrate a simple approach to error detection. Consider the binary source message

$$011001 \quad 001100 \quad 100110$$

which we have divided into three blocks. (Thus, each source symbol is a binary string of length six.) One possibility for detecting errors in the transmission of this message is to add what is known as an *even parity check digit* to each block. To do this, we simply adjoin to each block a single bit (0 or 1) to insure that the total number of 1s in the block is even. For clarity, we will underline the redundant bits

$$0110011 \quad 0011000 \quad 1001101$$

Now imagine that a single error should occur in say, the first block. For example, suppose the first block is received as 0100011. The receiver immediately counts the number of 1s in the received string, and finds that this number is odd. This tells the receiver that an error has occurred. In this way, the addition of an even parity check digit detects a single error (in each block.) Furthermore, in this encoding, we have only added one bit of redundant data for every 6 bits of source message. Thus, for a modest increase in redundancy, we get single error detection. If we are willing to allow more redundant data, we can encode the strings in such a way as to not only detect single errors, but to correct them as well.

We should mention in closing that the goals of encoding for efficiency and encoding for error detection/correction are distinct. If both goals are sought in a particular instance, the usual procedure is to first encode for efficiency and then encode for error detection/correction. (Encoding in the reverse order would make little sense.) As far as this author is aware, there is no single encoding scheme that achieves significant success in attaining both goals.

0

Preliminaries

In this chapter, we will briefly discuss various topics that will be used throughout the book. Many of these topics may already be familiar to you, but we suggest that you look through the material in any case, if only to set the terminology.

0.1 Miscellany

The Greek Alphabet

We will use only a few Greek letters in this text, but for reference we list the entire alphabet.

A	α	alpha	I	ι	iota	P	ρ	rho
B	β	beta	K	κ	kappa	Σ	σ	sigma
Γ	γ	gamma	Λ	λ	lambda	T	τ	tau
Δ	δ	delta	M	μ	mu	Υ	υ	upsilon
E	ϵ	epsilon	N	ν	nu	Φ	ϕ	phi
Z	ζ	zeta	Ξ	ξ	xi	X	χ	chi
H	η	eta	O	o	omicron	Ψ	ψ	psi
Θ	θ	theta	Π	π	pi	Ω	ω	omega

Sets

The size (number of elements) of a finite set S will be denoted by $|S|$. The set with no elements, called the **empty set**, is denoted by $\emptyset$.

The **union** $S \cup T$ of sets S and T is the set of all elements that are in *either* S or T, and the **intersection** $S \cap T$ is the set of all elements that are in *both* S and T. When $S \cap T = \emptyset$, we say that S and T are **disjoint**. In other words, two sets are disjoint if they have no elements in common.

If every element of a set E is also an element of S, then E is a **subset** of S, and we write $E \subseteq S$. We also say that S is a **superset** of E. Of course, $\emptyset \subseteq S$ and $S \subseteq S$, for all sets S. Subsets of S other than S itself are called **proper subsets**.

If $E \subseteq S$, the **complement** of E in S is denoted by $S - E$ and consists of the elements of S that are not in E. When no confusion can arise, the complement of E is written E^c.

The set $\mathbb{Z}_n = \{0, 1, 2, \ldots, n-1\}$ consisting of the first n non-negative integers will play a predominant role in this book. We will be especially interested in $\mathbb{Z}_n$ when n is a prime number. (A **prime number** is an integer $p \geq 2$ that has no positive divisors except 1 and p itself.)

Summation Notation

We will often use summation notation to indicate sums. If $s_1, s_2, \ldots, s_n$ are numbers or algebraic expressions, we denote their sum by

$$\sum_{k=1}^{n} s_k$$

For instance,

$$\sum_{k=1}^{20} k = 1 + 2 + 3 + \cdots + 20, \qquad \sum_{i=2}^{10} i^2 = 2^2 + 3^2 + 4^2 + \cdots + 10^2$$

and

$$\sum_{k=0}^{n} \frac{1}{(x+1)^k} = 1 + \frac{1}{x+1} + \frac{1}{(x+1)^2} + \cdots + \frac{1}{(x+1)^n}$$

Permutations

There are two equivalent ways to view permutations—as functions or as ordered arrangements. Using the functional point of view, a **permutation** of a finite set A is a bijective (one-to-one and onto) function $f : A \to A$. For instance, if $A = \{0, 1, 2, 3, 4\}$ then the function $f : A \to A$ defined by

$$f(0) = 2, \quad f(1) = 3, \quad f(2) = 0, \quad f(3) = 4, \quad f(4) = 1$$

is a permutation. Using the arrangement point of view, a **permutation** of a set A is simply an ordered arrangement of the elements of A. For instance, the permutation f may simply be written as 23041. We will feel free to use both representations of permutations in this book.

The number of permutations of a set of size n is

$$n! = 1 \cdot 2 \cdots n$$

The symbol $n!$ is read n factorial and is just the product of the first n positive integers. For instance, the number of permutations of the set A is $5! = 1 \cdot 2 \cdot 3 \cdot 4 \cdot 5 = 120$. It is also customary to set $0! = 1$.

If A is a set of size n, then a **permutation of size** $k \leq n$, taken from A, is just a permutation of k of the elements of A. For instance, taking the ordered arrangement point of view, if $A = \{0, 1, 2, 3, 4\}$ then the permutations of size 2, taken from A, are

$$01, 10, 02, 20, 03, 30, 04, 40,$$
$$12, 21, 13, 31, 14, 41,$$
$$23, 32, 24, 42,$$
$$34, 43$$

Notice that there are 20 such permutations.

In general, the number of permutations of size k, taken from a set of size n, is

$$\frac{n!}{(n - k)!}$$

For instance, the number of permutations of size 2 from the set A is

$$\frac{5!}{(5 - 2)!} = \frac{5!}{3!} = 20$$

as we have seen.

Binomial Coefficients

If $0 \leq k \leq n$, the **binomial coefficient** $\binom{n}{k}$ (read n choose k) is defined by

$$\binom{n}{k} = \frac{n!}{k!(n-k)!}$$

The following theorem shows why binomial coefficients are so important.

Theorem 0.1.1 *A set S of size n has precisely $\binom{n}{k}$ subsets of size k. Put another way, $\binom{n}{k}$ is the number of ways of choosing k elements from a set of size n.* □

For instance, the set $\mathbb{Z}_9 = \{0, 1, 2, 3, 4, 5, 6, 7, 8\}$ has

$$\binom{9}{3} = \frac{9!}{3!6!} = 84$$

subsets of size 3. There are 84 ways to choose 3 numbers from the set $\mathbb{Z}_9$.

There are a vast number of identities involving binomial coefficients, and we mention only two here

$$\binom{n}{k} = \binom{n}{n-k}$$
$$\binom{n}{k} = \binom{n-1}{k} + \binom{n-1}{k-1}$$

Binary Numbers

As you no doubt know, a decimal number is just a nonempty string over the alphabet $\{0, 1, 2, 3, 4, 5, 6, 7, 8, 9\}$. (We generally omit leading 0s, but that is not essential to the meaning of a number.) Each position in the string represents a power of 10. Reading from right to left, the positions represent

$$10^0 = 1, \quad 10^1 = 10, \quad 10^2 = 100, \quad 10^3 = 1000, \quad \dots$$

The digits of a decimal number are the coefficients of the corresponding powers of 10. For instance, in the decimal number 72305, the digit 3 is the coefficient of 10^2.

Binary numbers work in an entirely analogous way. A binary number is just a string over the set {0, 1}. (Leading 0s may be deleted.) In the case of binary numbers, each position in the string represents a power of 2. Reading from right to left, the positions represent

$$2^0 = 1, \quad 2^1 = 2, \quad 2^2 = 4, \quad 2^3 = 8, \quad \ldots$$

The bits (abbreviation for binary digits) are the coefficients of the corresponding powers of 2.

Of course, a number is a number—instead of saying binary [resp: decimal] number, we should really be saying number written in binary [resp: decimal] form, nevertheless, we will continue to abuse the terminology.

Using these facts, we can easily convert from binary notation to decimal notation. For instance, the binary number 11010 is equal to

$$1 \cdot 2^4 + 1 \cdot 2^3 + 0 \cdot 2^2 + 1 \cdot 2^1 + 0 \cdot 2^0 = 26$$

in decimal. This is sometimes written $11010_2 = 26_{10}$.

To convert from decimal notation to binary takes a bit more work. For example, consider the decimal number 298. The largest power of 2 that is less than or equal to 298 is $256 = 2^8$. Hence,

$$298 = 2^8 + 42$$

Now, the largest power of 2 less than or equal to 42 is $32 = 2^5$. Hence,

$$298 = 2^8 + 2^5 + 10 = 2^8 + 2^5 + 2^3 + 2$$

which tells us that the binary representation of 298_{10} is a string of bits with 1s in the ninth, sixth, fourth, and second positions and 0s elsewhere, that is,

$$298_{10} = 100101010_2$$

Arithmetic operations on binary numbers are easily performed by remembering that

$$0 + 0 = 0, \quad 0 + 1 = 1, \quad 1 + 0 = 1 \quad \text{and} \quad 1 + 1 = 10$$

and

$$0 \cdot 0 = 0, \quad 0 \cdot 1 = 0, \quad 1 \cdot 0 = 0 \quad \text{and} 1 \cdot 1 = 1$$

Note that addition and multiplication of bits looks exactly the same as if these bits were decimal digits, except that $1 + 1 = 10$. Thus, for instance,

$$
\begin{array}{r}
10111 \\
+\,1101 \\
\hline
100100
\end{array}
$$

Binary Trees

The following concepts will be encountered when we discuss Huffman encoding in Chapter 2. (You may postpone a reading of this subsection until then.)

Definition   A **graph** is a finite nonempty collection of points, called **nodes** or **vertices** together with a collection of line segments, called **edges**, connecting pairs of nodes. (This is a somewhat informal definition of the term graph.)   □

(Do not confuse this concept of a graph with that of the graph of a function. The two are not related, except by a common name.)

For example, Figure 0.1.1 below is a graph.

The following definition of a binary tree comes from computer science.

Definition   A **complete binary tree** is a graph with the following properties.

1. It is possible to draw a series of equidistant horizontal straight lines so that every node lies on one of these lines. These lines are not part of the graph, but indicate a level for each node, the top line being level 1.

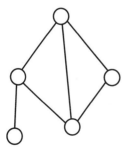

FIGURE 0.1.1

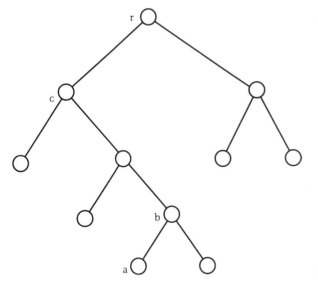

FIGURE 0.1.2

2. There is only one node on level 1. It is called the **root**.

3. If two nodes are connected by an edge, they must lie on adjacent levels. If node a on level i is connected to node b on level $i + 1$, then a is called the **parent** of b and b is a **child** of a. If there is a sequence of nodes $c = c_1, c_2, \ldots, c_n = d$ where c_i is a child of c_{i+1} for all $i = 1, \ldots, n - 1$, then c is called a **descendent** of d.

4. Every node is a descendent of the root; every node has either no or exactly two children and every node except the root has exactly one parent. A node with no children is called a **leaf**. Two nodes that have the same parent are called **siblings**.   □

Figure 0.1.2 is an example of a binary tree with five levels. Node r is the root, node a is a leaf and is a child of node b, which is the parent of a. Node a is also a descendent of node c (as well as of other nodes).

0.2   Some Probability

Probability plays an important role in information and coding theory. In this section, we will briefly discuss some aspects of the theory that will be useful to us. A few of our examples refer to the concept of a string. You may wish to refer to Section 1.1 for the relevant definitions.

Let us begin with a few simple definitions. By a process, we mean a procedure that produces outcomes from a certain set. For example, flipping a coin is a process, since it produces outcomes from the set {Heads, Tails}.

Definition The set of all possible outcomes of a process is called the **sample space** of the process. Any subset of the sample space, that is, any set of outcomes, is called an **event**.   □

In this book, we will deal exclusively with finite sample spaces, that is, sample spaces with only a finite number of elements.

Example 0.2.1

1. The sample space for the process of flipping a coin is the set $S = \{H, T\}$, where H represents heads and T represents tails.

2. The sample space for the process of rolling a pair of dice is the set S of all ordered pairs of the form (x, y), where x is the value on the first die, and y is the value on the second die. (We assume that the dice are distinguishable by some means.) Thus,

$$S = \{(1, 1), (1, 2), (1, 3), \ldots, (6, 4), (6, 5), (6, 6)\}$$

Note that $|S| = 6^2 = 36$. The event of getting a sum equal to 7 is the set

$$E = \{(1, 6), (2, 5), (3, 4), (4, 3), (5, 2), (6, 1)\}$$

3. Consider a "black box" that randomly emits elements from an alphabet $A = \{s_1, s_2, \ldots, s_n\}$. Suppose that one element is emitted each second for 6 seconds. The sample space in this case is the set S of all sequences of letters of length 6, taken from A. Hence, if no repeats are allowed, then

$$|S| = n(n - 1)(n - 2)(n - 3)(n - 4)(n - 5)$$

and if repeats are allowed, then $|S| = n^6$.   □

Once the sample space S for a given process has been determined, the next step is to assign probabilities to each of the possible outcomes.

Definition Let $S = \{s_1, \ldots, s_n\}$ be a sample space. Suppose that, for each element $s_i \in S$, we assign a real number denoted by p_i such that

1. $$0 \le p_i \le 1 \text{ for all } i$$

2. $$\sum_{i=1}^{n} p_i = 1.$$

Then the sequence $p_1, \ldots, p_n$ is called a **probability distribution** for S and the assignment $\mathcal{P} : S \to \mathbb{R}$ defined by $\mathcal{P}(s_i) = p_i$ is a **probability law** for S. □

The exact method of assigning probabilities to the outcomes of a process depends on the assumptions made about the process. For instance, the assumption that a coin is fair is equivalent to making the assignments

$$\mathcal{P}(H) = \frac{1}{2} \text{ and } \mathcal{P}(T) = \frac{1}{2}$$

Definition Let E be a nonempty event in a sample space S. Then the **probability** of E, denoted by $\mathcal{P}(E)$, is the sum of the probabilities of each outcome in the event. We also set $\mathcal{P}(\emptyset) = 0$. □

Example 0.2.2 Consider the process of rolling two fair dice, whose sample space consists of the 36 ordered pairs described in Example 0.2.1b. Since we are assuming that the dice are fair, the probabilities of each outcome must be equal, that is $p_i = \frac{1}{36}$, for all $i = 1, \ldots, 36$. Thus, using the description of E in Example 0.2.1b, we have

$$\begin{aligned}
\mathcal{P}(\text{getting a sum of 7}) &= \mathcal{P}(E) \\
&= \mathcal{P}((1,6)) + \mathcal{P}((2,5)) + \mathcal{P}((3,4)) + \mathcal{P}((4,3)) \\
&\quad + \mathcal{P}((5,2)) + \mathcal{P}((6,1)) \\
&= 6 \cdot \frac{1}{36} = \frac{1}{6}
\end{aligned}$$
□

It is not uncommon for each outcome of a process to be equally likely, in which case the probability of each outcome is $\frac{1}{|S|}$.

Definition The **uniform probability distribution** for a sample space S of size n is the sequence $\frac{1}{n}, \ldots, \frac{1}{n}$. □

Theorem 0.2.1 *Let S be a finite sample space with uniform probability law. If E is an event in S, then*

$$\mathcal{P}(E) = \frac{|E|}{|S|}$$
□

The basic properties of the probability function $\mathcal{P}$ are as follows.

Theorem 0.2.2   *Let S be a finite sample space. Then*

1. $\mathcal{P}(\emptyset) = 0$

2. $\mathcal{P}(S) = 1$

3. $0 \leq \mathcal{P}(E) \leq 1$, *for all events* $E \subseteq S$

4. $\mathcal{P}(E^c) = 1 - \mathcal{P}(E)$, *for all events* $E \subseteq S$

5. *If E and F are events in S, then*

$$E \cap F = \emptyset \text{ implies } \mathcal{P}(E \cup F) = \mathcal{P}(E) + \mathcal{P}(F) \qquad \square$$

It is important to keep in mind that part 5. of Theorem 0.2.2 holds only when the events E and F are disjoint. In probability theory disjoint events are said to be **mutually exclusive**. Part 5. can be extended to more than two events. In particular, if $E_1, \ldots, E_n$ are events in S that are pairwise mutually exclusive, that is $E_i \cap E_j = \emptyset$ for all $i \neq j$, then

$$\mathcal{P}(E_1 \cup \cdots \cup E_n) = \sum_{i=1}^{n} \mathcal{P}(E_i)$$

Example 0.2.3   A black box emits symbols from the set $A = \{0, 1\}$, at the rate of one per second. Each symbol is emitted with equal probability. After 4 seconds, what is the probability that the string emitted from the black box has an even number of 1s?

The sample space is the set S of binary strings of length 4. A string with k 1s is formed simply by choosing k of the 4 positions in which to place 1s, the remaining positions being filled by 0s. Hence, there are $\binom{4}{k}$ strings with exactly k 1s and the event E of getting an even number of 1s has size

$$\binom{4}{0} + \binom{4}{2} + \binom{4}{4} = 8$$

Thus,

$$\mathcal{P}(\text{even number of 1s}) = \frac{8}{2^4} = \frac{1}{2} \qquad \square$$

Example 0.2.4   Five cards are drawn, each with equal probability, from a deck of 52 cards. What is the probability of getting exactly 3 aces?

In this case, the sample space is the set of all possible 5-card hands, and this space has size

$$|S| = \binom{52}{5} = 2{,}598{,}960$$

Now we must compute the size of the event E of getting exactly 3 aces. Such a hand can be formed by first choosing 3 of the 4 aces, and this can be done in $\binom{4}{3}$ ways, and then choosing 2 of the remaining 48 (non-ace) cards, which can be done in $\binom{48}{2}$ ways. Thus,

$$|E| = \binom{4}{3}\binom{48}{2}$$

and

$$P(E) = \frac{|E|}{|S|} = \frac{\binom{4}{3}\binom{48}{2}}{\binom{52}{5}} = \frac{4 \cdot 1128}{2598960} \approx 0.0017 \qquad \square$$

Example 0.2.5   Five decimal digits $a_1, a_2, \ldots, a_5$ are chosen at random. What is the probability that a_5 is the same as one of the previous 4 digits?

The sample space is the set S of all strings of length 5 over $\{0, 1, 2, \ldots, 9\}$, and so $|S| = 10^5$. Let E be the event that a_5 is the same as one of a_1, a_2, a_3, or a_4. Determining the size of E is actually a bit awkward (since, in particular, some of a_1, a_2, a_3, or a_4 may be the same). In this case, it turns out to be easier to determine the size of the complement E^c of E in S.

To determine the size of E^c, we observe that for each of the 10 possibilities for a_5, the digits a_1, a_2, a_3, and a_4 must be taken from the 9 remaining digits. Thus $|E^c| = 9^4 \cdot 10$, and

$$P(E^c) = \frac{|E^c|}{|S|} = \frac{9^4 \cdot 10}{10^5} = 0.6561$$

Finally, we use part 4) of Theorem 0.2.2 to get

$$P(E) = 1 - P(E^c) = 0.3439 \qquad \square$$

Independent Events

Definition   Two events E and F from the same sample space are independent if

$$P(E \cap F) = P(E) \cdot P(F) \qquad \Box$$

The definition of independent events can be extended to more than two events, although the definition becomes a bit more involved then you might think at first.

Definition   Three events E, F, and G from the same sample space are independent if the following two conditions hold.

1. $P(E \cap F \cap G) = P(E) \cdot P(F) \cdot P(G)$

2. Each pair of events from among E, F and G is independent, that is,

$$P(E \cap F) = P(E) \cdot P(F), \qquad P(E \cap G) = P(E) \cdot P(G)$$
$$P(F \cap G) = P(F) \cdot P(G) \qquad \Box$$

Definition   The events $E_1, \ldots, E_n$ from the same sample space S are independent if the probability of the intersection of every subcollection of these events is equal to the product of the probabilities of these events. In symbols

$$P(E_{i_1} \cap \cdots \cap E_{i_k}) = P(E_{i_1}) \cdots P(E_{i_k})$$

for any $i_1, \ldots, i_k$. $\qquad \Box$

Example 0.2.6   Suppose we toss two fair coins in the air, one at a time. Let E be the event that the first toss results in heads, and let F be the event that the second toss results in heads. Then $P(E) = P(F) = \frac{1}{2}$. Since $E \cap F$ is the event that both tosses result in heads, we have $P(E \cap F) = \frac{1}{4}$. Thus $P(E \cap F) = P(E) \cdot P(F)$ and so the events E and F are independent. $\qquad \Box$

Example 0.2.7   A card is chosen at random from a deck of 52 cards. Let E be the event that the card is a 10 or a deuce, and let F be the event that the card has face value at most 6. (An ace has face value 1, face cards have no face value.) Then $P(E) = \frac{8}{52} = \frac{2}{13}$ and $P(F) = \frac{24}{52} = \frac{6}{13}$. But since $E \cap F$ is the event that the card chosen is a deuce, we have $P(E \cap F) = \frac{4}{52} = \frac{1}{13}$. Now

$$P(E) \cdot P(F) = \left(\frac{2}{13} \frac{6}{13} \right) = \frac{12}{169} \neq P(E \cap F)$$

and so the events are not independent.   □

Here are some examples that relate directly to the subject of this book.

Example 0.2.8 Suppose that binary strings of length 5 are sent over a noisy communication line, such as a telephone line. Assume that, because of the noise, the probability that a bit (0 or 1) is received correctly is 0.75. Assume also that the event that one bit is received correctly is independent of the event that another bit is received correctly.

1. What is the probability that a string will be received correctly?

2. What is the probability that exactly 3 of the 5 bits in a string are received correctly?

Solutions

1. If E_i is the event that the ith bit is received correctly, then $\mathcal{P}(E_i) = 0.75$, and since E_1, E_2, E_3, E_4, and E_5 are independent, we have

$$\begin{aligned}
\mathcal{P}(\text{string received correctly}) &= \mathcal{P}(E_1 \cap E_2 \cap E_3 \cap E_4 \cap E_5) \\
&= \mathcal{P}(E_1)\mathcal{P}(E_2)\mathcal{P}(E_3)\mathcal{P}(E_4)\mathcal{P}(E_5) \\
&= (0.75)^5 \approx 0.237
\end{aligned}$$

2. Consider the case where the first three bits are received correctly, and the other two are not. Since the probability that a bit is not received correctly is $1 - 0.75 = 0.25$, the probability of this occurring is

$$\begin{aligned}
\mathcal{P}(E_1 \cap E_2 \cap E_3 \cap E_4^c \cap E_5^c) &= \mathcal{P}(E_1)\mathcal{P}(E_2)\mathcal{P}(E_3)\mathcal{P}(E_4^c)\mathcal{P}(E_5^c) \\
&= (0.75)^3(0.25)^2
\end{aligned}$$

But this probability would be the same if any set of 3 bits were received correctly, and since there are $\binom{5}{3}$ possibilities for the positions of 3 correct bits, we have

$$\mathcal{P}(\text{exactly 3 bits received correctly}) = \binom{5}{3}(0.75)^3(0.25)^2 \approx 0.264$$

Thus, we see that it is more likely that exactly 2 errors are made in transmission than that no errors are made!   □

Since the previous example is very important, let us generalize it.

Example 0.2.9 Suppose that binary strings of length n are sent over a noisy communication line. Assume that, because of the noise, the probability that a bit (0 or 1) is received correctly is p. Assume also that the event that one bit is received correctly is independent of the event that another bit is received correctly.

1. What is the probability that an entire string will be received correctly?

2. What is the probability that a specified set of k bits (such as the first k bits) are received correctly, but that the remaining bits are incorrect?

3. What is the probability that exactly k bits (any k bits) are received correctly?

4. What is the probability that at least k bits are received correctly?

Solutions

1. If E_i is the event that the ith bit is received correctly, then $P(E_i) = p$, and since $E_1, \ldots, E_n$ are independent, we have

$$P(\text{string received correctly}) = P(E_1 \cap \cdots \cap E_n)$$
$$= P(E_1) \cdots P(E_n) = p^n$$

2. Consider the case where the first k bits are received correctly, and the others are not. Since the probability that a bit is not received correctly is $1 - p$, the probability of this occurring is

$$P(E_1 \cap \cdots \cap E_k \cap E_{k+1}^c \cap \cdots \cap E_n^c) = p^k(1 - p)^{n-k}$$

But this probability would be the same for any set of k bits, and so the probability that a specified set of k bits is received correctly (and the rest are received incorrectly) is $p^k(1 - p)^{n-k}$.

3. Since there are $\binom{n}{k}$ possibilities for the k correct bit positions, and since we saw in part b) that the probability that a specified set of k bits is received correctly is $p^k(1 - p)^{n-k}$, the probability that exactly k bits are received correctly is

$$P(\text{exactly } k \text{ correct bits}) = \binom{n}{k} p^k(1 - p)^{n-k}$$

4. From part c), we have

$$P(\text{at least } k \text{ bits are correct}) = \sum_{i=k}^{n} P(\text{exactly } i \text{ bits are correct})$$
$$= \sum_{i=k}^{n} \binom{n}{i} p^i(1 - p)^{n-i} \qquad \square$$

Conditional Probability

It is often the case that we have some additional knowledge about the outcome of a process and want to use this knowledge to obtain more accurate probabilities. This leads us to define the concept of conditional probability.

Definition   Let S be a sample space and suppose that E and F are events with $P(F) \neq 0$. The **conditional probability** of E, given that F has occurred, is

$$P(E \mid F) = \frac{P(E \cap F)}{P(F)} \qquad \square$$

Example 0.2.10   A certain operation results in complete recovery 60% of the time, partial recovery 30% of the time and death 10% of the time. What is the probability of complete recovery, given that a patient survives the operation?

Solution   Let E be the event of complete recovery and let F be the event of surviving the operation. We seek the conditional probability $P(E \mid F)$. Since F^c is the event of dying, we have

$$P(F) = 1 - P(F^c) = 1 - \frac{1}{10} = \frac{9}{10}$$

The intersection $E \cap F$ is the event that the patient recovers completely and survives, which is just E. Hence

$$P(E \cap F) = P(E) = \frac{6}{10}$$

Thus, we have

$$P(E \mid F) = \frac{P(E \cap F)}{P(F)} = \frac{P(E)}{P(F)} = \frac{6/10}{9/10} = \frac{2}{3} \approx 67\% \qquad \square$$

Note that the formula for conditional probability can be written in the sometimes useful form

$$P(E \cap F) = P(E \mid F) \cdot P(F)$$

Example 0.2.11   A company that manufactures computer chips uses two different manufacturing processes. Process 1 produces nondefective chips 98.5% of the time and process 2 produces nondefective chips 97.1%

of the time. Process 1 is used 60% of the time. What is the probability that a randomly chosen chip was produced by process 2 and is nondefective?

Solution   Let E be the event that the chip is nondefective and let F be the event that the chip was produced by process 2. We seek the probability $P(E \cap F)$. Since process 2 is used 40% of the time, the probability that a given chip was made by process 2 is

$$P(F) = 0.40$$

We are also given the conditional probability that the chip is nondefective, given that it was produced by process 2,

$$P(E \mid F) = 0.971$$

Hence,

$$P(E \cap F) = P(E \mid F) \cdot P(F) = (0.971)(0.40) = 0.3884 \qquad \square$$

We will have use for two very important formulas related to conditional probability. To state them, we first need a definition.

Definition   Let S be a sample space. The events $E_1, \ldots, E_n$ form a **partition** of S if

1. $P(E_i) > 0$ for all i,
2. the events are pairwise mutually exclusive, that is, $E_i \cap E_j = \emptyset$ for all $i \neq j$,
3. $E_1 \cup \cdots \cup E_n = S$ $\qquad \square$

Theorem 0.2.3 (Theorem on Total Probabilities)   *Let S be a sample space and let $E_1, \ldots, E_n$ form a partition of S. Provided that $P(E_k) \neq 0$ for all k, we have for any event A in S,*

$$P(A) = \sum_{k=1}^{n} P(A \mid E_k) P(E_k)$$

Proof   Since the events E_i form a partition of S, the events $A \cap E_i$ form a partition of A. That is, $A \cap E_1, \ldots, A \cap E_n$ are disjoint and their union is A. It follows that

$$P(A) = \sum_{k=1}^{n} P(A \cap E_k) = \sum_{k-1}^{n} P(A \mid E_k) P(E_k)$$

which is what we wanted to prove. $\qquad \blacksquare$

Theorem 0.2.4 (Bayes' Theorem)   *Let S be a sample space and let E_1, ..., E_n form a partition of S. For any event A in S with $\mathcal{P}(A) > 0$, we have*

$$\mathcal{P}(E_j \mid A) = \frac{\mathcal{P}(A \mid E_j)\mathcal{P}(E_j)}{\sum_{k=1}^{n} \mathcal{P}(A \mid E_k)\mathcal{P}(E_k)}$$

Proof   This follows directly from the definition of conditional probability and the Theorem on Total Probabilities

$$\mathcal{P}(E_j \mid A) = \frac{\mathcal{P}(E_j \cap A)}{\mathcal{P}(A)} = \frac{\mathcal{P}(A \mid E_j)\mathcal{P}(E_j)}{\sum_{k=1}^{n} \mathcal{P}(A \mid E_k)\mathcal{P}(E_k)} \qquad \blacksquare$$

0.3   Matrices

A **matrix** is a rectangular array of elements, such as

$$\begin{bmatrix} 1 & 4 \\ -2 & 0 \end{bmatrix}, \quad \begin{bmatrix} 0 & 1 & 2 & 3 \end{bmatrix}, \quad \begin{bmatrix} 3 & 5 & \frac{1}{2} \\ 8 & 0 & 0 \end{bmatrix}, \quad \begin{bmatrix} 1 \\ 0 \\ 0 \end{bmatrix}$$

Our interest will center on matrices whose elements, or **entries**, come from the set $\mathbb{Z}_p$. The entry in the ith row and jth column of a matrix is referred to as the $(\mathbf{i}, \mathbf{j})$**th entry**.

A matrix M is said to have **size $\mathbf{m} \times \mathbf{n}$** if it has m rows and n columns. We then say that M is an $\mathbf{m} \times \mathbf{n}$ **matrix**. For instance, the matrices above have size 2×2, 1×4, 2×3 and 3×1, respectively. A matrix with the same number of rows as columns (that is, a matrix of size $n \times n$), is called a **square matrix**. The first matrix above is square.

A matrix of size $1 \times n$, consisting of a single row, is often referred to as a **row matrix**. Similarly, a matrix of size $m \times 1$, consisting of a single column, is referred to as a **column matrix**. The second matrix above is a row matrix; the last matrix is a column matrix.

A matrix whose entries are all equal to 0 is called a **zero matrix**, and is denoted by 0. If we need to emphasize the size of the matrix, we will use the notation $0_{m,n}$.

A square matrix that has 1s along its main diagonal and 0s elsewhere is called an **identity matrix**. We use the symbol I_n to denote the identity matrix of size $n \times n$. For instance,

$$I_2 = \begin{bmatrix} 1 & 0 \\ 0 & 1 \end{bmatrix}, \quad I_3 = \begin{bmatrix} 1 & 0 & 0 \\ 0 & 1 & 0 \\ 0 & 0 & 1 \end{bmatrix}, \quad I_4 = \begin{bmatrix} 1 & 0 & 0 & 0 \\ 0 & 1 & 0 & 0 \\ 0 & 0 & 1 & 0 \\ 0 & 0 & 0 & 1 \end{bmatrix}$$

We define addition of matrices simply by adding corresponding entries. For instance,

$$\begin{bmatrix} 1 & 2 & 3 \\ 0 & 4 & 2 \end{bmatrix} + \begin{bmatrix} 2 & -2 & 0 \\ 3 & 1 & 1 \end{bmatrix} = \begin{bmatrix} 3 & 0 & 3 \\ 3 & 5 & 3 \end{bmatrix}$$

Similarly, we can multiply a matrix by a number by multiplying each entry by that number, for example,

$$5 \begin{bmatrix} 1 & 2 \\ -2 & 0 \end{bmatrix} = \begin{bmatrix} 5 & 10 \\ -10 & 0 \end{bmatrix}$$

The product of a row matrix of size $1 \times n$ and a column matrix of size $n \times 1$ is the 1×1 matrix defined as follows

$$\begin{bmatrix} u_1 & u_2 & \cdots & u_n \end{bmatrix} \begin{bmatrix} v_1 \\ v_2 \\ \vdots \\ v_n \end{bmatrix} = \begin{bmatrix} u_1 v_1 + u_2 v_2 + \cdots + u_n v_n \end{bmatrix}$$

Example 0.3.1   In $\mathbb{Z}_5$ (see Section 1.1), we have

$$\begin{bmatrix} 1 & 2 & 3 & 4 \end{bmatrix} \begin{bmatrix} 2 \\ 1 \\ 0 \\ 3 \end{bmatrix} = \begin{bmatrix} 1 \cdot 2 + 2 \cdot 1 + 3 \cdot 0 + 4 \cdot 3 \end{bmatrix} = \begin{bmatrix} 1 \end{bmatrix} \quad \square$$

This definition can be extended to the product of an $n \times m$ matrix and an $m \times k$ matrix. However, we shall only need the following special cases. The product of an $m \times n$ matrix and an $n \times 1$ matrix is an $m \times 1$ matrix formed by taking the products of the rows of the first matrix with the second (column) matrix, thus

$$\begin{bmatrix} u_{11} & u_{12} & \cdots & u_{1n} \\ u_{21} & u_{22} & \cdots & u_{2n} \\ & & \vdots & \\ u_{m1} & u_{m2} & \cdots & u_{mn} \end{bmatrix} \begin{bmatrix} v_1 \\ v_2 \\ \vdots \\ v_n \end{bmatrix} = \begin{bmatrix} u_{11}v_1 + u_{12}v_2 + \cdots + u_{1n}v_n \\ u_{21}v_1 + u_{22}v_2 + \cdots + u_{2n}v_n \\ \vdots \\ u_{m1}v_1 + u_{m2}v_2 + \cdots + u_{mn}v_n \end{bmatrix}$$

The product of a $1 \times m$ matrix and an $m \times n$ matrix is the $1 \times n$ matrix formed by taking the product of the first (row) matrix with the columns of the second matrix, thus

$$
\begin{bmatrix} v_1 & v_2 \cdots v_m \end{bmatrix}
\begin{bmatrix}
u_{11} & u_{12} & \cdots & u_{1n} \\
u_{21} & u_{22} & \cdots & u_{2n} \\
& & \vdots & \\
u_{m1} & u_{m2} & \cdots & u_{mn}
\end{bmatrix}
$$

$$
= [v_1 u_{11} + v_2 u_{21} + \cdots + v_m u_{m1}
$$
$$
v_1 u_{12} + v_2 u_{22} + \cdots + v_m u_{2m}
$$
$$
\cdots v_1 u_{1n} + v_2 u_{2n} + \cdots + v_m u_{mn}]
$$

Example 0.3.2   Over $\mathbb{Z}_2$ (see Section 1.1), we have

$$
\begin{bmatrix}
1 & 0 & 1 & 1 & 1 \\
0 & 0 & 0 & 1 & 1 \\
1 & 1 & 1 & 1 & 1
\end{bmatrix}
\begin{bmatrix}
1 \\ 1 \\ 0 \\ 0 \\ 1
\end{bmatrix}
$$

$$
= \begin{bmatrix}
1 \cdot 1 + 0 \cdot 1 + 1 \cdot 0 + 1 \cdot 0 + 1 \cdot 1 \\
0 \cdot 1 + 0 \cdot 1 + 0 \cdot 0 + 1 \cdot 0 + 1 \cdot 1 \\
1 \cdot 1 + 1 \cdot 1 + 1 \cdot 0 + 1 \cdot 0 + 1 \cdot 1
\end{bmatrix}
$$

$$
= \begin{bmatrix}
0 \\ 1 \\ 1
\end{bmatrix}
\qquad \square
$$

It is sometimes convenient to write a string $x = x_1 x_2 \cdots x_n$ in the form of a row matrix

$$
x = \begin{bmatrix} x_1 & x_2 & \cdots & x_n \end{bmatrix}
$$

or a column matrix

$$
x = \begin{bmatrix} x_1 \\ x_2 \\ \vdots \\ x_n \end{bmatrix}
$$

In this case, we will say that the string x is in row form, or in column form. We will use the same notation x for a string written in its usual form, and in column or row form.

Definition The **transpose** M^t of an $n \times m$ matrix M is the $m \times n$ matrix whose first column is the first row of M, whose second column is the second row of M, and so on. Thus, if

$$M = \begin{bmatrix} u_{11} & u_{12} & \cdots & u_{1n} \\ u_{21} & u_{22} & \cdots & u_{2n} \\ & & \vdots & \\ u_{m1} & u_{m2} & \cdots & u_{mn} \end{bmatrix}$$

then

$$M^t = \begin{bmatrix} u_{11} & u_{21} & \cdots & u_{m1} \\ u_{12} & u_{22} & \cdots & u_{m2} \\ & & \vdots & \\ u_{1n} & u_{2n} & \cdots & u_{mn} \end{bmatrix} \qquad \square$$

For example,

$$\text{if } M = \begin{bmatrix} 1 & 0 & 1 & 1 & 1 \\ 0 & 0 & 0 & 1 & 1 \\ 1 & 1 & 1 & 1 & 1 \end{bmatrix}, \text{ then } M^t = \begin{bmatrix} 1 & 0 & 1 \\ 0 & 0 & 1 \\ 1 & 0 & 1 \\ 1 & 1 & 1 \\ 1 & 1 & 1 \end{bmatrix}$$

A matrix M for which $M = M^t$ is said to be **symmetric**. For instance, the matrix

$$M = \begin{bmatrix} 1 & 2 & 3 \\ 2 & 4 & 5 \\ 3 & 5 & 0 \end{bmatrix}$$

is symmetric.

1

CHAPTER

An Introduction to Codes

1.1 Strings and Things

Modular Arithmetic in $\mathbb{Z}_n$

If n is a positive integer, we let $\mathbb{Z}_n$ denote the set $\{0, 1, \ldots, n-1\}$ consisting of the first n non-negative integers. Much of what we will do in this book (especially in the coding theory portion) involves this set.

If α and β are integers, the following three conditions are equivalent.

1. $\alpha - \beta$ is divisible by n

2. there exists an integer k for which $\alpha = \beta + kn$

3. α and β have the same remainder when divided by n.

If any (and hence all) of these conditions hold, we say that α and β are **congruent modulo** n, and write

$$\alpha \equiv \beta \, (\mathrm{mod}\, n)$$

Example 1.1.1

1. $5 \equiv 3 \, (\mathrm{mod}\, 2)$

2. $178 \equiv 17 \, (\mathrm{mod}\, 7)$

3. $-4 \equiv 12 \, (\mathrm{mod}\, 4)$

4. $15 \equiv 0 (\mathrm{mod}\, 5)$

5. $4 \not\equiv 6 (\mathrm{mod}\, 3)$

6. $14 \equiv 9 \equiv 4 \equiv -1 \equiv -6 (\mathrm{mod}\, 5)$ □

Given an integer α, there is exactly one integer k in $\mathbb{Z}_n$ that is congruent to α, that is, for which $k \equiv \alpha (\mathrm{mod}\, n)$. Put another way, in $\mathbb{Z}_n$, there exists a unique solution to the equation

$$x \equiv \alpha (\mathrm{mod}\, n)$$

This solution is simply the remainder obtained by dividing α by n and is referred to as the **residue** of α modulo n. It can also be found by adding, or subtracting, a suitable multiple of n in order to produce a number in $\mathbb{Z}_n$. For instance, the residue of 25 modulo 7 is found by subtracting $3 \cdot 7 = 21$ from 25, to get 4. That is, $4 \equiv 25 \ (\mathrm{mod}\ 7)$, where $4 \in \mathbb{Z}_7$.

We can define two algebraic operations, known as **addition modulo** n and **multiplication modulo** n, on the set $\mathbb{Z}_n$. If $x, y \in \mathbb{Z}_n$, we set

$$x \oplus_n y = \text{the remainder obtained by dividing } x + y \text{ by } n$$

and

$$x \otimes_n y = \text{the remainder obtained by dividing } xy \text{ by } n$$

Hence, $x \oplus_n y \equiv (x + y) \ (\mathrm{mod}\ n)$ and $x \otimes_n y \equiv xy \ (\mathrm{mod}\ n)$.

Example 1.1.2

1. In $\mathbb{Z}_5$,

$$3 \oplus_5 4 = 2 \text{ and } 3 \otimes_5 2 = 1$$

2. In $\mathbb{Z}_{10}$,

$$7 \oplus_{10} 8 = 5 \text{ and } 7 \otimes_{10} 8 = 6 \qquad\qquad □$$

The set $\mathbb{Z}_n$, together with the operations of addition and multiplication modulo n, is referred to as the **integers modulo** n. It is customary, whenever the context makes it clear, to use the ordinary symbols $+$ and $\cdot$, instead of $\oplus_n$ and $\otimes_n$, and we will also follow this custom.

In $\mathbb{Z}_2$, addition and multiplication modulo 2 are described by the following tables.

Addition modulo 2		
+	0	1
0	0	1
1	1	0

Multiplication modulo 2		
·	0	1
0	0	0
1	0	1

The Field $\mathbb{Z}_p$

Our main interest in the set $\mathbb{Z}_p$ is when p is a prime. The reason is that, when p is a prime, addition and multiplication modulo p have much nicer properties than in the nonprime case. The following theorem is a case in point.

Theorem 1.1.1 *The following property holds in $\mathbb{Z}_n$ if and only if n is a prime*

$$\alpha\beta = 0 \text{ implies } \alpha = 0 \text{ or } \beta = 0$$

Proof Let $n = p$ be a prime and suppose that $\alpha\beta = 0$ in $\mathbb{Z}_p$. This is equivalent to $\alpha\beta \equiv 0 \pmod{p}$, which holds if and only if p divides $\alpha\beta$. Since p is a prime, it divides the product $\alpha\beta$ if and only if it divides at least one of the factors. But if p divides α then $\alpha \equiv 0 \pmod{p}$; that is, $\alpha = 0$ in $\mathbb{Z}_p$. Similarly, if p divides β then $\beta \equiv 0 \pmod{p}$; that is, $\beta = 0$ in $\mathbb{Z}_p$.

To see that the property does not hold in $\mathbb{Z}_n$ when n is not a prime, observe that, if n is not prime, it has the form $n = \alpha\beta$, where $2 \leq \alpha, \beta \leq n - 1$. Hence, α and β are nonzero elements of $\mathbb{Z}_n$, and

$$\alpha\beta = n \equiv 0 \pmod{n}$$

that is, $\alpha\beta = 0$ in $\mathbb{Z}_n$. ∎

The set $\mathbb{Z}_p$, together with the operations of addition and multiplication modulo p, forms a field. Before giving a formal definition, we need to define the concept of a binary operation on a set.

Definition Let S be a nonempty set. A **binary operation** on S is a function $* : S \times S \rightarrow S$, from the set $S \times S$ of all ordered pairs of elements of S to S. We usually denote $*((\alpha, \beta))$ by $\alpha * \beta$. □

For example, ordinary addition of real numbers is a binary operation $+\mathbb{R} \times \mathbb{R} \to \mathbb{R}$, where $+(\alpha, \beta)$ is almost always written $\alpha + \beta$.

Now let us give a formal definition of a field.

Definition   A **field** is a nonempty set $\mathcal{F}$, together with two binary operations on $\mathcal{F}$, called **addition** (denoted by $+$) and **multiplication** (denoted by juxtaposition), satisfying the following properties.
Associative properties: For all α, β, and γ in $\mathcal{F}$,

$$\alpha + (\beta + \gamma) = (\alpha + \beta) + \gamma \text{ and } \alpha(\beta\gamma) = (\alpha\beta)\gamma$$

Commutative properties: For all α and β in $\mathcal{F}$,

$$\alpha + \beta = \beta + \alpha \text{ and } \alpha\beta = \beta\alpha$$

Distributive property: For all α, β and γ in $\mathcal{F}$,

$$\alpha(\beta + \gamma) = \alpha\beta + \alpha\gamma$$

Properties of 0 and 1: There exist two distinct special elements in $\mathcal{F}$, one called the **zero element** and denoted by 0 and the other called the **identity element** and denoted by 1, with the properties that, for all $\alpha \in \mathcal{F}$,

$$0 + \alpha = \alpha + 0 = \alpha \text{ and } 1 \cdot \alpha = \alpha \cdot 1 = \alpha$$

Inverse properties: For every $\alpha \in \mathcal{F}$, there exists another element in $\mathcal{F}$, denoted by $-\alpha$ and called the **negative** of α, for which

$$\alpha + (-\alpha) = (-\alpha) + \alpha = 0$$

For every nonzero $\alpha \in \mathcal{F}$, there exists another element in $\mathcal{F}$, denoted by α^{-1}, and called the **inverse** of α, for which

$$\alpha \cdot (\alpha^{-1}) = (\alpha^{-1}) \cdot \alpha = 1 \qquad \square$$

The most familiar fields are the sets $\mathbb{Q}$ of rational numbers, $\mathbb{R}$ of real numbers, and $\mathbb{C}$ of complex numbers, each with ordinary addition and multiplication. However, it happens that, for every prime power $q = p^n$, there is a field of size q. In fact, there is essentially only one field for each prime power. For $q = p$ a prime, this field is easy to describe, for it is just the integers modulo p.

Theorem 1.1.2   *The set $\mathbb{Z}_n$ of integers modulo n is a field if and only if n is a prime number.*

Proof Suppose first that $n = p$ is a prime. We will not establish all of the properties in the definition of field, but show only that every nonzero element α of $\mathbb{Z}_p$ has an inverse. Consider all possible products of α with the elements of $\mathbb{Z}_p$

$$\{\alpha\beta \mid \beta \in \mathbb{Z}_p\}$$

These p products are distinct, for if $\alpha\beta = \alpha\beta'$ then $\alpha(\beta - \beta') = 0$ and Theorem 0.1.2 gives $\beta - \beta' = 0$, whence $\beta = \beta'$. Since the p products are distinct, they represent every element in $\mathbb{Z}_p$. In particular, one of the products must equal 1, that is, $\alpha\beta = 1$ for some $\beta \in \mathbb{Z}_p$. By commutativity we also have $\beta\alpha = 1$ and so β is the inverse of α.

If n is not a prime, then $n = \alpha\beta$ for some nonzero $\alpha, \beta \in \mathbb{Z}_n$. It follows that $\alpha\beta = 0$ in $\mathbb{Z}_n$ and so α cannot have an inverse. For if α^{-1} did exist, then we would have

$$\beta = \alpha^{-1}(\alpha\beta) = \alpha^{-1}0 = 0$$

which is not the case. Since the nonzero element α does not have an inverse, the set $\mathbb{Z}_n$ cannot be a field. ∎

Example 1.1.3 Since $2 + 5 = 0$ in $\mathbb{Z}_7$, the negative of 2 is 5. In symbols, $-2 = 5$. Note that this is true only in $\mathbb{Z}_7$, using addition modulo 7. It is certainly not true in the familiar field of real numbers.

Since $2 \cdot 4 = 1$ in $\mathbb{Z}_7$, the inverse of 2 is 4, that is, $2^{-1} = 4$. Again, this is true only in $\mathbb{Z}_7$, and not in the field of real numbers. □

When p is not a prime, all of the properties in the definition of a field hold except that not all nonzero numbers have inverses. However, since $\mathbb{Z}_n$ is not a field if n is not prime, many other properties that hold in $\mathbb{Z}_p$ do not hold in $\mathbb{Z}_n$. Theorem 1.1.1 is a case in point.

The field $\mathbb{Z}_2$ of integers modulo 2 has an interesting property not shared by the other fields $\mathbb{Z}_p$ for $p > 2$. Since $1 + 1 = 0$ in $\mathbb{Z}_2$, we have $1 = -1$. Certainly $0 = -0$, and so, if e is either 0 or 1, then

$$e - e = e + (-e) = e + e$$

In other words, in $\mathbb{Z}_2$, each element is its own negative and subtraction is the same as addition. Note that this is not true in $\mathbb{Z}_p$, for $p \neq 2$.

We should caution against confusing addition modulo 2 in $\mathbb{Z}_2$ with addition of binary numbers. In $\mathbb{Z}_2$, we have $1 + 1 = 0$, but for addition of binary numbers, we have $1 + 1 = 10$.

Strings

The concept of a string is fundamental to the subjects of coding and information theory. Let $S = \{s_1, s_2, \ldots, s_n\}$ be a finite, nonempty set, which we refer to as an **alphabet**. A **string**, or **word**, over S is simply a finite sequence of elements of S. For instance, if $S = \{\alpha, \beta, 1, 2\}$ then

$$\beta, \alpha 1, 11\beta 2, \text{ and } 2222$$

are strings over S. A sequence does not have to be meaningful in some language to qualify as a word. For instance, if $S = \{a, b, c, \ldots, z\}$, then xyah is a word over S, even though it is not a word in the English language. The terms string and word are synonymous and will be used interchangeably.

Strings will be denoted by boldface letters, such as $\mathbf{x}$, $\mathbf{y}$ and $\mathbf{z}$. If $\mathbf{x} = x_1 x_2 \cdots x_k$ is a string over S, then each x_i in $\mathbf{x}$ is an **element** of $\mathbf{x}$. The **length** of a string $\mathbf{x}$, denoted by $\text{len}(\mathbf{x})$, is the number of elements in the string.

The **juxtaposition** of two strings $\mathbf{x}$ and $\mathbf{y}$ is the string $\mathbf{xy}$. For instance, the juxtaposition of $\mathbf{x} = 101$ and $\mathbf{y} = 1000$ is $\mathbf{xy} = 1011000$. If a string has the form $\mathbf{z} = \mathbf{xy}$, we say that $\mathbf{x}$ is a **prefix** of $\mathbf{z}$. For instance, 110 is a prefix of 1101010. It is clear that

$$\text{len}(\mathbf{xy}) = \text{len}(\mathbf{x}) + \text{len}(\mathbf{y})$$

The set of all strings over S is denoted by S^*. We also include in S^* the **empty string**, denoted by θ (the Greek letter theta), and defined to be the string with no elements. Thus, $\text{len}(\theta) = 0$. If n is a non-negative integer, the set of all strings over S of length n is denoted by S^n, and the set of all strings over S of length n or less is denoted by S_n. Thus, $S^n \subseteq S_n \subseteq S^*$. Note that $S^0 = S_0$ consists of just the empty string.

A string over $\mathbb{Z}_2 = \{0, 1\}$ is called a **binary string**. Each of the elements 0 and 1 is called a **bit**, which is a contraction of binary digit. For instance, 011101 is a binary string of length 6. The **complement** $\mathbf{x}^c$ of a binary string $\mathbf{x}$ is defined to be the string obtained by replacing all 0s by 1s and all 1s by 0s. For instance, $(11001)^c = 00110$. A string over $\mathbb{Z}_3 = \{0, 1, 2\}$ is called a **ternary string**.

If 0 is in the alphabet, then the **stringzero string** $00 \cdots 0$ is denoted by a boldface $\mathbf{0}$. If 1 is in the alphabet, the string consisting of all 1s is denoted by a boldface $\mathbf{1}$. For instance, in $\mathbb{Z}_2^5$, we have $\mathbf{0} = 00000$ and $\mathbf{1} = 11111$. If $0 \leq i \leq n$, the notation $\mathbf{e}_i$ is reserved strictly for the string all of whose elements are 0 except for a 1 in the ith position. For instance, in $\mathbb{Z}_2^4$,

$$\mathbf{e}_1 = 1000, \mathbf{e}_2 = 0100, \mathbf{e}_3 = 0010, \mathbf{e}_4 = 0001$$

In $\mathbb{Z}_5$, we have $\mathbf{e}_1 = 10000$. Thus, there is some ambiguity in this notation, since there is a different string $\mathbf{e}_i$ in each $\mathbb{Z}_p^n$ with $n \geq i$. However, this should not cause any problems, since the context will always resolve any ambiguity.

The next theorem tells us how many strings there are in S^n and S_n.

Theorem 1.1.3 *Let S be an alphabet of size $k > 1$. Then*

$$1)\ |S^n| = k^n \qquad\qquad 2)\ |S_n| = \frac{k^{n+1} - 1}{k - 1}$$

Proof For part 1), note that each string in S^n can be formed by picking one of the k symbols in S for the first position, one for the second position, one for the third position, and so on. Because there are k choices for each of the n positions, the number of strings that can be formed in this way is k^n. That is, $|S^n| = k^n$.

For part 2), we use the results of part 1),

$$|S_n| = |S^0| + |S^1| + \cdots + |S^n|$$
$$= 1 + k + k^2 + \cdots + k^n = \frac{k^{n+1} - 1}{k - 1} \qquad\blacksquare$$

Theorem 1.1.4
1. *In $\mathbb{Z}_2^n$, the number of strings with exactly k 0s is $\binom{n}{k}$.*
2. *In $\mathbb{Z}_r^n$, the number of strings with exactly k 0s is $\binom{n}{k}(r-1)^{n-k}$.*

Proof Part 1) follows from the fact that there is one such string for every way of choosing k of the n positions in which to place the 0s. Part 2) is similar, and left as an exercise. $\qquad\blacksquare$

Example 1.1.4
1. The set $\mathbb{Z}_2^8$ has size $2^8 = 256$. Furthermore, there are $\binom{8}{3} = 56$ binary strings in $\mathbb{Z}_2^8$ containing exactly three 0s.
2. The set $\mathbb{Z}_3^6$ has size $3^6 = 729$. There are $\binom{6}{2}2^4 = 240$ strings in $\mathbb{Z}_3^6$ that contain exactly two 0s.
3. The number of strings in $\mathbb{Z}_5^{10}$ that contain exactly three 0s and exactly two 1s is $\binom{10}{3}\binom{7}{2}3^5 = 612360$. This follows from the fact that there are

$\binom{10}{3}$ ways to choose the 3 positions for the 0s, then there are $\binom{7}{2}$ ways to choose 2 of the remaining 7 positions for the two 1s and, finally, there are 3^5 ways to fill in the remaining positions with any of the 3 other elements of $\mathbb{Z}_5$.

4. The number of strings in $\mathbb{Z}_7^{12}$ with at least nine 0s is

$$\binom{12}{9}6^3 + \binom{12}{10}6^2 + \binom{12}{11}6 + \binom{12}{12} = 49969 \qquad \square$$

Exercises

1. Write out the addition and multiplication modulo 3 tables for $\mathbb{Z}_3$.

2. Show that the cancellation law

$$\alpha\beta = \alpha\gamma, \alpha \neq 0 \text{ implies } \beta = \gamma$$

holds in $\mathbb{Z}_p$, p a prime. Does it hold in $\mathbb{Z}_n$ when n is not prime?

3. Find the following residues
   a) 23 (mod 11)
   b) -17 (mod 3)
   c) 1345 (mod 5)
   d) 1232456 (mod 2)
   e) 133 (mod 3)
   f) -1793 (mod 5)

4. Find the following inverses
   a) 3^{-1} in $\mathbb{Z}_5$
   b) 3^{-1} in $\mathbb{Z}_7$
   c) 8^{-1} in $\mathbb{Z}_{11}$

5. How many strings are there in $\mathbb{Z}_5^8$ that contain exactly four 0s? How many strings are there that contain exactly three 0s and two 1s?

6. Show that there are $\binom{n}{k}(r-1)^{n-k}$ strings in $\mathbb{Z}_r^n$ containing exactly k 0s.

7. How many strings are there in $\mathbb{Z}_3^6$ with at most two 0s?

8. How many strings are there in $\mathbb{Z}_2^{20}$ with at least two 0s?

9. How many strings are there in $\mathbb{Z}_r^n$ that contain exactly k nonzero elements?

10. Prove that $1 + k + k^2 + \cdots + k^n = \frac{1-k^{n+1}}{1-k}$.

11. In $\mathbb{Z}_2^n$, show that

    a) $(\mathbf{x} + \mathbf{y})^c = \mathbf{x} + \mathbf{y}^c$

    b) $\mathbf{x}^c + \mathbf{y}^c = \mathbf{x} + \mathbf{y}$.

For the next exercises, we use the following concept (which will be defined more formally later in the text). The distance between two strings of the same length is the number of positions in which these strings differ. For instance, $d(0011, 0110) = 2$, since these two strings differ in 2 places (the second and fourth).

12. Let $\mathbf{x}$ be a string in $\mathbb{Z}_r^n$. Show that there are $\binom{n}{k}(r-1)^k$ strings in $\mathbb{Z}_r^n$ that have distance k from $\mathbf{x}$.

13. If $\mathbf{x}$ and $\mathbf{y}$ are binary strings of length n, find expressions for

    a) $d(\mathbf{x}^c, \mathbf{y})$

    b) $d(\mathbf{x}^c, \mathbf{y}^c)$

    in terms of $d(\mathbf{x}, \mathbf{y})$.

14. Show that if $\mathbf{x}$, $\mathbf{y}$, and $\mathbf{z}$ are strings of the same length, then

$$d(\mathbf{x}, \mathbf{y}) \leq d(\mathbf{x}, \mathbf{z}) + d(\mathbf{z}, \mathbf{y})$$

This is known as the triangle inequality.

1.2   What Are Codes?

A code is nothing more than a set of strings over a certain alphabet. For example, the set

$$C = \{0, 10, 110, 1110\}$$

is a code over the alphabet $\mathbb{Z}_2$. Of course, codes are generally used to encode messages. For instance, we may use C to encode the first four letters of the alphabet, as follows

$$a \to 0$$
$$b \to 10$$
$$c \to 110$$
$$d \to 1110$$

Then we can encode words (or messages) built up from these letters. The word cab, for instance, is encoded as

$$\text{cab} \to 110010$$

These ideas lead us to make the following definitions.

Definition   Let $A = \{a_1, a_2, \ldots, a_r\}$ be a finite set, which we call a **code alphabet**. An **r-ary code** over A is a subset C of the set A^* of all words over A. The elements of C are called **codewords**. The number r is called the **radix** of the code. $\qquad\square$

The most commonly used (and studied) alphabet is the set $\mathbb{Z}_2$. Codes over $\mathbb{Z}_2$ are referred to as **binary codes**. Codes over the alphabet $\mathbb{Z}_3$ are referred to as **ternary codes**.

Definition   Let $S = \{s_1, s_2, \ldots, s_q\}$ be a finite set, which we refer to as a **source alphabet**. Let C be a code. An **encoding function** is a bijective function $f : S \to C$, from S onto C. If C is a code and $f : S \to C$ is an encoding function, we refer to the ordered pair (C, f) as an **encoding scheme** for S. $\qquad\square$

Because an encoding function is bijective (that is, both one-to-one and onto), it associates to each source symbol in the source alphabet one and only one codeword. Moreover, every codeword is associated to a source symbol. This makes it possible to decode any sequence of codewords.

Example 1.2.1   The 26 letters of the alphabet can be encoded as follows. Let the source alphabet be $S = \{a, b, c, \ldots, z\}$, let the code alphabet be $A = \{0, 1, \ldots, 9\}$, and let the code be $C = \{00, 01, 02, \ldots, 25\}$. Let $f : S \to C$ be defined by

$$f(a) = 00, f(b) = 01, \ f(c) = 02, f(d) = 03, f(e) = 04,$$
$$f(f) = 05, f(g) = 06, \ f(h) = 07, \ f(i) = 08, \ f(j) = 09,$$
$$f(k) = 10, \ f(l) = 11, \ f(m) = 12, f(n) = 13, f(o) = 14,$$
$$f(p) = 15, f(q) = 16, \ f(r) = 17, f(s) = 18, f(t) = 19,$$
$$f(u) = 20, f(v) = 21, \ f(w) = 22, f(x) = 23, f(y) = 24,$$
$$f(z) = 25$$

TABLE 1.2.1   The ASCII Encoding Scheme (Partial)

A	→	01000001	J	→	01001010	S	→	01010011
B	→	01000010	K	→	01001011	T	→	01010100
C	→	01000011	L	→	01001100	U	→	01010101
D	→	01000100	M	→	01001101	V	→	01010110
E	→	01000101	N	→	01001110	W	→	01010111
F	→	01000110	O	→	01001111	X	→	01011000
G	→	01000111	P	→	01010000	Y	→	01011001
H	→	01001000	Q	→	01010001	Z	→	01011010
I	→	01001001	R	→	01010010	space	→	00100000

This encoding function may be used to encode any message. For instance,

$$\text{math is fun} \rightarrow 1200190708180520130$$

We purposely used two-digit numbers for each codeword. If we had taken the code to be $C' = \{0, 1, 2, 3, 4, 5, 6, 7, 8, 9, 10, \ldots, 25\}$, then uniqueness problems would have arisen. For instance, the string 1019 could have resulted from several different messages,

$$\text{bat} \rightarrow 1019, \quad \text{babj} \rightarrow 1019, \quad \text{kt} \rightarrow 1019$$

We will discuss the problem of uniqueness later in this chapter.   □

Example 1.2.2   Table 1.2.1 shows a portion of a very commonly used code, known as the **ASCII code**. The acronym ASCII stands for American Standard Code for Information Interchange. This code is used by microcomputers to store characters in memory or on storage media. In this case, the complete source alphabet consists of all upper- and lower-case letters, punctuation marks, and various other symbols. The code consists of all binary numbers from 0000000 to 1111111, that is, from 0 to 127 decimal. (The extended ASCII code, adopted by many personal computers, consists of 8-bit binary numbers.)

For instance, the ASCII code for the upper-case letter A is 1000001 (or 65 decimal). The first thirty-three ASCII codes (not shown in the table) are used for control characters, that is, characters that control the operation of a monitor, printer, or other device. For instance, the decimal number 12 (00001100 binary) is the ASCII code for a form feed, and 7 (00000111 binary) is the ASCII code for tintinnabulation.   □

Codes can be divided into two general types as follows.

Definition   A **fixed length code**, or **block code**, is a code whose codewords all have the same length n. In this case, the number n is also called the **length** of the code. If a code C contains codewords of varying lengths, it is called a **variable length code**.   □

Fixed length codes have advantages and disadvantages over variable length codes. One advantage is that they never require a special symbol to separate the characters in the message being coded. For example, consider the encoded ASCII message

$$01000011010011110100010001000101$$

Because the (binary) ASCII code is a fixed length code whose codewords have length 8, we know that the first 8 bits represents the first character of the original message, which according to Table 1.2.1, is C. Similarly, the second set of 8 bits represents the second character in the message, namely O. Continuing in this way, we decode the message to get the word CODE.

Perhaps the main disadvantage of fixed length codes, such as the ASCII code, is that characters that are used frequently, such as the letter e, have codes as long as characters that are used infrequently, such as the space character. On the other hand, variable length codes, which can encode frequently used characters using shorter codewords, can save a great deal of time and space. We will discuss both types of codes in this book.

Exercises

1. Suppose you require a binary block code containing 126 codewords. What is the minimum possible length for this code?

2. Suppose you require a binary block code containing n codewords. What is the minimum possible length for this code?

3. How many encoding functions are possible from the source alphabet $S = \{a, b\}$ to the code $C = \{0, 1\}$? List them.

4. How many encoding functions are possible from the source alphabet $S = \{a, b, c\}$ to the code $C = \{00, 01, 11\}$? List them.

5. Find a formula for the number of encoding functions from a source alphabet of size n to a code of size n.

6. How many r-ary block codes of length n are there over an alphabet A? How many binary codes are there of length 5?

7. How many r-ary codes are there with maximum codeword length n over an alphabet A? What is this number for $r = 2$ (binary codes) and $n = 5$?

1.3   Uniquely Decipherable Codes

One of the most important properties that a code can possess is **unique decipherability**. Informally speaking, this means that any sequence of symbols can be interpreted in at most one way as a sequence of codewords. More formally, we have the following definition.

Definition   A code C over an alphabet A is **uniquely decipherable** if, for every string $x_1 x_2 \cdots x_n$ over A, there exists at most one sequence of codewords $\mathbf{c}_1 \mathbf{c}_2 \cdots \mathbf{c}_m$ for which

$$\mathbf{c}_1 \mathbf{c}_2 \cdots \mathbf{c}_m = x_1 x_2 \cdots x_n$$

Put another way, a code is uniquely decipherable if no two different sequences of codewords represents the same string over A, in symbols, if

$$\mathbf{c}_1 \mathbf{c}_2 \cdots \mathbf{c}_n = \mathbf{d}_1 \mathbf{d}_2 \cdots \mathbf{d}_m$$

for codewords $\mathbf{c}_i$ and $\mathbf{d}_j$, then $m = n$ and

$$\mathbf{c}_1 = \mathbf{d}_1, \mathbf{c}_2 = \mathbf{d}_2, \ldots, \mathbf{c}_n = \mathbf{d}_n \qquad \square$$

Example 1.3.1   Consider the following codes

$$C_1 = \{\mathbf{c}_1 = 0, \mathbf{c}_2 = 01, \mathbf{c}_3 = 001\}, \quad C_2 = \{\mathbf{d}_1 = 0, \mathbf{d}_2 = 10, \mathbf{d}_3 = 110\}$$

Code C_1 is not uniquely decipherable, since the string 001 represents either the single codeword $\mathbf{c}_3$ or the string $\mathbf{c}_1 \mathbf{c}_2$. On the other hand, C_2 is uniquely decipherable, since any string corresponds to at most one sequence of codewords. For instance, consider the string

$$1000110$$

Reading from left to right, we see that 1 is not, by itself, a codeword. But 10 is. Furthermore, only one codeword begins with 10 so 10 must be $\mathbf{d}_2$

$$10|00110$$
$$\mathbf{d}_2|$$

Next, we come to a 0, which must represent $\mathbf{d}_1$, since no other codeword begins with 0. Continuing in this way, we see that this string represents only one sequence of codewords.

To prove that C_2 is uniquely decipherable, we will be content with giving a set of observations that show that any given sequence $\mathbf{x}$ of codewords can be interpreted in only one way. In this case, we have the following observations, assuming that $\mathbf{x}$ is read from left to right.

1. If we encounter a 0, this must represent $\mathbf{d}_1$.

2. If we encounter a 1 followed by a 0, this must represent $\mathbf{d}_2$.

3. If we encounter a 1 followed by another 1, the next element must be a 0 and so this must represent $\mathbf{d}_3$. □

Speaking loosely, if a code is uniquely decipherable, then it cannot have very many short codewords. To illustrate this point, if the word 010011 of length 6 is a codeword, then the words 010 and 011 of length 3 cannot both be codewords. We can be more precise about codeword lengths in the following theorem, known as **McMillan's Theorem** (first published in 1956).

Theorem 1.3.1   *(McMillan's Theorem) Let $C = \{c_1, c_2, \ldots, c_q\}$ be an r-ary code and let $\ell_i = \mathrm{len}(c_i)$. If C is uniquely decipherable, then its codeword lengths $\ell_1, \ell_2, \ldots, \ell_q$ must satisfy*

$$\sum_{k=1}^{q} \frac{1}{r^{\ell_k}} \leq 1$$

Proof   The following proof is the usual one given for this theorem, although it is not particularly intuitive. Suppose that α_j is the number of codewords in C of length j. Then we have

$$\sum_{k=1}^{q} \frac{1}{r^{\ell_k}} = \sum_{j=1}^{m} \frac{\alpha_j}{r^j},$$

where $m = \max_i\{\ell_i\}$.

Now let u be a positive integer, and consider the quantity

$$\left(\sum_{j=1}^{m} \frac{\alpha_j}{r^j}\right)^u = \left(\frac{\alpha_1}{r} + \frac{\alpha_2}{r^2} + \cdots + \frac{\alpha_m}{r^m}\right)^u$$

Multiplying this out gives

$$= \sum_{\substack{1 \leq i_j \leq m \\ i_1, i_2, \ldots, i_u}} \frac{\alpha_{i_1}}{r^{i_1}} \cdots \frac{\alpha_{i_u}}{r^{i_u}} = \sum_{\substack{1 \leq i_j \leq m \\ i_1, i_2, \ldots, i_u}} \frac{\alpha_{i_1} \alpha_{i_2} \cdots \alpha_{i_u}}{r^{i_1 + \cdots + i_u}}$$

Now, since $1 \leq i_j \leq m$, each sum $i_1 + \cdots + i_u$ is at least u and at most um. Collecting terms with a common sum $i_1 + \cdots + i_u$, we get

$$= \sum_{k=u}^{um} \left(\sum_{i_1 + \cdots + i_u = k} \alpha_{i_1} \alpha_{i_2} \cdots \alpha_{i_u}\right) \frac{1}{r^k} = \sum_{k=u}^{um} \frac{N_k}{r^k}$$

where

$$N_k = \sum_{i_1 + \cdots + i_u = k} \alpha_{i_1} \alpha_{i_2} \cdots \alpha_{i_u}$$

Now we are ready to use the fact that the code is uniquely decipherable. Recalling that α_i is the number of codewords in C of length i, we see that

$$\alpha_{i_1} \alpha_{i_2} \cdots \alpha_{i_u}$$

is the number of possible strings of length $k = i_1 + \cdots + i_u$ consisting of a codeword of length i_1, followed by a codeword of length i_2, and so on, ending with a codeword of length i_u.

Hence, the sum N_k is the total number of strings $\mathbf{c}_1 \cdots \mathbf{c}_u$ of length k made up of exactly u codewords. Since C is uniquely decipherable, no two sequences of u codewords can yield the same string of length k and so there can be at most r^k such sequences of codewords, since r^k is the total number of strings of length k from an r-ary alphabet. In other words,

$$N_k \leq r^k$$

and so

$$\left(\sum_{k=1}^{m} \frac{\alpha_k}{r^k}\right)^u \leq \sum_{k=u}^{um} \frac{N_k}{r^k} \leq \sum_{k=u}^{um} 1 \leq um$$

Taking uth roots gives

$$\sum_{k=1}^{m} \frac{\alpha_k}{r^k} \leq u^{1/u} m^{1/u}$$

Since this holds for all positive integers u, we may let u approach ∞. But $u^{1/u} m^{1/u} \to 1$ as $u \to \infty$, and so we must have

$$\sum_{k=1}^{m} \frac{\alpha_k}{r^k} \leq 1 \qquad \blacksquare$$

The inequality in McMillan's Theorem is called **Kraft's Inequality**. McMillan's Theorem confirms that, for a uniquely decipherable code, the codeword lengths must be reasonably large. (The numbers ℓ_i must be large in order to make the terms $1/r^{\ell_i}$ small.)

Example 1.3.2 Suppose we desire a binary code consisting of six codewords, but we restrict the codeword lengths to a maximum of 2. (That is, $\ell_i \leq 2$ for $i = 1, 2, \ldots, 6$.) Since there are precisely six nontrivial strings over $\{0, 1\}$ of length at most 2, our code must consist of these six strings. That is, $C = \{0, 1, 00, 01, 10, 11\}$. But this code is not uniquely decipherable. (The string 01, for example, has two interpretations.)

In this case, the "shortness" of the codewords forces us to use codewords, such as 01, that are made up of smaller codewords (0 and 1). This prevents the code from being uniquely decipherable.

Of course, we could have used McMillan's Theorem to tell us that such a code could not be uniquely decipherable. For, in this case, $\ell_i \leq 2$ for all i. Hence, $r^{\ell_i} \leq r^2$ and so

$$\frac{1}{r^{\ell_i}} \geq \frac{1}{r^2}$$

Thus, since $r = 2$, we have

$$\sum_{k=1}^{6} \frac{1}{r^{\ell_k}} \geq \sum_{k=1}^{6} \frac{1}{r^2} = \sum_{k=1}^{6} \frac{1}{4} = \frac{6}{4} > 1$$

This tells us that Kraft's inequality does not hold for this code, and so it cannot be uniquely decipherable.   □

Note that McMillan's Theorem cannot tell us when a particular code is uniquely decipherable, but only when it is not. For the theorem does not say that any code whose codeword lengths satisfy Kraft's inequality must be uniquely decipherable. Rather, it says that if a code is known to be uniquely decipherable, then its word lengths must satisfy Kraft's inequality. Hence, if a code does not satisfy this inequality, we may conclude that it cannot be uniquely decipherable.

Exercises

1. Is the code $C = \{0, 10, 1100, 1101, 1110, 1111\}$ uniquely decipherable? Justify.

2. Is the code $C = \{0, 10, 110, 1110, 11110, 11111\}$ uniquely decipherable? Justify.

3. Is the code $C = \{0, 01, 011, 0111, 01111, 11111\}$ uniquely decipherable? Justify.

4. Is the code $C = \{0, 10, 110, 1110, 1111, 1101\}$ uniquely decipherable? Justify.

5. Is the code $C = \{0, 10, 1101, 1110, 1011, 110110\}$ uniquely decipherable? Justify.

6. Determine whether or not there is a uniquely decipherable binary code with codeword lengths 1,2,3,3. If so, construct such a code.

7. Determine whether or not there is a uniquely decipherable binary code with codeword lengths 1,3,3,3,4,5,5,5. If so, construct such a code.

8. Is it possible to construct a uniquely decipherable code, over the alphabet $\{0, 1, 2, \ldots, 9\}$, with nine codewords of length 1, nine codewords of length 2, ten codewords of length 3, and ten codewords of length 4?

9. For a given binary code, let $N(k)$ be the total number of sequences of codewords that contain exactly k bits. For instance, if

$$C = \{\mathbf{c}_1 = 0, \mathbf{c}_2 = 10, \mathbf{c}_3 = 11\}$$

then $N(3) = 5$, since the five codeword sequences

$$c_1 c_1 c_1, \quad c_1 c_2, \quad c_1 c_3, \quad c_2 c_1, \quad c_3 c_1$$

each contain exactly 3 bits, and no other codeword sequences contain exactly 3 bits.

a) Determine $N(1)$ and $N(2)$.

b) Show that $N(k) = N(k-1) + 2N(k-2)$, for all $k \geq 3$. Hint: a string of length $k \geq 3$ begins with either a codeword of length 1 or a codeword of length 2.

c) Solve the recurrence relation in part b). Hint: assume a solution of the form $N(k) = \alpha^k$ and solve for α. Get a general solution of the form $N(k) = a\alpha_1^k + b\alpha_2^k$ and determine the values of a and b.

d) For the code $D = \{0, 10, 110, 111\}$, compute $N(1)$, $N(2)$, $N(3)$, $N(4)$, $N(5)$. Show that these values are consistent with the formula

$$N(k) = \frac{4}{7} \cdot 2^k + \frac{3}{7} \cos \frac{2\pi}{3}k + \frac{\sqrt{3}}{21} \sin \frac{2\pi}{3}k$$

1.4   Instantaneous Codes and Kraft's Theorem

It is clear that unique decipherability is a very desirable property. However, even though a code may have this property, it may still not be possible to interpret codewords as soon as they are received. The following simple example will illustrate this.

Example 1.4.1   The code

$$C_3 = \{\mathbf{c}_1 = 0, \mathbf{c}_2 = 01\}$$

is easily seen to be uniquely decipherable (by reading strings backwards). Now suppose that the string 0001 is being transmitted. Just after receiving the first 0, we cannot tell whether it should be interpreted as the codeword $\mathbf{c}_1$ or the beginning of $\mathbf{c}_2$. Once we receive the second 0 in the message, we know that the first 0 must represent $\mathbf{c}_1$, but we don't know about the second 0. Thus, codewords cannot be interpreted as soon as they are received.

On the other hand, for the code

$$C_4 = \{\mathbf{d}_1 = 0, \mathbf{d}_2 = 10\}$$

individual codewords can be interpreted as soon as they are received. For instance, consider the string 00100. As soon as the first 0 is received, we know immediately that it must be $\mathbf{d}_1$, and similarly for the second 0.

When the 1 is received, we know that a 0 is coming next, and that the string 10 must be d_2. Thus, the message can be interpreted codeword by codeword. □

The previous example motivates the following definition.

Definition A code is said to be **instantaneous** if, whenever any sequence of codewords is transmitted, each codeword can be interpreted as soon as it is received. □

If a code is instantaneous, then it is also uniquely decipherable. However, the converse is not true. There are codes that are uniquely decipherable but not instantaneous, as illustrated by the code C_3 of Example 1.4.1.

There is a very simple way to tell when a code is instantaneous. First we need a definition.

Definition A code is said to have the **prefix property** if no codeword is a prefix of any other codeword. □

Example 1.4.2 The code C_2 in Example 1.3.1 has the prefix property. However, the code C_1 does not, since c_1 is a prefix of c_2. □

The importance of the prefix property comes from the following theorem.

Theorem 1.4.1 *A code C is instantaneous if and only if it has the prefix property.*

Proof Suppose that C has the prefix property. Then, as soon as the first codeword is received, we can decode it, since it cannot be a prefix of another codeword. The same reasoning applies to the second codeword and indeed to all other codewords in the sequence. Hence, the code C is instantaneous.

Conversely, suppose that C is instantaneous. If a codeword c is a prefix of a codeword d, then the first c in the sequence cc cannot be decoded until we receive at least one additional symbol, for when we have received only the first string c, we cannot tell whether or not this should be interpreted as a c or as the beginning of a d. Hence, no codeword is a prefix of another codeword and C has the prefix property. ∎

Example 1.4.3 Since the code C_2 of Example 1.3.1 has the prefix property, Theorem 1.4.1 tells us that it is instantaneous. □

Example 1.4.4   The code

$$C = \{0, 10, 110, 1110, 11110, 11111\}$$

is an example of a **comma code**. This terminology comes from the fact that the symbol 0 acts as a kind of comma, telling the receiver when a codeword ends. (The receiver can tell when the last codeword ends by its length.)

Comma codes have the prefix property, and so they are instantaneous. On the other hand, consider the code obtained by reversing the order of the bits

$$D = \{0, 01, 011, 0111, 01111, 11111\}$$

This code does not have the prefix property, and so it is not instantaneous. But it is uniquely decipherable, since any sequence of codewords can be deciphered by starting from the end of the message and first picking out all strings of 1s of length 5, which must represent the codeword 11111, then picking out strings of 1s of length 4, and so on.   □

Kraft's Theorem

Now we come to a theorem that tells us precisely when an instantaneous code exists with given codeword lengths $\ell_1, \ell_2, \ldots, \ell_q$. This theorem was first published by L. G. Kraft in 1949.

Theorem 1.4.2

1. *(Kraft's Theorem) There exists an instantaneous r-ary code $C = \{c_1, c_2, \ldots, c_q\}$, with codeword lengths $\ell_1, \ell_2, \ldots, \ell_q$, if and only if these lengths satisfy Kraft's inequality,*

$$\sum_{k=1}^{q} \frac{1}{r^{\ell_k}} \leq 1$$

2. *Let C be an instantaneous r-ary code. Then C is **maximal instantaneous**, that is, C is not contained in any strictly larger instantaneous code, if and only if equality holds in Kraft's inequality.*

3. *Suppose that C is an instantaneous code with maximum codeword length m. If C is not maximal, then it is possible to add a word of length m to C without destroying the property of being instantaneous.*

Proof Let C be a code with codeword lengths $\ell_1, \ldots, \ell_q$. We refer to the sum on the left side of Kraft's inequality as Kraft's sum. Let us begin by rewriting Kraft's inequality in a different form. Suppose that C has u_i codewords of length i for $i = 1, \ldots, m$, where m is the maximum codeword length in C. The the Kraft sum can be written

$$\sum_{k=1}^{q} \frac{1}{r^{\ell_k}} = \sum_{i=1}^{m} \frac{u_i}{r^i} = \frac{1}{r^m} \sum_{i=1}^{m} u_i r^{m-i}$$

and Kraft's inequality can be written in the form

$$\sum_{i=1}^{m} u_i r^{m-i} \leq r^m$$

or

$$u_m + \sum_{i=1}^{m-1} u_i r^{m-i} \leq r^m \qquad (1.4.1)$$

We can now prove part a). First, we show that an instantaneous code C must satisfy Kraft's inequality. Let $\mathbf{c} \in C$ have length $i \leq m - 1$. Since $\mathbf{c}$ cannot be a prefix of any other codeword in C, none of the words $\mathbf{cx}$, where $\mathbf{x}$ is a string of length $m - i$, can be in C. Since $\text{len}(\mathbf{x}) = m - i$, there are r^{m-i} words of the form $\mathbf{cx}$, all of which must be excluded from C. Moreover, if $\mathbf{d}$ is another codeword, say of length j, then there are an additional r^{m-j} words of length m that must also be excluded from C. (If $\mathbf{cx} = \mathbf{dy}$ for $\mathbf{c} \neq \mathbf{d}$, then one of $\mathbf{c}$ or $\mathbf{d}$ is a prefix of the other, which is not possible.) Thus, the total number of excluded words of length m is precisely equal to the summation on the left side of (1.4.1). Adding the number u_m of words of length m that are in C must result in a number that is no greater than the total number r^m of words of length m. Hence, (1.4.1) holds.

For the converse of part a), we must show that if $\ell_1, \ell_2, \ldots, \ell_q$ satisfy Kraft's inequality, then there is an instantaneous code C with these codeword lengths. This can be proved by induction on the number q. If q is less than or equal to the radix r, then taking distinct code symbols gives an instantaneous code of size q, all of whose codewords have length 1. Certainly, we can extend the length of each codeword to get codewords of lengths $\ell_1, \ldots, \ell_q$, whilst preserving the prefix property. Hence, the result is true for $q \leq r$. Now assume that it is true for all sets of q or fewer lengths, and let $\ell_1, \ldots, \ell_{q+1}$ be $q + 1$ lengths that satisfy Kraft's inequality. We assume, by renumbering if necessary, that $\ell_i \leq \ell_{i+1}$ for all $i = 1, \ldots, q$.

By the induction hypothesis, there is an instantaneous code C of size q with codeword lengths $\ell_1, \ldots, \ell_q$. Moreover, these numbers give strict inequality in Kraft's inequality (1.4.1). Hence, the reasoning that led to (1.4.1) shows that there is at least one word $\mathbf{d}$ of length ℓ_q that has not been included in C, but is also not excluded by virtue of having a codeword in C as prefix. It follows that we may include this word in C and still have an instantaneous code. Lengthening $\mathbf{d}$ (if necessary) by adjoining 0s to the right end will give a codeword of length ℓ_{q+1} without violating the prefix property and so we have an instantaneous code with codeword lengths $\ell_1, \ldots, \ell_{q+1}$.

For part b), suppose first that the codeword lengths of an instantaneous code C give equality in Kraft's inequality. If we add any word to C, the resulting code would not satisfy Kraft's inequality, which implies by what we have just proved above that it cannot be instantaneous. Hence, C is maximal instantaneous.

For the converse, we must show that if a code C is maximal instantaneous, then equality holds in Kraft's inequality. But this follows easily by looking at Kraft's inequality in the form (1.4.1). For if equality does not hold in Kraft's inequality, then the left side of (1.4.1) is less than r^m and so there is at least one word of length m that has not been included in C but is also not excluded by virtue of having a codeword as prefix. Hence, we may include this word in C and still have an instantaneous code. Thus, if C is maximal, equality must hold in Kraft's inequality. This finishes the proof of part b) and also proves part c). ∎

Note that Kraft's Theorem says that, if the lengths $\ell_1, \ell_2, \ldots, \ell_q$ satisfy Kraft's inequality, then there must exist some instantaneous code with these codeword lengths. It does not say that any code whose codeword lengths satisfy Kraft's inequality must be instantaneous. As we see in the next example, this is not necessarily the case.

Example 1.4.5 Consider the binary code $C = \{0, 11, 100, 110\}$. The codeword lengths are 1,2, and 3, and since $|A| = 2$, the left side of Kraft's inequality is

$$\frac{1}{2} + \frac{1}{2^2} + \frac{1}{2^3} + \frac{1}{2^3} = 1$$

Hence, these lengths do satisfy Kraft's inequality. Nonetheless, this code is not instantaneous, since the second codeword is a prefix of the fourth.   □

Parts b) and c) of Theorem 1.4.2 actually gives us a clue as to how to construct an instantaneous code with given codeword lengths $\ell_1, \ldots, \ell_q$. Suppose that these lengths are arranged so that $\ell_1 \le \ell_2 \le \cdots \le \ell_q$. If we have succeeded in finding $k < q$ codewords $c_1, \ldots, c_k$ with lengths $\ell_1, \ldots, \ell_k$, then the Kraft sum, using only these lengths, is strictly less than 1 and so, according to part b) of Theorem 1.4.2, the code $\{c_1, \ldots, c_k\}$ is not maximal. Hence, by part c), we may include an additional codeword c of length ℓ_k or greater, in particular, of length ℓ_{k+1}. The point is that we may add any codeword of length ℓ_{k+1} as long as it does not violate the prefix property, for then we can repeat the process until we have a code of size q. Here is an example.

Example 1.4.6 Let $A = \{0, 1, 2\}$ and let $\ell_1 = 1, \ell_2 = 1, \ell_3 = 2, \ell_4 = 4, \ell_5 = 4, \ell_6 = 5$. Kraft's inequality is satisfied

$$\frac{1}{3} + \frac{1}{3} + \frac{1}{3^2} + \frac{1}{3^4} + \frac{1}{3^4} + \frac{1}{3^5} = \frac{3^4 + 3^4 + 3^3 + 3 + 3 + 1}{3^5} = \frac{196}{243} < 1$$

and so there exists an instantaneous code C over A with these codeword lengths.

First, we choose the two codewords of the smallest length 1, say

$$c_1 = 0 \text{ and } c_2 = 1$$

Then we choose any codeword c_3 of the next smallest length 2 that does not cause the prefix property to be violated. Hence, c_2 cannot start with 0 or 1. Let us choose

$$c_3 = 20$$

Next we choose any two codewords of length 4 that begin with 2, but not with 20. Let us choose

$$c_4 = 2100 \text{ and } c_5 = 2101$$

Finally, we choose any codeword of length 5, not beginning with any previously chosen codeword. We may pick

$$c_6 = 21100$$

Thus, $C = \{0, 1, 20, 2100, 2101, 21100\}$. Of course, this process is by no means unique and there are other instantaneous codes with these codeword lengths. □

Kraft's Theorem and McMillan's Theorem together tell us something interesting about the relationship between uniquely decipherable codes and instantaneous codes. In particular, if there exists a uniquely decipherable code with codeword lengths $\ell_1, \ell_2, \ldots, \ell_n$, then according to McMillan's Theorem, these lengths must satisfy Kraft's inequality. But then we may apply Kraft's Theorem to conclude that there must also exist an instantaneous code with these lengths. In summary, we have the following remarkable theorem.

Theorem 1.4.3 *If a uniquely decipherable code exits with codeword lengths* $\ell_1, \ell_2, \ldots, \ell_n$, *then an instantaneous code must also exist with these same codeword lengths.* □

Our interest in this theorem will come later, when we turn to the question of finding desirable codes with the shortest possible codeword lengths. For it tells us that we lose nothing by considering only instantaneous codes (rather than all uniquely decipherable codes).

Let us conclude with another application of Kraft's Theorem.

Example 1.4.7 Let $A = \{0, 1\}$. Suppose that we want an instantaneous code C that contains the codewords 0, 10, and 110. How many additional codewords of length 5 could be added to this code?

Since $|A| = 2$, the three aforementioned codewords contribute

$$\frac{1}{2} + \frac{1}{2^2} + \frac{1}{2^3} = \frac{7}{8}$$

to the sum on the left side of Kraft's inequality. Thus, we have $\frac{1}{8}$ left to work with, so to speak. Now, a codeword of length 5 will contribute $\frac{1}{2^5} = \frac{1}{32}$ to the Kraft sum, and so we cannot add more than four such codewords, since $4 \cdot (\frac{1}{32}) = \frac{1}{8}$. (We may not be able to add as many as four codewords, but we cannot add more than 4.) Checking the possibilities shows that each codeword of length 4 must begin with 111. This leads us to the only possibilities, namely, 11100, 11101, 11110, and 11111. It is not hard to check that we may add these four codewords to our code, that is, that the code

$$\{0, 10, 110, 11100, 11101, 11110, 11111\}$$

is instantaneous.   □

Exercises

1. Show that if a code is instantaneous, then it is also uniquely decipherable.

2. Is the code $C = \{0, 10, 1100, 1101, 1110, 1111\}$ instantaneous?

3. Is the code $C = \{0, 10, 110, 1110, 1011, 1101\}$ instantaneous?

4. Can a block code fail to have the prefix property? Explain.

5. Can you construct an instantaneous binary code with codewords 0,10 and an additional nine codewords of length 5? Explain.

6. Find an example of a binary code that is uniquely decipherable but not instantaneous, different from any of the codes in the book.

7. How many prefixes does a word of length n have?

In Exercises 8 through 14, determine whether or not there is an instantaneous code with given radix r and codeword lengths. If so, construct such a code.

8. $r = 2$, lengths 1,2,3,3

9. $r = 2$, lengths 1,2,2,3,3

10. $r = 2$, lengths 1,3,3,3,4,4

11. $r = 2$, lengths 2,2,3,3,4,4,5,5

12. $r = 3$, lengths 1,1,2,2,3,3,3

13. $r = 5$, lengths 1,1,1,1,2,2,2,2,3,3,3,4,4,4

14. $r = 5$, lengths 1,1,1,1,1,8,9

15. Suppose that we want an instantaneous binary code that contains the codewords 0, 10, and 1100. How many additional codewords of length 6 could be added to this code? Construct a code with these additional codewords?

16. Suppose that $\ell_1, \ldots, \ell_q$, and r give equality in Kraft's inequality. Let C be an instantaneous r-ary code with these codeword lengths. If $L = \max\{\ell_i\}$, show that C must contain at least two codewords of maximum length L.

I

PART

Information Theory

2 | Efficient Encoding

2.1 Information Sources; Average Codeword Length

In order to achieve unique decipherability, McMillan's Theorem tells us that we must allow reasonably long codewords. Unfortunately, this tends to reduce the efficiency of a code, by requiring longer strings to encode a given amount of data.

On the other hand, it is often the case that not all source symbols occur with the same frequency within a given class of messages. Thus, it makes sense to assign the longer codewords to the less frequently used source symbols, thereby reducing the average number of code symbols per source symbol, and improving the efficiency of the code.

Our plan in this chapter is to construct a certain class of instantaneous encoding schemes that are the most efficient possible among all instantaneous encoding schemes, in a sense that we shall now make precise. (An encoding scheme is instantaneous if the corresponding code is instantaneous.) To this end, we will assume that each source symbol has associated to it a probability of occurrence. This leads us to make the following definition.

Definition An **information source** (or simply **source**) is an ordered pair $\mathcal{S} = (S, \mathcal{P})$, where $S = \{s_1, s_2, \ldots, s_q\}$ is a source alphabet, and $\mathcal{P}$ is a probability law that assigns to each element s_i of S a probability $\mathcal{P}(s_i)$. The sequence $\mathcal{P}(s_1), \ldots, \mathcal{P}(s_q)$ is the **probability distribution for S**.   □

Often we will be interested only in the probability distribution of the source, which we may simply write in the form $P = \{p_1, \ldots, p_q\}$.

A source can be thought of as a "black box" that emits source symbols, one at a time, to form a message. We will assume that the emission of source symbols is independent of time. In other words, the fact that a given source symbol is emitted at a given instant has no effect on which source symbol will be emitted at any other instant.

As a measure of the efficiency of an encoding scheme, we use the average codeword length.

Definition   Let $\mathcal{S} = (S, \mathcal{P})$ be an information source, and let (C, f) be an encoding scheme for $S = \{s_1, \ldots, s_q\}$. The **average codeword length** of (C, f) is

$$\sum_{i=1}^{q} \mathrm{len}(f(s_i))\mathcal{P}(s_i)$$ □

Example 2.1.1   Consider the source alphabet $S = \{a,b,c,d\}$, with probabilities of occurrence

$$\mathcal{P}(a) = \frac{2}{17}, \quad \mathcal{P}(b) = \frac{2}{17}, \quad \mathcal{P}(c) = \frac{8}{17}, \quad \mathcal{P}(d) = \frac{5}{17}$$

Consider also the two encoding schemes shown below

Scheme 1	**Scheme 2**
a → 11	a → 01010
b → 0	b → 00
c → 100	c → 10
d → 1010	d → 11

We have

$$\text{Average length for scheme 1} = 2 \cdot \frac{2}{17} + 1 \cdot \frac{2}{17} + 3 \cdot \frac{8}{17} + 4 \cdot \frac{5}{17} = \frac{50}{17}$$

and

$$\text{Average length for scheme 2} = 5 \cdot \frac{2}{17} + 2 \cdot \frac{2}{17} + 2 \cdot \frac{8}{17} + 2 \cdot \frac{5}{17} = \frac{40}{17}$$

Thus, encoding scheme 2 has a smaller average codeword length. In this sense, it is more efficient than scheme 1.   □

Example 2.1.2 Table 2.1.1 (see next page) shows the letters of the alphabet and the space character, along with approximate probabilities of occurrence in the English language, based on statistical data. The last three columns of the table show three different encoding schemes.

The first scheme is a simple fixed length code, using the first 27 binary numbers. In this case, the codewords all have length 5, which is the minimum possible codeword length for a fixed length code (since $2^4 < 27 \leq 2^5$). Thus, the average codeword length of this scheme is 5.

The second scheme uses a comma code, discussed in Section 1.4. We have not written out all of the codewords, since their lengths become rather large. (The last two codewords have length 26.) Calculation gives an average codeword length of approximately 7.0607 for this scheme. Hence, the fixed length code is more efficient.

The third scheme is the Huffman encoding scheme. As we will see, Huffman encoding produces the most efficient scheme, in the sense of having the smallest average codeword length, among all instantaneous codes. In this case, a computation shows that the average codeword length is approximately 4.1195, a savings of approximately 18% over the fixed length code.

It is worth noting that the comma code is less efficient than the fixed length code because the probabilities of occurrence are all fairly close to each other. Had the probability of occurrence of the space character, for instance, been much larger compared to the other probabilities, then the comma code would have been more efficient than the fixed length code (but not the Huffman code).   □

Exercises

1. Let $S = \{a,b,c,d,e\}$ and $\mathcal{P}(a) = \frac{1}{5}$, $\mathcal{P}(b) = \frac{1}{5}$, $\mathcal{P}(c) = \frac{3}{10}$, $\mathcal{P}(d) = \frac{1}{10}$, $\mathcal{P}(e) = \frac{1}{5}$. Which scheme is more efficient

   (a) $a \rightarrow 0$, $b \rightarrow 10$, $c \rightarrow 110$, $d \rightarrow 1110$, $e \rightarrow 11110$, or

   (b) $a \rightarrow 000$, $b \rightarrow 001$, $c \rightarrow 010$, $d \rightarrow 011$, $e \rightarrow 100$?

2. Let $S = \{a,b,c,d,e,f\}$ and $\mathcal{P}(a) = 0.2$, $\mathcal{P}(b) = 0.2$, $\mathcal{P}(c) = 0.3$, $\mathcal{P}(d) = 0.1$, $\mathcal{P}(e) = 0.1$, $\mathcal{P}(f) = 0.1$. Which scheme is more efficient

TABLE 2.1.1

Symbol	Probability	Block Code	Comma code	Huffman code
(Space)	0.1859	00000	0	111
E	0.1031	00001	10	010
T	0.0796	00010	110	1101
A	0.0642	00011	1110	1011
O	0.0632	00100	11110	1001
I	0.0575	00101	111110	0111
N	0.0574	00110	.	0110
S	0.0514	00111	.	0011
R	0.0484	01000	.	0010
H	0.0467	01001		0001
L	0.0321	01010		10101
D	0.0317	01011		10100
U	0.0228	01100		00001
C	0.0218	01101		00000
F	0.0208	01110		110011
M	0.0198	01111		110010
W	0.0175	10000		110001
Y	0.0164	10001		100011
P	0.0152	10010		100010
G	0.0152	10011		100001
B	0.0127	10100		100000
V	0.0083	10101		1100000
K	0.0049	10110		11000011
X	0.0013	10111		1100001011
Q	0.0008	11000		1100001010
J	0.0008	11001	$111\cdots 10$	1100001001
Z	0.0005	11010	$111\cdots 11$	1100001000

(a) $a \to 0$, $b \to 10$, $c \to 110$, $d \to 1110$, $e \to 11110$, $f \to 111110$, or

(b) $a \to 000$, $b \to 001$, $c \to 010$, $d \to 011$, $e \to 100$, $f \to 101$?

3. Assuming a source with a uniform probability distribution, what is the average codeword length of a comma code with ten codewords?

4. Assuming a source with a uniform probability distribution, what is the average codeword length of a comma code with n codewords?

5. How do we minimize the average codeword length of an encoding scheme for a source with a uniform probability distribution?

2.2 Huffman Encoding

In 1952, D. A. Huffman published a method for constructing highly efficient instantaneous encoding schemes. This method is now known as *Huffman encoding*. Before giving an example of Huffman encoding, let us state the reason why this type of encoding is so important. (A proof will be given in the next section.)

Theorem 2.2.1 *Let* $S = (S, \mathcal{P})$ *be an information source. Then all Huffman encoding schemes for* S *are instantaneous. Furthermore, Huffman encoding schemes have the smallest average codeword length among all instantaneous encoding schemes for* S. $\square$

The minimum average codeword length, taken over all uniquely decipherable r-ary encoding schemes for S will be denoted by MinAveCode-Len$_r(S)$. According to Theorem 1.4.3, this is the same as the minimum, taken over all instantaneous encoding schemes. We also denote the average codeword length of any r-ary Huffman encoding scheme for S by AveCodeLenHuff$_r(S)$. Then Theorem 2.2.1 can be summarized by writing

$$\text{AveCodeLenHuff}_r(S) = \text{MinAveCodeLen}_r(S)$$

Now let us give an example of Huffman encoding. Although r-ary Huffman encoding schemes can be constructed for all $r \geq 2$, we will restrict attention to binary Huffman codes. (For information on nonbinary Huffman encoding, we refer the reader to *Coding and Information Theory*, by this author. See also the exercises for this chapter.)

Before reading this example, you should familiarize yourself with the terminology on binary trees in Section 0.1.

Example 2.2.1 Consider the source alphabet and probabilities shown below.

Symbol	Probability
a	0.35
b	0.10
c	0.19
d	0.25
1	0.06
2	0.05

The Huffman encoding scheme is constructed by constructing a complete binary tree as follows.

Step 1   Place each symbol inside a node. Then label each node with the probability of occurrence of the symbol and arrange the nodes in order of increasing probability of occurrence.

Step 2   Connect the two leftmost nodes to a new node, as shown below. Label the new node with the sum of the probabilities associated to the original nodes. Lower this portion of the figure so that the new node is at the top row.

Step 3   Repeat the process of arranging the figure so that the nodes on the top level are in increasing order of probabilities, and then connecting the two leftmost nodes, until only one node remains on the top row. Here are the steps required in this case.

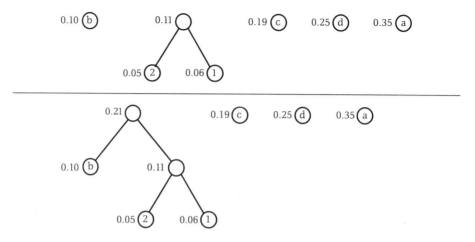

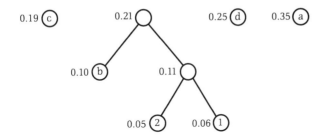

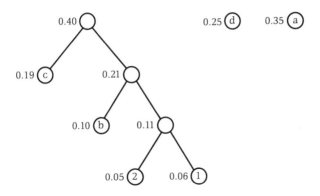

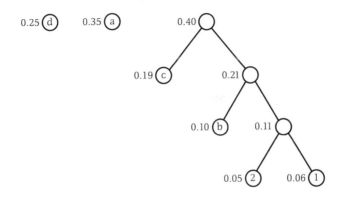

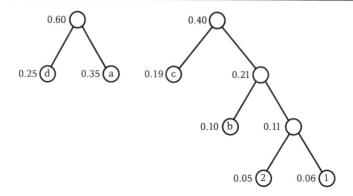

Step 4   Discard all of the probabilities, and label each line segment that slants up (from left to right) with a 0 and each line segment that slants down (from left to right) with a 1. This is done in Figure 2.2.2. The result is referred to as a **Huffman tree**.

To determine the codeword associated to each source symbol, start at the root and write down the sequence of bits encountered en route to the

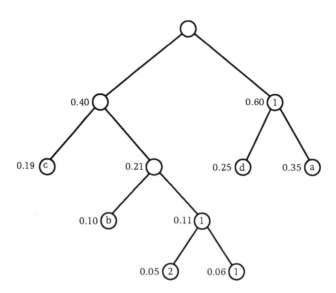

FIGURE 2.2.1

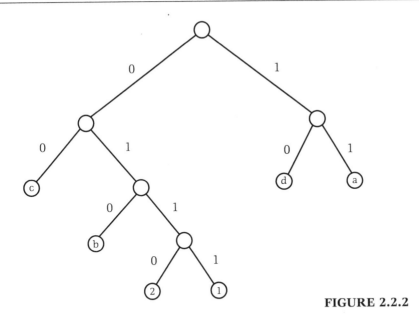

FIGURE 2.2.2

source symbol. In this case, the Huffman encoding is

Source Symbol	Code
a	11
b	010
c	00
d	10
1	0111
2	0110

We will leave it as an exercise to verify that this is an instantaneous code, whose average codeword length is 2.32.

Notice that a binary fixed length code would require codewords of length at least 3 to encode 6 source symbols ($2^2 < 6 \leq 2^3$). Hence, the average codeword length of a fixed length code is 3, and the Huffman code reduces the average codeword length by 22.7%.   □

We should remark that the Huffman encoding scheme need not be unique for a given source S. This is due to the ambiguity that occurs when two nodes in the top row have the same probability. Nevertheless, all Huffman encoding schemes for S have the same average codeword length which, according to Theorem 2.2.1, is the smallest among all instantaneous encoding schemes for S.

Exercises

In Exercises 1–6, find a Huffman encoding of the given probability distribution, using the source symbols $A, B, C, \ldots$ (in this order). Determine the savings over the most efficient fixed length code.

1. $P = \{0.1, 0.2, 0.4, 0.2, 0.1\}$

2. $P = \{0.25, 0.25, 0.25, 0.24, 0.01\}$

3. $P = \{0.1, 0.2, 0.4, 0.1, 0.1, 0.1\}$

4. $P = \{0.05, 0.1, 0.55, 0.05, 0.1, 0.1, 0.05\}$

5. $P = \{0.1, \ldots, 0.1\}$

6. $P = \{0.9, 0.09, 0.009, 0.0009, 0.0001\}$

7. Write a computer program to implement Huffman encoding.

8. State a condition in terms of the sizes of the probabilities that guarantee uniqueness (up to switching 0s and 1s) in Huffman encoding.

9. Determine all source probability distributions $\{p_1, p_2, p_3, p_4\}$ that have $\{00, 01, 10, 11\}$ as Huffman codewords. Hint: think about the Huffman tree.

10. Let (C, f) be a binary Huffman encoding and suppose that the codeword c_i has length ℓ_i for $i = 1, \ldots, k$. Prove that equality holds in Kraft's inequality, that is,

$$\sum \frac{1}{2^{\ell_i}} = 1$$

Hint: Show that there does not exist an instantaneous code D whose codeword lengths m_i satisfy $m_i \leq \ell_i$ for all i and $m_j < \ell_j$ for some j. How does this cause a problem if the Kraft sum is strictly less than 1?

11. (**Huffman codes of radix >2.**) When the radix is greater than 2, Huffman encoding proceeds in a manner entirely analogous to the case $r = 2$, with one exception. In each step, we want to group those r nodes on the top level with smallest probabilities together into a single node, which is labeled with the sum of these r probabilities. However, at the penultimate step, we want exactly r nodes, before combining them into the root node. Thus, the first step may require that we combine fewer than r nodes. Determine the correct number

of nodes to combine into a single node on the first step, so that we may combine r nodes into one on each of the subsequent steps and have exactly r nodes at the penultimate step. How many reduction steps are necessary to complete the Huffman tree?

12. Let (C, f) be a binary Huffman encoding. Let $L = max\{\ell_i\}$. Show that C must contain two codewords $\mathbf{c}$ and $\mathbf{d}$ of maximum length L with the property that they differ only in their last positions.

13. Let (C, f) be a binary Huffman encoding for the uniform probability distribution $P = \{1/n, \ldots, 1/n\}$, and suppose that the codeword lengths of C are ℓ_i.

    (a) Show that (C, f) has minimum total codeword length $T = \sum \ell_i$ among all instantaneous encodings for P.

    (b) Show that C contains two codewords $\mathbf{c}$ and $\mathbf{d}$ of maximum codeword length and that $\mathbf{c}$ and $\mathbf{d}$ differ only in their last positions.

    (c) Show that $\ell_i = L$ or $\ell_i = L - 1$ for all i.

    (d) Let $n = \alpha 2^k$, where $1 < \alpha \leq 2$. Let $\mathbf{u}$ be the number of codewords of length $L - 1$ and let v be the number of codewords of length L. Determine u, v, and L in terms of α and k.

    (e) Find $\text{MinAveCodeLen}_2(\frac{1}{n}, \ldots, \frac{1}{n})$.

14. Given n source symbols and thinking in terms of using frequencies in place of probabilities (which does not affect the results of the Huffman algorithm), what are the minimum possible frequencies (frequencies must be positive integers) to produce a Huffman code with the largest possible maximum codeword length? What is this largest length?

2.3 The Proof that Huffman Encoding Is the Most Efficient

We are now ready to prove the following theorem, first stated in the previous section.

Theorem 2.3.1 *Let S be an information source. Then all Huffman encoding schemes for S are instantaneous. Furthermore, Huffman encoding schemes have the smallest average codeword length among all instantaneous encoding* *schemes for S. In symbols,*

$$\text{AveCodeLenHuff}_r(S) = \text{MinAveCodeLen}_r(S) \qquad \square$$

Proof   Again, we restrict our attention to binary ($r = 2$) codes. The fact that Huffman coding schemes are instantaneous can be seen most easily by considering the Huffman tree. If one codeword was the prefix of another, then it would be possible to get to the second codeword by traveling from the root to the first codeword, and then continuing down the tree from there. But codewords come only at the end nodes (the leaves) of the tree, and so this is not possible. Hence, Huffman encoding schemes have the prefix property, and so they are instantaneous.

Now we must show that Huffman encoding schemes have the smallest average codeword length among all instantaneous encoding schemes. Let (H, f) be a binary Huffman encoding scheme for S, and let (C, g) be any other instantaneous binary encoding scheme. Let us denote the average codeword lengths of these schemes by $\text{AveLen}(H, f)$ and $\text{AveLen}(C, g)$. Thus, we want to prove that

$$\text{AveLen}(H, f) \leq \text{AveLen}(C, g)$$

We begin by making several observations. Table 2.3.1 will set the notation.

Of course, we may assume by reordering the source symbols if necessary that

$$p_1 \geq p_2 \geq \cdots \geq p_q$$

Observation 1   We may assume that the codeword lengths for C satisfy

$$m_1 \leq m_2 \leq \cdots \leq m_q$$

TABLE 2.3.1

Source Symbol	Probability	Huffman Codeword	Huffman Length	Other Codeword	Other Length
s_1	p_1	$\mathbf{h}_1$	ℓ_1	$\mathbf{c}_1$	m_1
s_2	p_2	$\mathbf{h}_2$	ℓ_2	$\mathbf{c}_2$	m_2
s_3	p_3	$\mathbf{h}_3$	ℓ_3	$\mathbf{c}_3$	m_3
$\vdots$	$\vdots$	$\vdots$	$\vdots$	$\vdots$	$\vdots$
s_q	p_q	$\mathbf{h}_q$	ℓ_q	$\mathbf{c}_q$	m_q

For if not, then we may interchange two codewords and produce an encoding scheme with smaller average codeword length, which can be used in place of (C, g).

We may also assume that the last two codewords $\mathbf{c}_{q-1}$ and $\mathbf{c}_q$ of C are of equal length, that is, $m_{q-1} = m_q$, and differ only in their last bits. For if not, we may again replace (C, g) by an encoding scheme with smaller average codeword length. To see this, suppose that $m_{q-1} < m_q$, and consider the codeword $\mathbf{c}'_q$, obtained by removing the last bit from $\mathbf{c}_q$. Since C is instantaneous, $\mathbf{c}'_q$ is not a codeword in C and is too long to be a prefix of a codeword in C. Hence, we may replace $\mathbf{c}_q$ by the shorter word $\mathbf{c}'_q$, which will reduce the average codeword length.

Observation 2   During the initial step of Huffman's algorithm, the relative order in which we place source symbols with the same probability of occurrence has no effect on the average codeword length of the encoding scheme, for it amounts to nothing more than a relabeling of source symbols with the same probability of occurrence. Hence, we may arrange it so that the source symbols s_q and s_{q-1} occupy the first two positions on the left (in that order).

It is also clear that, since s_q and s_{q-1} are siblings in the Huffman tree, the codewords $\mathbf{c}_q$ and $\mathbf{c}_{q-1}$ have the same length and, in fact, differ only in their last bits. Hence $\ell_q = \ell_{q-1}$. (See the previous exercise set.)

With these observations in mind, a proof can be constructed using induction on the number q of source symbols. If $q = 2$, then the Huffman encoding scheme has average codeword length 1, and so

$$\text{AveLen}(H, f) \leq \text{AveLen}(C, g)$$

Let us assume that the result is true for all source alphabets of size $q - 1$, and then prove that it is also true for source alphabets of size q.

Consider the Huffman scheme (H, f) for the source $\mathcal{S}$. For purposes of induction, we form a new source $\mathcal{S}'$ by replacing s_q and s_{q-1} with a single source symbol s, with probability of occurrence $p_q + p_{q-1}$. This gives us a source alphabet $\mathcal{S}' = \{s_1, s_2, \ldots, s_{q-2}, s\}$ of size $q - 1$. To determine the average codeword length for a Huffman encoding (H', f') of this source, observe that, at the second step in the encoding of the original source $\mathcal{S}$, we connect the two leftmost nodes, to get

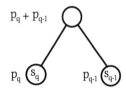

From this point on, we will get the same tree by replacing this "mini-tree" with a single node labeled s, with probability $p_q + p_{q-1}$ and then putting the minitree back at the end. Hence, we can encode the new source $\mathcal{S}'$ by encoding the original source and then replacing the minitree above by the single node s.

Removing this minitree causes the removal of the two nodes for s_q and s_{q-1}, and subtracts

$$\ell_{q-1} p_{q-1} + \ell_q p_q$$

from the average codeword length. But the inclusion of the node for s adds

$$(\ell_{q-1} - 1)(p_{q-1} + p_q)$$

to the average codeword length. Hence, there is a net change of (since $\ell_q = \ell_{q-1}$)

$$(\ell_{q-1} - 1)(p_{q-1} + p_q) - (\ell_{q-1} p_{q-1} + \ell_q p_q) = -(p_{q-1} + p_q)$$

Thus

$$\text{AveLen}(H', f') = \text{AveLen}(H, f) - (p_{q-1} + p_q) \qquad (2.3.1)$$

Now consider the encoding scheme (C, g). The last two codewords of C have the form

$$\mathbf{c}_q = x_1 x_2 \cdots x_u 0$$

and

$$\mathbf{c}_{q-1} = x_1 x_2 \cdots x_u 1$$

Since C is instantaneous, the string $\mathbf{c} = x_1 x_2 \cdots x_u$ is not a codeword in C and so we may encode the new source $\mathcal{S}'$ using $\mathbf{c}$ as the codeword for the source symbol s. This results in a net change in average codeword length of (because $m_{q-1} = m_q$)

$$(m_{q-1} - 1)(p_{q-1} + p_q) - (m_{q-1} p_{q-1} + m_q p_q) = -(p_{q-1} + p_q)$$

which is the same as in the Huffman case. Hence,

$$\text{AveLen}(C', g') = \text{AveLen}(C, g) - (p_{q-1} + p_q) \qquad (2.3.2)$$

But, the induction hypothesis implies that

$$\text{AveLen}(H', f') \leq \text{AveLen}(C', g')$$

and this, together with (2.3.1) and (2.3.2), shows that

$$\text{AveLen}(H, f) \leq \text{AveLen}(C, g)$$

as desired.   ∎

We should conclude by making a few remarks on how Huffman encoding is implemented. Given a message to encode, one does not usually know ahead of time the proper probabilities of occurrence. Using Table 2.1.1 on the relative frequencies of letters in the English language (compiled statistically) may be far from ideal for a given message, and can even lead to a lengthening of some messages.

Accordingly, a static approach is to first scan the message and compile a table of frequencies for each source symbol. Since the Huffman algorithm is not affected by a proportional scaling of the probabilities, these frequences can be used in place of probabilities. (Scaling the frequencies by dividing by their sum would, of course, yield a probability distribution but, in practice, these numbers must be stored in a computer and roundoff errors may actually change the shape of the tree.)

The static approach provides a probability distribution (or set of frequencies) that statistically models the given message, but has the disadvantage that the table of frequencies (or the code itself) must also be transmitted along with the message, for decoding purposes.

A more efficient approach is to use dynamic (also called adaptive) Huffman encoding, which involves scanning the message only once and constantly updating the frequency information after each symbol is encoded. Thus, the probability model (via frequencies) is constantly changing. It is important to emphasize that this approach is entirely outside the theoretical scope of our discussion (both previous and forthcoming), where we assume that a fixed probability distribution is given.

Exercises

1. A complete binary tree is said to be **weighted** if a) each node has a number associated with it, called a **weight**, and b) the weight of a node that is not a leaf is the sum of the weights of its two children. If we divide each weight in a weighted binary tree by the sum of all of the weights, the result is a weighted binary tree the sum of whose weights is 1. Let us refer to such a binary tree as a **normalized weighted complete binary tree**. Prove that a graph G is a Huffman tree (that is, comes from an application of the Huffman algorithm applied to some source) if and only if

   (a) G is a normalized weighted complete binary tree, and

   (b) as we scan the weights of the nodes, going from left to right and starting on the bottom level and proceeding upward through the levels, the weights fall in increasing order by size.

3

CHAPTER

Noiseless Coding

3.1 Entropy

The results of the previous chapter show that Huffman encoding schemes are the most efficient, in the sense of having the smallest average codeword length, among all instantaneous encoding schemes. Our goal in this chapter is to determine just how efficient such an encoding scheme can be. We will see that, to every source $\mathcal{S}$, there is a number, called the *entropy* of $\mathcal{S}$, that has the property that the average codeword length of any instantaneous encoding scheme for $\mathcal{S}$ must be greater than or equal to the entropy of $\mathcal{S}$. In other words, the entropy provides a *lower bound* on the average codeword length of any instantaneous encoding scheme.

The entropy of a source is intended to measure in a precise way the amount of "information" in the source. In order to motivate the concept of the amount of information obtained from a source symbol, let us imagine that a contest is taking place. Each of two contestants has a "black box" that emits source symbols from a source $\mathcal{S}$ with source alphabet $S = \{s_1, s_2\}$, and probabilities $p_1 = \frac{99}{100}$, $p_2 = \frac{1}{100}$. The winner of the contest is the first one to name both source symbols, that is, the first one to have complete information about the set S. (We assume that neither contestant has seen the source symbols beforehand.)

Now, suppose that on the first round, the first contestant gets source symbol s_1, while the second contestant gets s_2. At this point, which contestant is more likely to win the contest?

Since the first contestant still needs to receive source symbol s_2, whose probability of occurrence is $\frac{1}{100}$, whereas the second contestant needs to receive s_1, whose probability of occurrence is $\frac{99}{100}$, it is clear that the second contestant is more likely to win than the first. In some sense then, the second contestant has received more *information* about S from the source symbol s_2, with the smaller probability of occurrence, than did the first contestant. This motivates the statement that, however we decide to define the information obtained from a source symbol, it should have the property that *the less likely a source symbol is to occur, the more information we obtain from an occurrence of that symbol, and conversely.*

Because the information obtained from a source symbol is not a function of the symbol itself, but rather of the symbol's probability of occurrence p, we use the notation $I(p)$ to denote the information obtained from a source symbol with probability of occurrence p. We will make the following reasonable assumptions about the function $I(p)$, defined for all $0 < p \leq 1$.

Assumption 1 $I(p) \geq 0$

Assumption 2   The function $I(p)$ is continuous in p.

Since we assume that the events of s_i and s_j occurring (on different transmissions) are independent, the information obtained from the knowledge that both s_i and s_j have occurred should be the sum of $I(p_i)$ and $I(p_j)$. Since the probability of both events occurring is the product $p_i p_j$, we get

Assumption 3 $I(p_i p_j) = I(p_i) + I(p_j)$

The remarkable fact about these three assumptions is that there is essentially only one function that satisfies them.

Theorem 3.1.1   *A function $I(p)$, defined for all $0 < p \leq 1$, satisfies the previous three assumptions if and only if it has the form*

$$I(p) = C \lg \frac{1}{p}$$

where C is a positive constant and $\lg$ is the logarithm base 2. $\square$

Proof   We leave it as an exercise to show that any function of this form satisfies all three assumptions. For the converse, observe first that, by assumption 3,

$$I(p^2) = I(p \cdot p) = I(p) + I(p) = 2I(p)$$

and similarly,

$$I(p^3) = I(p^2 \cdot p) = I(p^2) + I(p) = 3I(p)$$

In general, for any positive integer n,

$$I(p^n) = nI(p) \tag{3.1.1}$$

a statement that can be proved formally by induction. Replacing p by $p^{1/n}$ gives

$$I(p^{1/n}) = \frac{1}{n}I(p) \tag{3.1.2}$$

Since (3.1.1) and (3.1.2) hold for all positive integers n, we have

$$I(p^{n/m}) = \frac{1}{m}I(p^n) = \frac{n}{m}I(p)$$

that is,

$$I(p^q) = qI(p)$$

for all positive rational numbers q.

Since, for any positive real number r, there is a sequence of positive rational numbers q_n for which $\lim_{n \to \infty} q_n = r$, and thus $\lim_{n \to \infty} p^{q_n} = p^r$, the continuity of $I(p)$ implies that

$$I(p^r) = I(\lim_{n \to \infty} p^{q_n}) = \lim_{n \to \infty} I(p^{q_n}) = I(p) \lim_{n \to \infty} q_n = rI(p)$$

Now let us fix a value of p for which $0 < p < 1$. Since any q satisfying $0 < q < 1$ can be written in the form $q = p^{\log_p q}$, we have

$$I(q) = I(p^{\log_p q}) = I(p) \log_p q = C \lg \frac{1}{q}$$

for some constant $C > 0$. Finally, the continuity of the information function gives $I(1) = 0$.   ∎

Since the arbitrary multiplicative constant can be absorbed in the units of measurement of information, the previous theorem justifies the following definition.

Definition   The **information** $I(p)$ obtained from a source symbol s with probability of occurrence $p > 0$, is given by

$$I(p) = \lg \frac{1}{p}$$

where lg is the base 2 logarithm.   □

When it is convenient, we will also use the notation $I(s)$ for the information obtained from the source symbol s. However, it is important to keep in mind that the information $I(s)$ depends only on the probability of occurrence of s.

The unit of measurement of information is the bit, which is a contraction of binary unit. The connection between the binary unit and the binary digit (also abbreviated bit) comes from the following observation. If the source is $S = \{0, 1\}$, $P = \{\frac{1}{2}, \frac{1}{2}\}$, then the information given by either source symbol is $I(\frac{1}{2}) = \lg 2 = 1$. In other words, if the source randomly emits 1 binary digit (bit), then the information obtained by a single emission is 1 binary unit (bit).

Example 3.1.1   A personal computer monitor is capable of displaying pictures made up of pixels at a resolution of 1024 columns by 768 rows (and higher). Hence, if each pixel can be in any one of $256 = 2^8$ colors, there are a total of $2^{8 \times 1024 \times 768} = 2^{6291456}$ different pictures. If each of these pictures is considered to be equally likely, the probability of a given picture occurring is $2^{-6291456}$, and so the information obtained from a single picture is

$$I = \lg 2^{6291456} = 6,291,456 \text{ bits}$$

On the other hand, let us estimate the information obtained from a random speech of 1000 words. (While it is true that most people do not speak in random sequences of words, politicians often do, for example.) A 10,000 word vocabulary would be considered quite excellent (in fact, quite amazing), and the probability of speaking a given sequence of 1000 words from such a vocabulary is 10000^{-1000}. Hence, the amount of information obtained by such a speech is

$$I = \lg 10000^{1000} = 1000 \lg 10000 < 14,000 \text{ bits}$$

This proves that a picture is worth more than a thousand words!   □

We can now define the concept of entropy.

Definition Let $\mathcal{S} = (S, \mathcal{P})$ be a source, with probability distribution $P = \{p_1, \ldots, p_q\}$. The average information obtained from a single sample from $\mathcal{S}$ is

$$H(\mathcal{S}) = \sum_{i=1}^{q} p_i I(p_i) = \sum_{i=1}^{q} p_i \lg \frac{1}{p_i} = -\sum_{i=1}^{q} p_i \lg p_i$$

The quantity $H(\mathcal{S})$ is called the **entropy** of the source. (When $p_i = 0$, we set $p_i \lg \frac{1}{p_i} = 0$.) Since this quantity depends only on the probability distribution P (and not on the source alphabet S), we also use the notations $H(P)$ and $H(p_1, \ldots, p_n)$ for the entropy. $\square$

Example 3.1.2 Consider a source $\mathcal{S}_1 = (S_1, \mathcal{P}_1)$ for which each source symbol is equally likely to occur, that is, for which $\mathcal{P}_1(s_i) = p_i = 1/q$, for all $i = 1, 2, \ldots, q$. Then

$$H(\mathcal{S}_1) = H(\frac{1}{q}, \ldots, \frac{1}{q}) = \sum_{i=1}^{q} p_i \lg \frac{1}{p_i} = \sum_{i=1}^{q} \frac{1}{q} \lg q = \lg q$$

On the other hand, for a source $\mathcal{S}_2 = (S_2, \mathcal{P}_2)$, where $p_1 = 1$ and $p_i = 0$ for all $i = 2, 3, \ldots, q$, we have

$$H(\mathcal{S}_2) = H(1, 0, \ldots, 0) = p_1 \lg \frac{1}{p_1} = 0 \qquad \square$$

The previous example illustrates why the entropy of a source is often thought of as a measure of the amount of uncertainty in the source. The source $\mathcal{S}_1$, which emits all symbols with equal probability, is in a much greater state of uncertainty than the source $\mathcal{S}_2$, which always emits the same symbol. Thus, the greater the entropy, the greater the uncertainty in each sample and the more information is obtained from the sample. (The term disorder is also used in this context.)

Example 3.1.3 If $\mathcal{S} = (S, \mathcal{P})$, where $S = \{s_1, s_2, s_3\}$ and

$$\mathcal{P}(s_1) = \frac{1}{4}, \mathcal{P}(s_2) = \frac{1}{4}, \mathcal{P}(s_3) = \frac{1}{2}$$

then the entropy is

$$H(\mathcal{S}) = H(\frac{1}{4}, \frac{1}{4}, \frac{1}{2}) = \frac{1}{4} \lg 4 + \frac{1}{4} \lg 4 + \frac{1}{2} \lg 2 = 1.5$$

This is compared to an entropy of $\lg 3 = 1.585$ for a source of size 3 where each source symbol is equally likely. $\square$

Example 3.1.4 The entropy of the source $S = (S, \mathcal{P})$ where $S = \{s_1, s_2, s_3\}$ and

$$\mathcal{P}(s_1) = \frac{1}{2}, \mathcal{P}(s_2) = \frac{1}{2}, \mathcal{P}(s_3) = 0$$

is

$$H(S) = H(\frac{1}{2}, \frac{1}{2}, 0) = \frac{1}{2}\lg 2 + \frac{1}{2}\lg 2 + 0 = \lg 2 = 1$$

This is the same as the entropy of a two-symbol source, each of whose symbols is equally likely. This example illustrates the fact that the addition of a source symbol (or symbols) that cannot occur does not effect the amount of information obtained from a sampling of the source.　□

Example 3.1.5 The first two columns of Table 2.1.1 show the information source associated with the letters of the alphabet used in the English language. A computation shows that the entropy for this source is approximately 4.07991. Thus, one gets an average of 4.07991 bits of information by sampling a single letter from English text.

Note that the average codeword length for the Huffman encoding scheme in Table 2.1.1 is approximately 4.1195 bits and so there is a small amount of additional information in the Huffman code beyond what is contained in the source itself. Recall also that no other instantaneous binary code can do better in terms of average codeword length.　□

Exercises

1. Compute the entropy of the probability distribution $\{\frac{1}{3}, \frac{2}{3}\}$.

2. Compute the entropy of the probability distribution $\{\frac{1}{8}, \frac{1}{8}, \frac{3}{4}\}$.

3. Compute the entropy of the probability distribution $\{\frac{1}{a}, \frac{1}{a}, \ldots, \frac{2}{a}, \frac{2}{a}\}$ where $a \geq 5$ is an integer.

4. Show that any function of the form $I(p) = C\lg\frac{1}{p}$ satisfies all three assumptions for the entropy function.

5. When is the entropy $H(S)$ of a source equal to 0?

6. Suppose a fair coin is tossed and if the outcome is a heads, we toss it again. How much information do we get if the final outcome is a heads? A tails? How much uncertainty is there in the final outcome?

7. Suppose we toss a fair coin and roll a fair die. Do we get more information (on the average) from this experiment or from the experiment of tossing three fair coins? Four fair coins?

8. How much information do we get (on the average) by sampling from a deck of cards if

   (a) each card is equally likely to be drawn?

   (b) the black cards are twice as likely to be drawn as the red cards?

9. Suppose that we roll a fair die that has two faces numbered 1, two faces numbered 2, and two faces numbered 3. Then we toss a fair coin the number of times indicated by the number on the die and count the number of heads. How much information is obtained (on the average) by this procedure?

3.2 Properties of Entropy

In Example 3.1.2, we saw that the entropy of a source S_1 of size q with uniform probability distribution is equal to $\lg q$, and that the entropy of a source S_2 where one symbol has probability of occurrence 1 is equal to 0. These are the two "extreme" cases for the value of the entropy of any source. In other words, the entropy satisfies $0 \le H(S) \le \lg q$ for all sources of size q.

In order to prove this fact, we must first establish some preliminary results concerning logarithms, whose proofs are left as exercises.

Lemma 3.2.1

1. *If* $\ln$ *denotes the natural logarithm then, for all* $x > 0$,

$$\ln x \le x - 1$$

2. *If* $\lg$ *denotes the logarithm base 2 then, for all* $x > 0$,

$$\lg x \le \frac{x - 1}{\ln 2}$$

In both cases, equality holds if and only if $x = 1$. $\qquad\square$

Lemma 3.2.2 *Let* $P = \{p_1, p_2, \ldots, p_q\}$ *be a probability distribution. Let* $R = \{r_1, r_2, \ldots, r_q\}$ *have the property that* $0 \leq r_i \leq 1$ *for all i, and*

$$\sum_{i=1}^{q} r_i \leq 1$$

(Note the inequality here.) Then

$$\sum_{i=1}^{q} p_i \lg \frac{1}{p_i} \leq \sum_{i=1}^{q} p_i \lg \frac{1}{r_i}$$

with equality holding if and only if $p_i = r_i$ *for all i.* $\qquad\qquad \square$

Proof   According to Lemma 3.2.1,

$$\sum_{i=1}^{q} p_i \lg \frac{r_i}{p_i} \leq \frac{1}{\ln 2} \sum_{i=1}^{q} p_i \left(\frac{r_i}{p_i} - 1 \right)$$

$$= \frac{1}{\ln 2} \sum_{i=1}^{q} (r_i - p_i)$$

$$= \frac{1}{\ln 2} \left(\sum_{i=1}^{q} r_i - \sum_{i=1}^{q} p_i \right)$$

$$= \frac{1}{\ln 2} \left(\sum_{i=1}^{q} r_i - 1 \right) \leq 0$$

Thus

$$\sum_{i=1}^{q} p_i \lg \frac{r_i}{q_i} \leq 0$$

Writing $\lg(r_i/p_i) = \lg(1/p_i) - \lg(1/r_i)$ and rearranging gives

$$\sum_{i=1}^{q} p_i \lg \frac{1}{p_i} \leq \sum_{i=1}^{q} p_i \lg \frac{1}{r_i}$$

Finally, equality holds here if and only if it holds in Lemma 3.2.1, which happens if and only if $r_i/p_i = 1$ for all i. $\blacksquare$

With these lemmas at our disposal, we can prove the main result of this section.

Theorem 3.2.3   *For a source* $\mathcal{S} = (S, \mathcal{P})$ *of size q, the entropy* $H(\mathcal{S})$ *satisfies*

$$0 \leq H(\mathcal{S}) \leq \lg q$$

Furthermore, $H(S) = \lg q$ if and only if all of the source symbols are equally likely to occur, and $H(S) = 0$ if and only if one of the source symbols has probability 1 of occurring. □

Proof Let $P = \{p_1, \ldots, p_q\}$ be the probability distribution of S and let $R = \{1/q, 1/q, \ldots, 1/q\}$ be the uniform distribution. Applying Lemma 3.2.2 to P and R gives

$$H(S) = \sum_{i=1}^{q} p_i \lg \frac{1}{p_i} \leq \sum_{i=1}^{q} p_i \lg \frac{1}{1/q}$$

$$= \sum_{i=1}^{q} p_i \lg q = (\lg q) \sum_{i=1}^{q} p_i = \lg q$$

Thus, $H(S) \leq \lg q$. As for equality, this happens precisely when equality holds in Lemma 3.2.2, that is, when $p_i = 1/q$ for all i. Proof of the final statement is left as an exercise. ∎

Theorem 3.2.3 confirms the fact that, on the average, the most information is obtained from sources for which each source symbol is equally likely to occur.

Let us examine a bit more closely the entropy of the special **binary source** $S = \{0, 1\}$, with probability distribution of the form $P = \{p, 1-p\}$. Thus, the entropy of a binary source is

$$H(S) = p \lg \frac{1}{p} + (1-p) \lg \frac{1}{1-p}$$

The function on the right is often denoted by $H(p)$

$$H(p) = p \lg \frac{1}{p} + (1-p) \lg \frac{1}{1-p} \tag{3.2.1}$$

and called the *entropy function*. Its graph is shown in Figure 3.2.1. (Note that $p \lg(\frac{1}{p})$ is defined to be 0 when $p = 0$.) As expected, the entropy function reaches its maximum value when $p = 1 - p = 1/2$.

A final note. The definition of entropy involves base 2 logarithms, but it is sometimes convenient to use logarithms to other bases. Accordingly, for any positive integer r, we define the r-**ary entropy of a source** S by

$$H_r(S) = \sum_{i=1}^{q} p_i \log_r \frac{1}{p_i}$$

Thus, the entropy $H(S) = H_2(S)$ is the binary entropy.

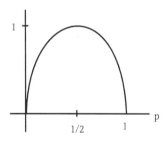

FIGURE 3.2.1   The entropy function $H(p)$

Exercises

1. Prove Lemma 3.2.1.

2. Compute the derivative of the entropy function $H(p)$ given in (3.2.1).

3. Prove that the entropy function $H(p)$ in (3.2.1) is symmetric about the line $x = \frac{1}{2}$.

4. Show that

$$H_r(\mathcal{S}) = \frac{H(\mathcal{S})}{\lg r}$$

5. Find a relationship between $H_r(\mathcal{S})$ and $H_s(\mathcal{S})$.

6. Let $P = \{p_1, \ldots, p_n\}$ be a probability distribution. Suppose that ϵ is a positive real number and that $p_1 - \epsilon > p_2 + \epsilon \geq 0$. Thus, $\{p_1 - \epsilon, p_2 + \epsilon, \ldots, p_n\}$ is also a probability distribution. Interpret the inequality

$$H(p_1, \ldots, p_n) < H(p_1 - \epsilon, p_2 + \epsilon, p_3, \ldots, p_n)$$

in words. Verify this inequality.

7. Use Lemma 3.2.2 to prove that, if $\{p_1, \ldots, p_n\}$ is a probability distribution, then

$$x_1^{p_1} \cdots x_n^{p_n} \leq p_1 x_1 + \cdots + p_n x_n$$

where $x_1, \ldots, x_n$ are positive real numbers. This says that the geometric mean of the x_i is less than or equal to the arithmetic mean. Prove that equality holds if and only if the x_i are all equal. Hint: consider the expressions $r_i = p_i x_i / \sum_j p_j x_j$.

3.3 Extensions of an Information Source

Consider the binary source $S = \{s_1, s_2\}$, with probabilities

$$\mathcal{P}(s_1) = p_1 = 0.25, \mathcal{P}(s_2) = p_2 = 0.75$$

A Huffman encoding for this source is

$$s_1 \to 0$$
$$s_2 \to 1$$

with average codeword length 1.

Rather than encoding each symbol from S, suppose we encode all strings of length two over S. In other words, consider the source with alphabet

$$S^2 = \{s_1 s_1, s_1 s_2, s_2 s_1, s_2 s_2\}$$

where the probabilities of occurrence are determined by multiplication,

$$\mathcal{P}(s_1 s_1) = p_1 p_1 = (0.25)(0.25) = 0.0625$$
$$\mathcal{P}(s_1 s_2) = p_1 p_2 = (0.25)(0.75) = 0.1875$$
$$\mathcal{P}(s_2 s_1) = p_2 p_1 = (0.75)(0.25) = 0.1875$$
$$\mathcal{P}(s_2 s_2) = p_2 p_2 = (0.75)(0.75) = 0.5625$$

The Huffman algorithm gives the encoding

$$s_1 s_1 \to 010$$
$$s_1 s_2 \to 011$$
$$s_2 s_1 \to 00$$
$$s_2 s_2 \to 1$$

This scheme has average codeword length

$$(0.0625) \cdot 3 + (0.1875) \cdot 3 + (0.1875) \cdot 2 + (0.5625) \cdot 1 = 1.6875$$

But since each codeword represents two source symbols, the average codeword length per original source symbol is $1.6875/2 = 0.84375$, which is an improvement over encoding the original source. Continuing this theme, let S^3 be the source alphabet consisting of strings of length 3 over S. Each source symbol in S^3 is assigned a probability as before. For instance,

$$\mathcal{P}(s_1 s_2 s_1) = p_1 p_2 p_1 = (0.25)(0.75)(0.25) = 0.046875$$

A Huffman encoding of this source is

$$s_1 s_1 s_1 \twoheadrightarrow 11100$$
$$s_1 s_1 s_2 \twoheadrightarrow 11101$$
$$s_1 s_2 s_1 \twoheadrightarrow 11110$$
$$s_1 s_2 s_2 \twoheadrightarrow 100$$
$$s_2 s_1 s_1 \twoheadrightarrow 11111$$
$$s_2 s_1 s_2 \twoheadrightarrow 101$$
$$s_2 s_2 s_1 \twoheadrightarrow 110$$
$$s_2 s_2 s_2 \twoheadrightarrow 0$$

which has an average codeword length of 2.46875, or an average codeword length per original source symbol of 2.46875/3 = 0.82292, which is an additional improvement over the original encoding.

From these examples, we see that it may be possible to improve the average codeword length per original source symbol by grouping source symbols to form a new source. While it is true that, in some cases, this method does not result in improvements, the method is important, and does lead, as we shall see, to significant theoretical results. This leads us to make the following definition.

Definition   Let $\mathcal{S} = (S, \mathcal{P})$ be an information source. The **nth extension** of $\mathcal{S}$ is the source $\mathcal{S}^n = (S^n, \mathcal{P}^n)$, where S^n is the set of all words of length n over S, and $\mathcal{P}^n$ is the probability distribution defined as follows. If $s = s_{i_1} s_{i_2} \cdots s_{i_n}$ is a word in S^n, then

$$\mathcal{P}^n(s) = \mathcal{P}(s_{i_1} s_{i_2} \cdots s_{i_n}) = p_{i_1} p_{i_2} \cdots p_{i_n} \qquad \square$$

The entropy of an extension $\mathcal{S}^n$ is related to the entropy of $\mathcal{S}$ in a very simple way. In fact, when we think of entropy as the average amount of information obtained per symbol, it seems intuitively clear that, since we get n times as much information from a word of length n as from a single character, the entropy of $\mathcal{S}^n$ should be n times the entropy of $\mathcal{S}$. The following theorem confirms this.

Theorem 3.3.1   *Let $\mathcal{S}$ be an information source, and let $\mathcal{S}^n$ be its nth extension. Then $H(\mathcal{S}^n) = nH(\mathcal{S})$.* $\qquad \square$

Proof The entropy of the nth extension is

$$H(\mathcal{S}^n) = \sum_{\substack{i_1,i_2,\dots,i_n \\ 0 \le i_k \le q}} p_{i_1} p_{i_2} \cdots p_{i_n} \lg \frac{1}{p_{i_1} p_{i_2} \cdots p_{i_n}}$$

The properties of logarithms give

$$H(\mathcal{S}^n) = \sum_{i_1,i_2,\dots,i_n} p_{i_1} p_{i_2} \cdots p_{i_n} \lg \frac{1}{p_{i_1}} \qquad (3.3.1)$$

$$+ \sum_{i_1,i_2,\dots,i_n} p_{i_1} p_{i_2} \cdots p_{i_n} \lg \frac{1}{p_{i_2}}$$

$$+ \cdots + \sum_{i_1,i_2,\dots,i_n} p_{i_1} p_{i_2} \cdots p_{i_n} \lg \frac{1}{p_{i_n}}$$

Now, let us look at the first of these summations

$$\sum_{i_1,i_2,\dots,i_n} p_{i_1} p_{i_2} \cdots p_{i_n} \lg \frac{1}{p_{i_1}} = \sum_{i_1=1}^{q} p_{i_1} \lg \frac{1}{p_{i_1}} \times \sum_{i_2=1}^{q} p_{i_2} \times \cdots \times \sum_{i_n=1}^{q} p_{i_n}$$

Since the sum of the probabilities p_j equals 1, this equals

$$\sum_{i_1=1}^{q} p_{i_1} \lg \frac{1}{p_{i_1}} = H(\mathcal{S})$$

Since each of the other sums in the expression (3.3.4) for $H(\mathcal{S}^n)$ is also equal to $H(\mathcal{S})$, and since there are n such sums, we get

$$H(\mathcal{S}^n) = H(\mathcal{S}) + H(\mathcal{S}) + \cdots + H(\mathcal{S}) = nH(\mathcal{S}) \qquad \blacksquare$$

Example 3.3.1 The entropy of the binary source $S = \{s_1, s_2\}$, $p_1 = 0.25$, $p_2 = 0.75$, is

$$H(\mathcal{S}) = 0.25 \lg \frac{1}{0.25} + 0.75 \lg \frac{1}{0.75} = 0.81128$$

Hence, the entropy of the nth extension $\mathcal{S}^n$ is

$$H(\mathcal{S}^n) = nH(\mathcal{S}) = 0.81128n$$

As we will see in the next section, there is a simple relationship between the entropy of a source (or its extensions) and the average codeword length of any Huffman encoding of that source (or its extensions.) □

Exercises

1. Consider the S source with alphabet $S = \{a, b\}$ and probability distribution $\mathcal{P}(a) = \frac{1}{4}, \mathcal{P}(b) = \frac{3}{4}$. Construct a Huffman encoding scheme for S, S^2, and S^3 and find the average codeword lengths per source symbol.

2. Repeat the previous exercise with a uniform probability distribution on S.

3.4   The Noiseless Coding Theorem

When encoding a source S, it certainly seems reasonable that we will need at least as many bits of information in the encoding as there is in the source. (For efficient encoding, we also want as few extra bits in the encoding as possible.) Since the entropy of S measures the amount of information in S, it should come as no surprise that the minimum average codeword length of any encoding of S should be at least as great as the entropy of S. In symbols,

$$H_r(S) \leq \text{MinAveCodeLen}_r(S)$$

This is the content of part of the famous Noiseless Coding Theorem, first proved by Claude Shannon in 1948. (Noise refers to the introduction of errors in the code.)

Theorem 3.4.1 (The Noiseless Coding Theorem—Version 1)   *Let S be an information source. Then*

$$H_r(S) \leq \text{MinAveCodeLen}_r(S)$$

where $\text{MinAveCodeLen}_r(S)$ *denotes the minimum average codeword length among all uniquely decipherable r-ary encoding schemes for S.*   □

Proof   Denote the probability distribution of the source S by $P = \{p_1, p_2, \ldots, p_q\}$. Let (C, f) be a uniquely decipherable r-ary encoding scheme for S, with codeword lengths $\ell_1, \ell_2, \ldots, \ell_q$ and consider the numbers

$$r_i = \frac{1}{r^{\ell_i}}$$

The r_i satisfy $0 \le r_i \le 1$. Furthermore, since C is uniquely decipherable, McMillan's Theorem tells us that

$$\sum_{i=1}^{q} r_i = \sum_{i=1}^{q} \frac{1}{r^{\ell_i}} \le 1$$

Thus, Lemma 3.2.2 implies that

$$H(\mathcal{S}) = \sum_{i=1}^{q} p_i \lg \frac{1}{p_i} \le \sum_{i=1}^{q} p_i \lg \frac{1}{r_i} = \sum_{i=1}^{q} p_i \lg r^{\ell_i} = \sum_{i=1}^{q} p_i \ell_i \lg r$$

$$= \lg r \sum_{i=1}^{q} p_i \ell_i = (\lg r) \text{AveCodeLen}(C, f)$$

Dividing by $\lg r$, and noting that $H_r(\mathcal{S}) = H(\mathcal{S})/\lg r$, we get

$$H_r(\mathcal{S}) \le \text{AveCodeLen}(C, f)$$

Since this holds for any uniquely decipherable r-ary encoding scheme for $\mathcal{S}$, the result follows. ∎

Example 3.4.1 Consider the source $\mathcal{S} = (S, \mathcal{P})$, where $S = \{0, 1, \ldots, 9\}$ and $\mathcal{P}$ is uniform. The entropy of this source is $\lg 10$. According to the Noiseless Coding Theorem, the average codeword length of any uniquely decipherable *ternary* encoding scheme (alphabet of size 3) must be at least

$$H_3(\mathcal{S}) = \frac{H(\mathcal{S})}{\lg 3} = 2.0959 \qquad (3.4.1) \quad \square$$

Example 3.4.2 Table 2.1.1 contains an information source corresponding to the letters of the English language. In Example 3.1.5, we noted that the entropy of this source is approximately 4.07991, and so the Noiseless Coding Theorem tell us that any uniquely decipherable encoding scheme must have average codeword length of at least 4.07991.

Table 2.1.1 also shows a Huffman encoding scheme for this source. In Example 2.1.2, we mentioned that the average codeword length of this Huffman encoding scheme is approximately 4.1195, which is quite close to the minimum possible. $\square$

The first version of the Noiseless Coding Theorem says that the entropy $H_r(\mathcal{S})$ provides a *lower bound* on $\text{MinAveCodeLen}_r(\mathcal{S})$. Let us now turn to the issue of finding an *upper bound* on $\text{MinAveCodeLen}_r(\mathcal{S})$. For this, we wish to construct an instantaneous encoding of $\mathcal{S}$ with small codeword lengths. Recall that if the lengths $\ell_1, \ldots, \ell_q$ satisfy Kraft's inequality

$$\sum_{i=1}^{q} \frac{1}{r^{\ell_i}} \leq 1$$

then there is an instantaneous code with these codeword lengths. If $P = \{p_1, \ldots, p_q\}$ is the probability distribution for S, then Kraft's inequality can be written in the form

$$\sum_{i=1}^{q} \frac{1}{r^{\ell_i}} \leq \sum_{i=1}^{q} p_i$$

Thus, if

$$\frac{1}{r^{\ell_i}} \leq p_i$$

for all i, Kraft's inequality will be satisfied. This can be rewritten in the form

$$\log_r \frac{1}{p_i} \leq \ell_i$$

so let us choose ℓ_i to be the *smallest* integer satisfying this inequality. In other words, if the integers ℓ_i are chosen to satisfy

$$\log_r \frac{1}{p_i} \leq \ell_i < \log_r \frac{1}{p_i} + 1 \tag{3.4.1}$$

for all i, then there is an instantaneous encoding with these codeword lengths. An encoding scheme whose codeword lengths ℓ_i satisfy (3.4.2) is referred to as a **Shannon-Fano encoding scheme**. Moreover,

$$\text{AveCodeLen}_r(S) = \sum_{i=1}^{q} p_i \ell_i < \sum_{i=1}^{q} p_i \left(\log_r \frac{1}{p_i} + 1 \right)$$

$$= \sum_{i=1}^{q} p_i \log_r \frac{1}{p_i} + \sum_{i=1}^{q} p_i = H_r(S) + 1$$

Hence,

$$\text{AveCodeLen}_r(S) < H_r(S) + 1$$

from which it follows that

$$\text{MinAveCodeLen}_r(S) < H_r(S) + 1$$

Combining this upper bound with the first version of the Noiseless Coding Theorem gives the second version of this theorem.

Theorem 3.4.2 (The Noiseless Coding Theorem—Version 2) *Let S be an information source. Then*

$$H_r(S) \leq \text{MinAveCodeLen}_r(S) < H_r(S) + 1$$

where $\text{MinAveCodeLen}_r(S)$ is the minimum average codeword length among all uniquely decipherable r-ary encoding schemes for S.   □

This theorem tells us that $\text{MinAveCodeLen}_r(S)$ lies between $H_r(S)$ and $H_r(S) + 1$. However, the difference between these bounds is 1 r-ary unit per source symbol, and this is still quite a lot from a practical standpoint. Fortunately, better results can be achieved by considering the encoding of extensions of the source S.

In particular, since the nth extension S^n of S is a source in its own right, we may apply the Noiseless Coding Theorem to S^n, to get

$$H_r(S^n) \leq \text{MinAveCodeLen}_r(S^n) < H_r(S^n) + 1$$

But $H_r(S^n) = nH_r(S)$, and so

$$nH_r(S) \leq \text{MinAveCodeLen}_r(S^n) < nH_r(S) + 1$$

Dividing by n gives the final version of the Noiseless Coding Theorem.

Theorem 3.4.3 (The Noiseless Coding Theorem—Final Version) *Let S be an information source, and let S^n be its nth extension. Then*

$$H_r(S) \leq \frac{\text{MinAveCodeLen}_r(S^n)}{n} < H_r(S) + \frac{1}{n}$$

where $\text{MinAveCodeLen}_r(S^n)$ is the minimum average codeword length among all uniquely decipherable r-ary encoding schemes for S^n.   □

Since each codeword in the nth extension S^n encodes n source symbols from S, the number

$$\frac{\text{MinAveCodeLen}_r(S^n)}{n}$$

is the minimum average codeword length per source symbol of S, taken over all uniquely decipherable r-ary encodings of S^n. Furthermore, since $1/n$ tends to 0 as n gets large, the upper bound $H_r(S) + 1/n$ approaches the lower bound $H_r(S)$, and so, according to the Noiseless Coding Theorem, the number $\text{MinAveCodeLen}_r(\mathcal{S}^n)/n$ can be made as close to $H_r(\mathcal{S}^n)$ as desired by taking n large enough.

In other words, by encoding extensions of $\mathcal{S}$, that is, blocks of source symbols rather than individual source symbols, we can reduce the average codeword length per source symbol to as close to the entropy $H_r(\mathcal{S})$ as desired. This is the real essence of the Noiseless Coding Theorem. The penalty for doing so is that, since $|S^n| = q^n$, the number of codewords required to encode the nth extension $\mathcal{S}^n$ grows exceedingly large as n gets large. As a result, achieving the desired "closeness" to the entropy may be a practical impossibility.

Exercises

1. Consider the source $\mathcal{S} = (S, \mathcal{P})$, where $S = \{a, b, c\}$ and $\mathcal{P}(a) = 1/2$, $\mathcal{P}(b) = 1/4$, $\mathcal{P}(c) = 1/4$. What is the binary entropy of this source? Can we achieve a minimum average codeword length equal to the entropy for this source?

2. Consider the source $\mathcal{S} = (S, \mathcal{P})$, where $S = \{a, b, c\}$ and $\mathcal{P}(a) = 2/3$, $\mathcal{P}(b) = 1/6$, $\mathcal{P}(c) = 1/6$. What is the binary entropy of this source? Can we achieve a minimum average codeword length equal to the entropy for this source?

3. Let $\mathcal{S}$ be a binary source (thus $S = \{0, 1\}$). In order to guarantee that the average codeword length, per source symbol of $\mathcal{S}$, is at most 0.01 greater than the entropy of $\mathcal{S}$, which extension of $\mathcal{S}$ should we encode? How many codewords would we need?

4. In this exercise, we construct an entirely different type of encoding scheme. Rather than encoding each source symbol with a fixed codeword (as in Huffman encoding), source symbols are encoded in groups in a way that depends on each symbol's relationship to other source symbols in the source message. To be specific, let the source alphabet be $S = \{0, 1\}$ and suppose that $\mathcal{P}(0) = p$ and $\mathcal{P}(1) = 1 - p$. To encode a string of source symbols, we count the number of 0s occurring in the string before the appearance of a 1. The two encoding rules are

   (a) if eight 0s appear in a row, encode these 0's as a 0, that is,

$$00000000 \to 0$$

(b) if fewer than eight 0s appear (say k 0s) before the next 1, then determine the 3-bit binary representation of k (say $e_1 e_2 e_3$) and encode the string of k 0s followed by the 1 as the codeword $1 e_1 e_2 e_3$. For instance, the source string 0001 (which is three 0s followed by a 1) is encoded as 1011 since 011 is the binary representation of the number three.

    i. Show that the resulting code is instantaneous.

    ii. What is the probability that the source will emit k 0s followed by a 1? (Recall that we assume independence of the source emissions.)

    iii. Define an event as the construction of a codeword. Find the average codeword length per event.

    iv. Find the average number of source bits per event.

    v. For each event, compute the number of codeword bits needed per source bit. Then compute the average of these numbers.

    vi. For $p = 0.9$, determine the average codeword length per source bit for a Huffman encoding of the fourth extension $\mathcal{S}^4$. How does this number compare to the number in part v)? What significance does this have for the optimality of Huffman encoding? Does this violate the Noiseless Coding Theorem?

5. Let $\mathcal{S}$ be a source and let $\mathcal{S}^2$ be its second extension. Is the second extension of $\mathcal{S}^2$ equal to the fourth extension of $\mathcal{S}$? In symbols, is $(\mathcal{S}^2)^2 = \mathcal{S}^4$?

II

PART

Coding Theory

4 The Main Coding Theory Problem

4.1 Communications Channels

Up to now, we have been concerned with how to encode an information source in the most efficient way, in order to keep the average codeword length as small as possible. The Noiseless Coding Theorem tells us that, at least theoretically, we can achieve a level of efficiency as close to the value of the entropy as desired. Let us now turn to the question of how to encode source data in order to detect and perhaps correct errors in transmission or storage. Let us begin by setting the stage and defining the terms for our discussion.

Definition Let $A = \{a_1, \ldots, a_r\}$ be a set of size r, which we refer to as a **code alphabet** and whose elements are called **code symbols**. An **r-ary block code** C over A is a nonempty subset of A^n, the set of all strings over A of length n. The elements of **c** are referred to as **codewords**.

The number n is the **length** of the code and the number of codewords in **c** is the size of the code. A code of length n and size M is called an **(n, M)-code**. A code over the alphabet $\mathbb{Z}_2 = \{0, 1\}$ is called a **binary code**. A code over the alphabet $\mathbb{Z}_3 = \{0, 1, 2\}$ is called a **ternary code**. □

As before, boldface letters are used to denote codewords. Since we will restrict our attention to block codes from now on, we will simply refer to them as codes.

Our model for communications is that of a communications channel, which can be thought of simply as a "black box" that accepts individual code symbols as input and produces as output one code symbol per input symbol.

The idea is that we encode the source symbols in our message into codewords and then input the resulting sequence of codewords, one code symbol at a time, into the channel. Hopefully, the channel will reproduce these symbols accurately as output. The problem is that channels make mistakes and occassionally alter an input symbol. When the output symbol differs from the corresponding input symbol, a **symbol error** has occurred. If one or more symbol errors result in the alteration of a codeword, then a **word error** has occurred.

This model of a communications channel as a black box applies not only to the transmission of data, such as over telephone lines or through space in the form of electromagnetic radiation, but also to the storage of data, such as on computer disk, tape, CD ROM, and so on. In this case, the input corresponds to the writing of the data and the output to the subsequent reading of that data.

The issue before us is how to design a procedure for determining whether word errors have occurred and, if so, how to correct those errors. (There is no way to correct an individual symbol error by itself.) A procedure that substitutes a codeword (hopefully the codeword transmitted) for a given received word (when it feels that this is necessary), or else simply declares an error, is called a decision rule.

Note that when a received word is a codeword, it is not possible to tell, without some additional information, whether or not a word error has occurred (changing one codeword into another). Of course, sometimes the context of the received message will reveal that errors have occurred, but this type of error detection is not relevant to our discussion. On the other hand, when a received word is not a codeword, then a word error has occurred.

The choice of which decision rule to use depends on the nature of the communications channel. To illustrate this, let us consider a simple example. Suppose that a certain channel has the property that the probability of a symbol error is the same for each position in the string being transmitted and is considerably less than $\frac{1}{2}$. If the code is

$$C = \{00000000, 11111111\}$$

and the received word is y = 11101111, then the decision rule should decide that the codeword sent was 11111111, and not 00000000. Of course, this may be wrong in a particular case, but since it is far more likely to be 11111111 than 00000000, this decision procedure will be right a vast majority of the time.

On the other hand, suppose that the channel has the property that the first bit sent is always incorrect. Then if the string 111011111 is received, the decision rule should identify the codeword sent as 00000000, and not 11111111. Thus, the nature of the channel can have a profound effect on the choice of decision rule.

Consider a communications channel that accepts symbols from a code alphabet $A = \{a_1, \ldots, a_r\}$, which we also refer to as the channel alphabet. A given input symbol a_i produces an output symbol a_j. At a different time, the same input a_i might produce a different output a_k. Thus, the key feature of a communications channel is the set of conditional probabilities $\mathcal{P}(a_j$ received$| a_i$ sent). These probabilities are called the channel probabilities, or transition probabilities. We can now give a formal definition of a communications channel.

Definition A **communications channel** consists of a finite **channel alphabet** $A = \{a_1, \ldots, a_r\}$ and a set of (**forward**) **channel probabilities**, or **transition probabilities**, $\mathcal{P}(a_j$ received$| a_i$ sent), satisfying

$$\sum_{j=1}^{r} \mathcal{P}(a_j \text{ received } | a_i \text{ sent}) = 1$$

for all i. As the notation indicates, we think of $\mathcal{P}(a_j$ received $| a_i$ sent) as the probability that the code symbol a_j is received, given that a_i is sent through the channel. Furthermore, as the notation suggests, we assume that this probability does not change with time. The equation above simply says that, given that a_i was sent, some code symbol must be received. $\square$

It is important not to confuse the forward channel probability $\mathcal{P}(a_j$ received $| a_i$ sent) with the so-called **backward channel probability** $\mathcal{P}(a_i$ sent $| a_j$ received). In the forward probabilities, we assume a certain codeword was sent. In the backward probabilities, we assume a certain word is received. The backward channel probabilities will be discussed a bit later in this section.

We will deal only with channels that have no memory, in the following sense.

Definition A communications channel is said to be **memoryless** if the outcome of any one transmission is independent of the outcome of the previous transmission. Put more formally, if $\mathbf{c} = c_1 \cdots c_n$ and $\mathbf{x} = x_1 \cdots x_n$ are words of length n over the alphabet A, the probability $\mathcal{P}(\mathbf{x}$ received $\mid \mathbf{c}$ sent) that $\mathbf{x}$ is received, given that $\mathbf{c}$ is sent (one symbol at a time), is just the product

$$\mathcal{P}(\mathbf{x} \text{ received } \mid \mathbf{c} \text{ sent}) = \prod_{i=1}^{n} \mathcal{P}(x_i \text{ received} \mid c_i \text{ sent})$$

We will also refer to the probabilities $\mathcal{P}(\mathbf{x}$ received $\mid \mathbf{c}$ sent) as **forward channel probabilities**.   □

Communication channels can be described in a variety of ways. One way is to construct a graph as shown in Figure 4.1.1. There is one node on the left for each code symbol, and similarly on the right. Each node on the left is connected with each node on the right by an edge that is labeled with the corresponding channel probability. (In Figure 4.1.1, we have drawn only one edge as an illustration.)

Example 4.1.1 Perhaps the most important memoryless channel is the binary symmetric channel, which has channel alphabet $\{0,1\}$ and channel probabilities

$$\mathcal{P}(1 \text{ received } \mid 0 \text{ sent}) = \mathcal{P}(0 \text{ received } \mid 1 \text{ sent}) = p$$
$$\mathcal{P}(0 \text{ received } \mid 0 \text{ sent}) = \mathcal{P}(1 \text{ received } \mid 1 \text{ sent}) = 1 - p$$

FIGURE 4.1.1 A communications channel.

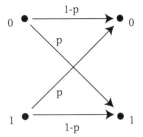

FIGURE 4.1.2   A binary symmetric channel.

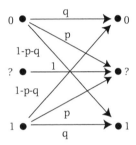

FIGURE 4.1.3   A binary erasure channel.

Thus, the probability of a symbol (bit) error, also called the **crossover probability**, is p. This channel is pictured in Figure 4.1.2.   □

Example 4.1.2   Another important memoryless channel is the **binary erasure channel**, whose alphabet is $\{0, 1, ?\}$, as shown in Figure 4.1.3. (For clarity, we have left out two arrows with 0 probability. Also, the intention for this channel is never to send the symbol "?". However, this symbol may be received and so it is included in the channel alphabet.) Can you explain why this is called an erasure channel?   □

Exercises

1. Draw the graph of a binary erasure channel that either transmits a bit correctly, or else erases the bit.

2. Explain why a communications channel whose diagram has the following form

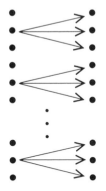

(all edges not shown have 0 probability) is called a **lossless channel**.

3. Explain why a communications channel whose diagram has the following form

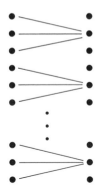

(all edges not shown have 0 probability) is called a **deterministic channel**.

4. Explain why a communications channel whose diagram has the following form

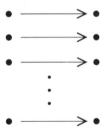

(all edges not shown have 0 probability) is called a **noiseless channel**.

5. Explain why a communications channel whose diagram has the following form

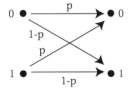

   is called a **useless channel**.

6. Draw a diagram of a channel, with channel alphabet $A = \{0, 1, 2\}$ and with the following two properties a) the probability that any symbol is transmitted correctly is independent of the symbol being sent, and b) when a symbol error is made, the received symbol is equally likely to be any of the symbols that were not sent. If the probability that a 0 is transmitted correctly is $\frac{9}{10}$, what is the probability that a 0 is changed into a 1?

7. What should one do with a binary symmetric channel whose crossover probability is significantly greater than $\frac{1}{2}$? Hint: it is not correct to answer "chuck it!"

4.2 Decision Rules

As mentioned previously, a procedure that substitutes a codeword for a given received word, or declares a decoding error, is called a decision rule. The concept of a decision rule can be defined more precisely as a function from the set of all words to the set of codewords, together with an additional symbol representing a decoding error.

Definition Let C be an (n, M)-code over a code alphabet A. Assume that **c** does not contain the symbol "?" as a codeword (otherwise replace "?" by some other symbol). A **decision rule** for C is a function $f : A^n \to C \cup \{?\}$. The process of applying a decision rule is referred to as **decoding**. If **x** is a (received) word in A^n, then the decision rule f **decodes x** as the codeword $f(\mathbf{x})$ if $f(\mathbf{x}) \in C$ or else declares a **decoding error** if $f(\mathbf{x}) = ?$.   □

Our goal is to find a decision rule that maximizes the probability of correct decoding, that is, the probability that $f(\mathbf{x})$ is actually the codeword that was sent. (It is probably worth remarking, however, that the receiver has no way of knowing whether or not the decoding process has resulted in the correct codeword.)

The probability of correct decoding can be expressed in a variety of ways. For instance, using the Theorem on Total Probabilities (Chapter 0), we can easily derive two formulas for this probability. Conditioning on the codeword sent gives

$$\mathcal{P}(\text{correct decoding}) = \sum_{c \in C} \mathcal{P}(\text{correct decoding} \mid \mathbf{c} \text{ sent})\mathcal{P}(\mathbf{c} \text{ sent})$$

(4.2.1)

Conditioning instead on the word received gives

$$\mathcal{P}(\text{correct decoding}) = \sum_{\mathbf{x} \in A^n} \mathcal{P}(\text{correct decoding} \mid \mathbf{x} \text{ received})\mathcal{P}(\mathbf{x} \text{ received})$$

(4.2.2)

where the sum now runs over all words in A^n.

Note that formula (4.2.1) explicitly involves the probabilities $\mathcal{P}(\mathbf{c}$ sent) that the various codewords $\mathbf{c}$ are sent through the channel. These probabilites

$$\{\mathcal{P}(\mathbf{c}_1 \text{ sent}), \dots, \mathcal{P}(\mathbf{c}_M \text{ sent})\}$$

form the so-called **input distribution** for the channel. The input distribution is not part of the channel and causes unfortunate complications, since it may vary depending on what type of messages are being sent. (If you are sending credit card numbers, for instance, there will be considerably higher probability that a codeword representing a digit is being sent then if you are sending the text of *Henry IV, Part I*.) Of course, formula (4.2.2) also involves the input distribution, since the probability that a given word $\mathbf{x}$ is received usually depends on which codeword was sent.

Let f be a decision rule for the code C. If the codeword $\mathbf{c}$ is sent, then a correct decoding takes place provided that the received word $\mathbf{x}$ satisfies $f(\mathbf{x}) = \mathbf{c}$. Hence,

$$\mathcal{P}(\text{correct decoding} \mid \mathbf{c} \text{ sent}) = \sum_{\mathbf{x}:f(\mathbf{x})=\mathbf{c}} \mathcal{P}(\mathbf{x} \text{ received} \mid \mathbf{c} \text{ sent})$$

Substituting into (4.2.1) gives

$$P(\text{correct decoding}) = \sum_{\mathbf{c}\in C} \sum_{\mathbf{x}:f(\mathbf{x})=\mathbf{c}} P(\mathbf{x} \text{ received} \mid \mathbf{c} \text{ sent})P(\mathbf{c} \text{ sent})$$

The double sum on the right is a bit nasty, however.

Fortunately, formula (4.2.2) provides us with a better handle on how to obtain a good decision rule. Under the decision rule f, a received word $\mathbf{x}$ is decoded correctly if the codeword sent was actually $f(\mathbf{x})$. Thus,

$$P(\text{correct decoding} \mid \mathbf{x} \text{ received}) = P(f(\mathbf{x}) \text{ sent} \mid \mathbf{x} \text{ received})$$

(no sum this time). Substituting into formula (4.2.2) gives

$$P(\text{correct decoding}) = \sum_{\mathbf{x}\in A^n} P(f(\mathbf{x}) \text{ sent} \mid \mathbf{x} \text{ received})P(\mathbf{x} \text{ received})$$

Hence, the probability of correct decoding can be maximized by choosing the decision rule that maximizes each of the conditional probabilities

$$P(f(\mathbf{x}) \text{ sent} \mid \mathbf{x} \text{ received})$$

In other words, given that $\mathbf{x}$ is received, we decide that the codeword sent is the one most likely to have been sent! This is accomplished by looking at the backward channel probabilities

$$P(\mathbf{c}_1 \text{ sent} \mid \mathbf{x} \text{ received}), \ldots, P(\mathbf{c}_M \text{ sent} \mid \mathbf{x} \text{ received})$$

and choosing the codeword $\mathbf{c}_i$ with the largest probability. (The issue of ties will be discussed in a moment.) This motivates the following definition.

Definition Any decision scheme f for which $f(\mathbf{x})$ has the property that

$$P(f(\mathbf{x}) \text{ sent} \mid \mathbf{x} \text{ received}) = \max_{\mathbf{c}\in C} P(\mathbf{c} \text{ sent} \mid \mathbf{x} \text{ received})$$

for all possible received words $\mathbf{x}$, is called an **ideal observer**. In words, an ideal observer is one for which $f(\mathbf{x})$ is a codeword most likely to have been sent, given that $\mathbf{x}$ is received. □

Our deliberations thus far have proven the following theorem.

Theorem 4.2.1 *An ideal observer decision rule maximizes the probability of the correct decoding of received words.* □

An ideal observer certainly lives up to its name in the sense of maximizing the probability of correct decoding. However, the problem with ideal observers is that they depend on the input distribution, as can be seen explicity by using Bayes' Theorem (see Chapter 0)

$$P(\mathbf{c} \text{ sent} \mid \mathbf{x} \text{ received}) = \frac{P(\mathbf{x} \text{ received} \mid \mathbf{c} \text{ sent})P(\mathbf{c} \text{ sent})}{\sum_{k=1}^{M} P(\mathbf{x} \text{ received} \mid \mathbf{c}_k \text{ sent})P(\mathbf{c}_k \text{ sent})}$$
(4.2.3)

Thus, in order to function, an ideal observer needs to know how likely it is that various codewords are sent through the channel, that is, it must know something about the messages being sent!

This is not so ideal.

One way around this is simply to assume that the input probability distribution is uniform

$$P(\mathbf{c} \text{ sent}) = \frac{1}{M}$$

where M is the size of the code. Thus, (4.2.3) becomes

$$P(\mathbf{c} \text{ sent} \mid \mathbf{x} \text{ received}) = \frac{P(\mathbf{x} \text{ received} \mid \mathbf{c} \text{ sent})}{\sum_{k=1}^{M} P(\mathbf{x} \text{ received} \mid \mathbf{c}_k \text{ sent})}$$

Now, the denominator on the right is a sum of forward channel probabilities and thus depends only on the channel. Thus, maximizing the left side of this equation is equivalent to maximizing the numerator on the right side; that is, the forward channel probabilities $P(\mathbf{x}$ received $\mid \mathbf{c}$ sent). This leads to the following definition and theorem.

Definition Any decision rule f for which $f(\mathbf{x})$ maximizes the forward channel probabilities, that is, for which

$$P(\mathbf{x} \text{ received} \mid f(\mathbf{x}) \text{ sent}) = \max_{\mathbf{c} \in C} P(\mathbf{x} \text{ received} \mid \mathbf{c} \text{ sent})$$

for all possible received words $\mathbf{x}$, is called a **maximum likelihood decision rule**. In words, $f(\mathbf{x})$ is a codeword with the property that for no other codeword would it be more likely that the given output $\mathbf{x}$ was received. □

Applying the maximum likelihood decision rule is called **maximum likelihood decoding**.

Theorem 4.2.2 *For the uniform input distribution, an ideal observer is one who applies maximum likelihood decoding.* □

Since the forward channel probabilities are given as part of the definition of the channel, it is relatively easy to apply maximum likelihood decoding.

Example 4.2.1   Suppose that codewords from the code $\{000, 111\}$ are being sent over a binary symmetric channel with crossover probability $p = 0.01$. Thus, the probability that a symbol is received correctly is $1 - p = 0.99$. Suppose that the string 100 is received. The forward channel probabilities are

$$\begin{aligned} \mathcal{P}(100 \text{ received} \mid 000 \text{ sent}) &= \mathcal{P}(1 \text{ received} \mid 0 \text{ sent})\mathcal{P}(0 \text{ received} \mid 0 \text{ sent})^2 \\ &= (0.01)(0.99)^2 = 0.009801 \end{aligned}$$

and

$$\begin{aligned} \mathcal{P}(100 \text{ received} \mid 111 \text{ sent}) &= \mathcal{P}(1 \text{ received} \mid 1 \text{ sent})\mathcal{P}(0 \text{ received} \mid 1 \text{ sent})^2 \\ &= (0.99)(0.01)^2 = 0.000099 \end{aligned}$$

Since the first probability is larger than the second, the maximum likelihood decision rule decodes 100 as 000. $\square$

In view of Theorem 4.2.2, from now on we will assume a uniform input probability distribution.

Before continuing, we should make a remark about ties. When two or more codewords, say c_1 and c_2, give the maximum forward channel probability, then maximum likelihood decoding results in a tie. In practice, the procedure for handling ties usually depends on the seriousness of making a decoding error. In some cases, we may wish to choose randomly from among the candidates. In other cases, it might be more desirable simply to admit a decoding error, thereby reducing the chance of getting an undetected error. The term **complete decoding** refers to the case where all received words are decoded (one way or another), and the term **incomplete decoding** refers to the case where we prefer occasionally to simply admit an error, rather than always decode.

Exercises

1. Verify formula (4.2.3).

2. Suppose that codewords from the code $\{0000, 1111\}$ are being sent over a binary symmetric channel with crossover probability $p = 0.01$. Use the maximum likelihood decision rule to decode the received words a) 0000, b) 0010, c) 1010.

3. Suppose that codewords from the code $\{000, 001, 111\}$ are being sent over a binary symmetric channel with crossover probability $p = 0.01$. Use the maximum likelihood decision rule to decode the received words a) 010, b) 110.

4. Consider a binary channel with channel probabilities

$$\mathcal{P}(0 \text{ received} \mid 0 \text{ sent}) = \frac{3}{4}, \mathcal{P}(1 \text{ received} \mid 1 \text{ sent}) = \frac{7}{8}$$

If codewords from the code $\{000, 001, 111\}$ are being sent over this channel, use maximum likelihood decoding to decode the received words a) 010, b) 110.

5. Consider a binary erasure channel, as described in Example 4.1.2, with probability $p = 0.008$, $q = 0.99$. If codewords from the code $\{000, 001, 111\}$ are being sent over this channel, use maximum likelihood decoding to decode the received word a) 010, b) 10?, c) ??0.

6. In this exercise, we compute the probability of a decoding error for three different channels. In each case, the code C consists of the 8 binary strings of length 3. Also, in each case we use the "no brainer" decision rule—decide that the output is correct! Denote an input codeword by $i_1 i_2 i_3$ and a received word by $o_1 o_2 o_3$. Finally, let BSC denote a binary symmetric channel with crossover probability p.

   (a) The first channel works as follows: send i_1 through the BSC to get o_1. Then choose o_2 and o_3 randomly (flip a fair coin for each).

   (b) The second channel works as follows: choose o_1 to be the majority bit among i_1, i_2 and i_3. Then set $o_2 = o_3 = o_1$.

   (c) The third channel works as follows: send i_1 through the BSC to get o_1, send i_2 through the BSC to get o_2, and send i_3 through the BSC to get o_3.

   Compute the probability of correct decoding for each of these channels, assuming a uniform input distribution. Which channel is best for $p = 0.001$? Hint: use formula (4.2.1).

4.3 Nearest Neighbor Decoding

By far the most thoroughly studied communications channel is the binary symmetric channel and we wish to examine maximum likelihood decoding more closely for this channel. (It turns out that what we learn for this channel applies to other channels as well.)

It is reasonable to assume that the crossover probability of a binary symmetric channel satisfies $p < \frac{1}{2}$ (in practice, it should be much less than $\frac{1}{2}$). Since the probability that a code symbol is received correctly is $1 - p$, the probability that a codeword of length n is received correctly is

$$\mathcal{P}(\text{no word error}) = (1 - p)^n$$

The probability that a single symbol error occurs in a specified place in the codeword is

$$\mathcal{P}(\text{one symbol error in a specified place}) = p(1 - p)^{n-1}$$

The probability that exactly two errors occur in specified places is

$$\mathcal{P}(\text{two symbol errors in specified places}) = p^2(1 - p)^{n-2}$$

More generally, the probability of exactly k errors occurring in k specified places is

$$\mathcal{P}(k \text{ symbol errors in specified places}) = p^k(1 - p)^{n-k}$$

(see Example 0.2.9).

As a result, if a codeword **c** differs from a word **x** in exactly k positions, then

$$\mathcal{P}(\mathbf{x} \text{ received} \mid \mathbf{c} \text{ sent}) = p^k(1 - p)^{n-k}$$

Since $p < \frac{1}{2}$, it follows that $1 - p > p$, and so this probability is larger for larger values of the exponent $n - k$, that is for smaller values of k. Hence, the probability $\mathcal{P}(\mathbf{x} \text{ received} \mid \mathbf{c} \text{ sent})$ is maximized by choosing a codeword for which k is as small as possible, that is, a codeword that has the fewest symbol differences with the received word **x**.

We can summarize as follows.

Theorem 4.3.1 *For a binary symmetric channel, the maximum likelihood decision rule is to choose a codeword that differs in the fewest places with the received word* **x**. □

We have come a long way from having to find maximum backward channel probabilities. Now all we need to do is compare bits in the received word with bits in the various codewords. A codeword with the fewest differences is a maximum likelihood codeword. This decision rule can be phrased a bit more elegantly by introducing a measure of the closeness of two strings.

Definition   Let $\mathbf{x}$ and $\mathbf{y}$ be strings of length n over an alphabet A. The **Hamming distance** from $\mathbf{x}$ to $\mathbf{y}$, denoted by $d(\mathbf{x}, \mathbf{y})$, is defined to be the number of places in which $\mathbf{x}$ and $\mathbf{y}$ differ.   □

Example 4.3.1

(a) Since the strings $\mathbf{x} = 11001$ and $\mathbf{y} = 11010$ differ in exactly two places, their distance apart is $d(\mathbf{x}, \mathbf{y}) = 2$.

(b) If $\mathbf{x} = 1234$ and $\mathbf{y} = 1423$, then $d(\mathbf{x}, \mathbf{y}) = 3$.   □

Since Hamming distance is the only type of distance that we will consider in this book, let us simply refer to it as distance.

Now suppose that C is a code of length n over A. The codewords that are closest (in Hamming distance) to a given received word $\mathbf{x}$ are referred to as **nearest neighbor codewords**. The **nearest neighbor decision rule** is the rule that decodes a received word $\mathbf{x}$ as a nearest neighbor codeword. Applying the nearest neighbor decision rule is referred to as **nearest neighbor decoding**, or **minimum distance decoding**. Notice that the distance $d(\mathbf{c}, \mathbf{x})$ between the codeword sent and the received word $\mathbf{x}$ is just the number of symbol errors that occurred in the transmission.

Theorem 4.3.2 *For a binary symmetric channel, maximum likelihood decoding is the same as nearest neighbor decoding.*   □

Example 4.3.2   Suppose that codewords from the code

$$C = \{0000, 0011, 1000, 1100\}$$

are being sent over a binary symmetric channel. Assuming $\mathbf{x} = 0111$ is received, then

$$d(0111, 0000) = 3$$
$$d(0111, 0011) = 1$$
$$d(0111, 1000) = 4$$
$$d(0111, 1100) = 3$$

Hence, the nearest neighbor codeword is 0011.   □

Nearest neighbor decoding is certainly easy to implement, at least in principle. However, for large codes, there may be practical problems in taking the "brute force" approach of computing the distance to each codeword. As we will see in Chapter 5, for a large class of very important codes, there are more efficient methods for implementing nearest neighbor decoding.

Let us conclude this section by stating some properties of the Hamming distance function. Proof of the next theorem is left for the exercises.

Theorem 4.3.3 *The Hamming distance function is a* **metric** *on the set* A^n, *that is, for all* **x**, **y**, *and* **z** *in* A^n, *we have*

1. *(***positive definiteness***)*

$$d(x, y) \geq 0, \text{ and } d(x, y) = 0 \text{ if and only if } x = y$$

2. *(***symmetry***)*

$$d(x, y) = d(y, x)$$

3. *(***triangle inequality***)*

$$d(x, z) \leq d(x, y) + d(y, z) \qquad \Box$$

Since Hamming distance **d** is a metric, the pair (A^n, d) is referred to as a **metric space**. The most famous metric space is the pair $(\mathbb{R}, r)$ where $\mathbb{R}$ is the set of real numbers and r is the ordinary Euclidean distance function $r(x, y) = |x - y|$.

Exercises

1. Compute the Hamming distances a) d(0110101, 1101011) b) d(123456789, 987654321)

2. For the code $C = \{11100, 01001, 10010, 00111\}$ use nearest neighbor decoding to decode the following received words a) 10000, b) 01100, c) 00100, and d) 01001.

3. For the ternary code $C = \{11200, 01221, 20012, 00111\}$ use nearest neighbor decoding to decode the following received words a) 12200, b) 21102, c) 00100, and d) 01201.

4. Prove that the Hamming distance function is a metric on A^n.

5. Consider a channel with channel alphabet $A = \{a_1, \ldots, a_r\}$. Assume that the probability that any given symbol is sent correctly is $p > \frac{1}{2}$ and that the probability that a symbol is changed into another symbol is the same for all other symbols. What is that probability? Show that, for such a channel, maximum likelihood decoding is equivalent to nearest neighbor decoding.

6. Show that maximum likelihood decoding is equivalent to nearest neighbor decoding for the binary erasure channel with the property that $\mathcal{P}(1 \text{ received} \mid 0 \text{ sent}) = \mathcal{P}(0 \text{ received} \mid 1 \text{ sent}) = 0$, assuming that we only consider codes over $\{0, 1\}$, that is, excluding ?.

7. Construct a binary channel (channel alphabet $\{0, 1\}$) for which maximum likelihood decoding is not the same as nearest neighbor decoding. Hint: the probability that a 0 is changed into a 1 must be different from the probability that a 1 is changed into a 0.

4.4 The Minimum Distance of a Code

If one or two symbol errors should occur in transmitting a codeword from the code $C = \{000, 111\}$, the resulting string cannot be another codeword and therefore such errors can always be detected. For this reason, C is referred to as a *double-error-detecting code*. Furthermore, if a single error should occur in transmission, the resulting string will be closer to one of the codewords than to the other and so nearest neighbor decoding will correct this error. Hence, C is a *single-error-correcting code*. Let us generalize these concepts in a definition.

Definition   Let u be a positive integer. A code C is ***u*-error-detecting** if, whenever a codeword incurs at least one but at most u errors, the resulting string is not a codeword. A code C is **exactly *u*-error-detecting** if it is u-error detecting but not $(u + 1)$-error-detecting.   □

We will assume that, in order to detect errors in transmission, the receiver checks the received string against a list of all codewords. If the string is not on the list, the receiver knows that at least one error has occurred, but he has no way of knowing exactly how many errors have occurred.

Since this list checking can be done by computer, we may imagine that the receiver has a red light on his desk. This light goes on whenever the received string is not a codeword. For a u-error-detecting code, the light will go on if at least one but at most u errors have occurred. However, the light may or may not go on if more than u errors have occurred. Thus, if the light goes on, the receiver can be certain only that at least one error has occurred. Furthermore, if the light does not go on, the receiver cannot conclude that no errors have occurred, for it is possible that "so many" errors occurred as to change the codeword sent into another codeword.

Example 4.4.1 The code C = {000000, 111000, 111111} is double-error-detecting, since changing any codeword in one or two positions does not result in another codeword. In fact, C is exactly double-error-detecting. □

Definition Let v be a positive integer. A code C is **v-error-correcting** if nearest neighbor decoding is able to correct v or fewer errors, assuming that if a tie occurs in the decoding process, a decoding error is reported. A code is **exactly v-error-correcting** if it is v-error-correcting but not $(v+1)$-error-correcting. □

It should be kept in mind that, as long as the received word is not a codeword, nearest neighbor decoding will decode it as some codeword, but the receiver has no way of knowing whether that codeword is the one that was actually sent. The receiver knows only that, under a v-error-correcting code, if no more than v errors were introduced, then nearest neighbor decoding will produce the codeword that was sent.

Example 4.4.2 If one or two errors should occur in the transmission of a codeword from the code C = {0000000000, 1111100000, 1111111111}, the resulting word is closer to the codeword sent than to any other codeword. Hence, **c** is double-error-correcting.

If more than two errors occur, nearest neighbor decoding may produce the wrong codeword. For instance, if 0000000000 is sent, but 1110000000 is received, the closest codeword is 1111100000, and a decoding error will result. Hence, C is exactly double-error-correcting. □

Example 4.4.3   The code $\text{Rep}_2(3) = \{000, 111\}$ is called a **binary repetition code** of length 3. Similarly, $\text{Rep}_2(4) = \{0000, 1111\}$ is a **binary repetition code** of length 4. More generally, the **r-ary repetition code** of length n is

$$\text{Rep}_r(n) = \{0\cdots 0, 1\cdots 1, 2\cdots 2, \ldots, (r-1)\cdots(r-1)\}$$

consisting of r codewords, each of length n.

The r-ary repetition code of length n can detect up to $n-1$ errors in transmission, and so it is $(n-1)$-error-detecting. Furthermore, it is exactly $\lfloor \frac{n-1}{2} \rfloor$-error-correcting.   □

Example 4.4.4   The set of all ten-digit telephone numbers (including area codes) is an example of a block code of length 10. There are currently about 115 million telephones in the United States, and a natural question arises as to whether it is possible to assign numbers to these telephones so that a single error in dialing can be corrected. In the parlance of coding theory, we ask if there is a single-error-correcting 10-ary code of length 10 and size 115,000,000. It is customary to refer to 10-ary codes as *decimal codes*.

At this time, it is not known what the maximum size is for a single-error-correcting decimal code of length 10. However, it is known, and we shall be able to prove it later, that this maximum is no bigger than 100,000,000. Therefore, it is not possible to encode all telephone numbers in this country with single error correction.

However, we will see later that there is a single-error-correcting decimal code of length 10 with size 82,644,629. Therefore, single error correction would be possible in many other countries.   □

The error detecting/correcting value of a code can be expressed more elegantly in terms of the minimum distance between codewords.

Definition   Let C be a code with at least two codewords. The **minimum distance** $d(C)$ of C is the smallest distance between distinct codewords. In symbols,

$$d(C) = \min\{d(\mathbf{c}, \mathbf{d}) \mid \mathbf{c}, \mathbf{d} \in C, \mathbf{c} \neq \mathbf{d}\} \tag{4.4.1}$$

Since $\mathbf{c} \neq \mathbf{d}$ implies that $d(\mathbf{c}, \mathbf{d}) \geq 1$, the minimum distance of a code must be at least 1.

Example 4.4.5

(a) If $C_1 = \{000, 010, 011\}$, then since $d(000, 010) = 1$, we have $d(C_1) = 1$.

(b) Let $C_2 = \{00011, 00101, 11101, 11000\}$. The following table shows all distances between distinct codewords of C_2.

	00011	00101	11101	11000
00011	*	2	4	4
00101	2	*	2	4
11101	4	2	*	2
11000	4	4	2	*

Hence, $d(C_2) = 2$.

(c) For the r-ary repetition code

$$\text{Rep}_r(n) = \{00 \cdots 0, 11 \cdots 1, \ldots, (r-1)(r-1) \cdots (r-1)\}$$

of length n, we have $d(\mathbf{c}, \mathbf{d}) = n$ for all $\mathbf{c}, \mathbf{d} \in \text{Rep}_n(n)$ and so $d(C) = n$.

(d) The double-error-detecting code in Example 4.4.2 has minimum distance $d(C) = 5$. $\square$

The next theorem is essentially just a restatement of the definition of u-error-detecting in terms of minimum distance.

Theorem 4.4.1 *A code C is u-error-detecting if and only if $d(C) \geq u + 1$.* $\square$

Here is the analog for error correction.

Theorem 4.4.2 *A code C is v-error-correcting if and only if $d(C) \geq 2v + 1$* $\square$

Proof Suppose first that $d(C) \geq 2v + 1$. If a codeword $\mathbf{c}$ suffers between 1 and v errors, the resulting string $\mathbf{x}$ satisfies $1 \leq d(\mathbf{c}, \mathbf{x}) \leq v$. This implies that $\mathbf{x}$ is closer to $\mathbf{c}$ than to any other codeword in $\mathbf{c}$. For if $d(\mathbf{x}, \mathbf{d}) \leq v$ for any codeword $d \neq c$, then the triangle inequality gives

$$d(\mathbf{c}, \mathbf{d}) \leq d(\mathbf{c}, \mathbf{x}) + d(\mathbf{x}, \mathbf{d}) \leq v + v = 2v$$

which contradicts the fact that $d(C) = 2v + 1 > 2v$. Hence, nearest neighbor decoding will correct v or fewer errors.

For the converse, suppose that C is v-error-correcting, but that there are distinct codewords $\mathbf{c}$ and $\mathbf{d}$ with $d(\mathbf{c}, \mathbf{d}) = d(C) \leq 2v$. We wish to show that, assuming $\mathbf{c}$ is sent and at most v errors occur, it is nevertheless possible that nearest neighbor decoding will either report a tie (which is considered a decoding error) or else decode the received word incorrectly as **d**. This contradiction to C being v-error-correcting shows that $d(C) \geq 2v + 1$.

First, we observe that $d(\mathbf{c}, \mathbf{d}) = d(C) \geq v + 1$, for otherwise **c** could be turned into **d** by suffering at most v errors, which would then go uncorrected. Thus, we may assume, for the sake of argument, that **c** and **d** differ in exactly the first $k = d(C)$ positions, where $v + 1 \leq k \leq 2v$. (This assumption can be justified in terms of reordering coordinates.) Consider the received word **x** that agrees with **c** in the first $k - v$ positions, agrees with **d** in the next v positions and agrees with both **c** and **d** in the last $n - k$ positions, as shown below

$$x = \underbrace{x_1 \cdots x_{k-v}}_{\text{agree with } \mathbf{c}} \underbrace{x_{k-v+1} \cdots x_k}_{\text{agree with } \mathbf{d}} \underbrace{x_{k+1} \cdots x_n}_{\text{agree with both}}$$

Since $d(\mathbf{c}, \mathbf{x}) = v$ and $d(\mathbf{d}, \mathbf{x}) = k - v \leq v$, it follows that either $d(\mathbf{c}, \mathbf{x}) = d(\mathbf{d}, \mathbf{x})$, in which case there is a tie, or $d(\mathbf{c}, \mathbf{x}) \geq d(\mathbf{d}, \mathbf{x})$, in which case **x** is decoded incorrectly as **d**. ∎

Since the concept of minimum distance is clearly important, the following notation has been devised.

Definition A code of size M, length n, and minimum distance d, is referred to as an **(n,M,d)-code**. The numbers n, M and d are called the **parameters** of the code. □

Example 4.4.6

(a) The code C_1 in Example 4.4.5 is a (3,3,1)-code.

(b) The code C_2 in Example 4.4.5 is a (5,4,2)-code.

(c) The code $C = \{0000, 1100\}$ is a (4,2,2)-code.

(d) The code $C = \{00, 01, 10, 11\}$ is a (2,4,1)-code.

(e) The r-ary repetition code $\text{Rep}_r(n)$ is an (n, r, n)-code. □

Example 4.4.7 In January 1979, the Mariner 9 spacecraft took black and white photographs of Mars. A grid of size 600×600 was placed over each photograph, and each of the resulting 360,000 grid components was assigned one of 64 shades of gray. Thus, the source information consisted of 64 different source symbols. Each source symbol was then encoded using a particular binary (32,64,16)-code, known as a *Reed-Muller code*. Since the minimum distance of this code is 16, Corollary 4.4.3 tells us that it is exactly seven-error-correcting. (We will study Reed-Muller codes later in the book.)   □

Example 4.4.8 In the period from 1979 through 1981, the Voyager spacecrafts took color photographs of Jupiter and Saturn. This required a source alphabet of size 4096 to represent various shades of color. The source information was then encoded using a particular binary (24,4096,8)-code, known as a Golay code. By Corollary 4.4.3, this code is exactly three-error-correcting. (We will study Golay codes later in the book.)   □

We can characterize being exactly v-error-correcting in terms of minimum distance. Proof of the following corollary is left as an exercise.

Corollary 4.4.3

1. *An (n, M, d)-code C is exactly v-error-correcting if and only if $d = 2v + 1$ or $d = 2v + 2$.*

2. *A code C has $d(C) = u$ if and only if it is exactly $\lfloor \frac{u-1}{2} \rfloor$-error-correcting.*   □

Mixed Error Detection and Error Correction

There is a somewhat subtle point that should be made about error detection and correction. Namely, both cannot take place at the same time and at maximum levels. To be more specific, suppose that a code C has minimum distance d. Thus, it is $(d - 1)$-error detecting and $\lfloor \frac{d-1}{2} \rfloor$-error-correcting.

If we use C for error detection only, it can detect up to $d - 1$ errors. On the other hand, if we want C to also correct errors whenever possible, then it can correct up to $\lfloor \frac{d-1}{2} \rfloor$ errors, but may no longer be able to detect a situation where more than $\lfloor \frac{d-1}{2} \rfloor$ but less than d errors have occurred. For if more than $\lfloor \frac{d-1}{2} \rfloor$ errors are made, nearest neighbor decoding might "correct" the received word to the wrong codeword, and thus the errors will go undetected. (In a sense, employing v-error-correction turns each codeword into a "magnet" that attracts any received word that is within a distance of v, even if the received word "came from" a more distant codeword.)

This issue is important since, in practice, it is not uncommon to use a mixed strategy of both error correction, which may need to be restricted because it can be expensive in both time and money, and error detection. Accordingly, we have the following for mixed strategies.

Definition   Consider the following strategy for error correction/detection. Let v be a positive integer. If a word $\mathbf{x}$ is received and if the closest codeword $\mathbf{c}$ to $\mathbf{x}$ is at a distance of at most v, and there is only one such codeword, then decode $\mathbf{x}$ as $\mathbf{c}$. If there is more than one codeword at minimum distance to $\mathbf{x}$, or if the closest codeword has distance greater than v, then simply declare a word error.

A code C is **simultaneously v-error-correcting and u-error-detecting** if, whenever at least one but at most v errors are made, the strategy described above will correct these errors and if, whenever at least $v + 1$ but at most $v + u$ errors are made, the strategy above simply reports a word error.   □

Theorem 4.4.4   *A code C is simultaneously v-error-correcting and u-error-detecting if and only if $d(C) \geq 2v + u + 1$.*   □

Proof   We leave proof of the fact that if $d(C) \geq 2v + u + 1$, then C is simultaneously v-error-correcting and u-error-detecting, as an exercise.

For the converse, suppose that C has the desired error detection/correction properties but that $d(C) \leq 2v + u$. We will follow a line similar to that in the proof of Theorem 4.4.2. Since C is v-error-correcting, Theorem 4.4.2 implies that $2v + 1 \leq d(C)$. Let $\mathbf{c}$ and $\mathbf{d}$ be codewords in C with distance $k = d(C) = d(\mathbf{c}, \mathbf{d})$. Thus, $\mathbf{c}$ and $\mathbf{d}$ disagree in exactly $k = d(\mathbf{c}, \mathbf{d})$ positions, with $2v + 1 \leq k \leq 2v + u$ and we may assume as before that these positions are the first k positions. Consider the received word $\mathbf{x}$ that agrees with $\mathbf{c}$ in the first v positions, agrees with $\mathbf{d}$ in the next $k - v$ positions and agrees with both in the remaining positions

$$\mathbf{x} = \underbrace{x_1 \cdots x_v}_{\text{agree with } \mathbf{c}} \underbrace{x_{v+1} \cdots x_k}_{\text{agree with } \mathbf{d}} \underbrace{x_{k+1} \cdots x_n}_{\text{agree with both}}$$

Notice that $d(\mathbf{c}, \mathbf{x}) = k - v$ and $d(\mathbf{d}, \mathbf{x}) = v$, and $v + 1 \leq k - v \leq v + u$.

If $\mathbf{c}$ is sent and $\mathbf{x}$ is received, then error detection will fail since error "correction" will ensue. More specifically, since the number of errors in transmission lies between $v + 1$ and $v + u$, the prescribed strategy should simply report a decoding error. But instead, since $d(\mathbf{x}, \mathbf{d}) = v$, the prescribed strategy will decode $\mathbf{x}$ incorrectly as $\mathbf{d}$.   ■

The Probability of a Decoding Error

Let us take another look at the probability of a decoding error, using minimum distance decoding, with a binary symmetric channel.

The binary code $C = \{000000, 111111, 111000\}$ can easily be made more efficient by adding the codeword 000111, which would add one additional codeword at no cost to its minimum distance. This leads us to make the following definition.

Definition   An (n, M, d)-code is said to be **maximal** if it is not contained in any larger code with the same minimum distance, that is, if it is not contained in any $(n, M + 1, d)$-code.   □

It is intuitively clear that, given any code C, we may continually add new codewords to it until we get a maximal code C' containing C. The following characterization of maximal codes, whose proof is left to the reader, makes this even more obvious.

Theorem 4.4.5   *An (n, M, d)-code C code is maximal if and only if, for all words* $\mathbf{x}$*, there is a codeword* $\mathbf{c}$ *with the property that* $d(\mathbf{x}, \mathbf{c}) < d$.   □

It is interesting to note that, if a particular (n, M, d)-code C is not maximal, there are advantages and disadvantages to enlarging it to a maximal code C'. On the one hand, C' is still $\lfloor \frac{d-1}{2} \rfloor$-error-correcting, which is good, and C' can encode more source symbols, which is also good. On the other hand, while C may be able to correct more than $\lfloor \frac{d-1}{2} \rfloor$ errors on occasion, C' will never be able to correct more than $\lfloor \frac{d-1}{2} \rfloor$ errors. As a simple example, consider the code $C = \{00000, 11000\}$, of minimum distance 2. This code is single-error-correcting, but will also correct many other errors. For instance, if 00000 is sent but 00111 is received, then 00000 is closer to the received word than 11000 and so correct decoding will take place, even though three errors were made. However, if we fill out C to a maximal code, then three symbol errors in a single word will always be decoded incorrectly. Thus, maximal codes are best when we are concerned only with a code's "guaranteed" error correction capability.

As a result, it is more difficult to get a handle on the probability of decoding error with nonmaximal codes. For maximal codes, we can be a bit more precise.

In the case of a maximal code, if a codeword $\mathbf{c}$ is transmitted, and if d or more symbol errors are made, so that the received word $\mathbf{x}$ has the property that $d(\mathbf{x}, \mathbf{c}) \geq d$, then $\mathbf{x}$ will be closer to a different codeword, and so minimum distance decoding will definitely result in a decoding error. Since the probability of making exactly k symbol errors in a binary symmetric channel with crossover probability p is

$$\binom{n}{k} p^k (1-p)^{n-k}$$

we get the following lower bound on the probability of a decoding error

$$\mathcal{P} \text{ (decode error)} \geq \sum_{k=d}^{n} \binom{n}{k} p^k (1-p)^{n-k}$$

On the other hand, any (n, M, d)-code C (maximal or not) is $\lfloor \frac{d-1}{2} \rfloor$-error-correcting. Hence, the probability of correct decoding is at least as large as the probability of making $\lfloor \frac{d-1}{2} \rfloor$ or fewer errors, that is,

$$\mathcal{P} \text{ (correct decoding)} \geq \sum_{k=0}^{\lfloor \frac{d-1}{2} \rfloor} \binom{n}{k} p^k (1-p)^{n-k}$$

Hence,

$$\mathcal{P} \text{ (decode error)} = 1 - \mathcal{P} \text{ (correct decoding)} \leq 1 - \sum_{k=0}^{\lfloor \frac{d-1}{2} \rfloor} \binom{n}{k} p^k (1-p)^{n-k}$$

We thus have both an upper bound and a lower bound on the probability of a decoding error. Let us summarize in a theorem.

Theorem 4.4.6 *For the binary symmetric channel using minimum distance decoding, the probability of a decoding error for a maximal (n, M, d)-code satisfies*

$$\sum_{k=d}^{n} \binom{n}{k} p^k (1-p)^{n-k} \leq \mathcal{P} \text{ (decode error)} \leq 1 - \sum_{k=0}^{\lfloor \frac{d-1}{2} \rfloor} \binom{n}{k} p^k (1-p)^{n-k}$$

For a nonmaximal code, the upper bound still holds, but the lower bound may not. □

Example 4.4.9 Consider the binary repetition code $\text{Rep}_2(4) = \{0000, 1111\}$ and a binary symmetric channel with crossover probability $p = 0.001$. Since this code is a maximal $(4,2,4)$-code, Theorem 4.4.6 gives the bounds

$$(0.001)^4 \leq \mathcal{P} \text{ (decode error)} \leq 1 - (0.999)^4 - 4(0.001)(0.999)^3$$

or

$$10^{-12} \leq \mathcal{P} \text{ (decode error)} \leq 5.992 \times 10^{-6}$$

As you can see, this is quite good. Unfortunately, the code $\text{Rep}_2(4)$ contains only two codewords and so it is not very useful.   □

Exercises

1. Consider the binary code $C = \{11100, 01001, 10010, 00111\}$.

   (a) Compute the minimum distance of C.

   (b) Decode the words 10000, 01100, and 00100 using minimum distance decoding.

2. In each case, construct a binary (n, M, d)-code with the given parameters (n, M, d) or prove that no such code can exist.

   (a) $(8,2,8)$

   (b) $(8,3,8)$

   (c) $(3,9,1)$

   (d) $(4,8,2)$

   (e) $(5,3,4)$

3. Consider the code C consisting of all words in $\mathbb{Z}_2^n$ that have an even number of 1s. What is the length, size, and minimum distance of C?

4. Construct an explicit example to illustrate that simultaneous error detection and correction can reduce the error detecting capabilites of a code (Theorem 4.4.4.)

5. Prove Corollary 4.4.3.

6. Estimate the probability of a decoding error using the binary repetition code of length 5 under a binary symmetric channel with crossover probability $p = 0.001$. (Assume minimum distance decoding.)

7. As we will see later in this chapter, one of the Hamming codes H has parameters $(15, 2^{11}, 3)$. Find an upper bound for the probability of a decoding error for this code, using a binary symmetric channel with crossover probability $p = 0.001$. (Assume minimum distance decoding.)

8. Does a binary (8,4,5)-code exist? Justify your answer.

9. Does a binary (7,3,5)-code exist? Justify your answer.

10. With reference to Theorem 4.4.4, prove that if $d(C) \geq 2v + u + 1$, then C is simultaneously v-error-correcting and u-error-detecting.

11. Prove that an (n, M, d)-code C code is maximal if and only if, for all words $\mathbf{x} \notin C$, there is a codeword $\mathbf{c}$ with the property that $d(\mathbf{x}, \mathbf{c}) < d$.

4.5 Perfect Codes and the Sphere-Packing Condition

Some Examples of Codes

Before continuing with the theory of codes, we should stop to describe some of the more famous families of codes, so that we will have something to use as examples. We will discuss these codes (and others) in detail in later chapters. The term *perfect* (as applied in these examples) is defined later in the section.

The Repetition Codes

We have already discussed the repetition codes, but let us include them here for completeness. The **r-ary repetition code** of length n is the (n, r, n)-code

$$Rep_r(n) = \{00\cdots 0, 11\cdots 1, 22\cdots 2, \ldots, (r-1)(r-1)\cdots(r-1)\}$$

consisting of r codewords, each of length n. These codes are perfect $\left\lfloor \frac{n-1}{2} \right\rfloor$-error-correcting codes.

The Hamming Codes

The family of Hamming codes $\mathcal{H}_r(h)$ is probably the most famous of all error-correcting codes. These codes were discovered independently by Marcel Golay in 1949 and Richard Hamming in 1950. The Hamming codes are perfect codes and have the advantage of being very easy to decode. However, these codes are only single-error-correcting, having minimum distance 3.

Specifically, for each prime power r and for each $h > 0$, the Hamming code $\mathcal{H}_r(h)$ is an r-ary (n, M, d)-code with parameters

$$n = \frac{r^h - 1}{r - 1}, M = r^{n-h}, d = 3$$

The most common case by far is the binary case, where

$$n = 2^h - 1, M = 2^{n-h}, d = 3$$

Golay Codes

In 1948, Marcel Golay introduced some very special codes, denoted by $\mathcal{G}_{23}$, $\mathcal{G}_{24}$, $\mathcal{G}_{11}$, and $\mathcal{G}_{12}$ that are now called Golay codes. The code $\mathcal{G}_{24}$ is a binary (24,4096,8)-code which, as we mentioned before, was used by the Voyager spacecraft to transmit color photographs of Jupiter and Saturn. The related code $\mathcal{G}_{23}$ is a perfect binary (23,4096,7)-code.

The code $\mathcal{G}_{11}$ is a perfect ternary (11,729,5)-code, and $\mathcal{G}_{12}$ is a ternary (12,729,6)-code. The binary Golay codes are among the most important codes, for both practical and theoretical reasons.

Reed-Muller Codes

The Reed-Muller codes are a family of binary codes that have good practical value and nice decoding properties. For each positive integer m, and each integer r for which $0 \le r \le m$, the r-th order Reed-Muller code $\mathcal{R}(r, m)$ has parameters

$$n = 2^m, M = 2^{1+\binom{m}{1}+\cdots+\binom{m}{r}}, d = 2^{m-r}$$

The first order Reed-Muller codes $\mathcal{R}(1, m)$ are perfect $(2^m, 2^{m+1}, 2^{m-1})$-codes. The code $\mathcal{R}(1,5)$ was used by Mariner 9 to transmit black and white photographs of Mars in 1972.

Spheres in $\mathbb{Z}_r^n$

Now let us return to a discussion of error correction. We can get a more geometric view of the error correction properties of a code by defining the concept of a sphere in $\mathbb{Z}_r^n$.

Definition   Let $\mathbf{x}$ be a string in $\mathbb{Z}_r^n$ and let $\rho \geq 0$. (ρ is the Greek letter rho.) The **sphere** $S_r^n(\mathbf{x}, \rho)$ in $\mathbb{Z}_r^n$ with center $\mathbf{x}$ and radius ρ is the set of all strings in $\mathbb{Z}_r^n$ whose distance from $\mathbf{x}$ is at most ρ. In symbols,

$$S_r^n(\mathbf{x},\rho) = \{\mathbf{y} \in \mathbb{Z}_r^n \mid d(\mathbf{x}, \mathbf{y}) \leq \rho\} \qquad \square$$

Example 4.5.1   The sphere $S_2^3(101,2)$ in $\mathbb{Z}_2^3$ consists of all binary strings of length 3 whose distance from 101 is at most 2. Thus,

$$S_2^3(101, 2) = \{101, 001, 111, 100, 011, 000, 110\} \qquad \square$$

Example 4.5.2   Figure 4.5.1 shows the set $\mathbb{Z}_2^3$. The words that lie in the sphere $S_2^3(111,1)$ of radius 1 about the word 111 are shown as solid dots in this figure. (Note that a sphere need not look very round!) $\square$

The Volume of a Sphere

To determine the number of strings contained in a sphere $S_r^n(\mathbf{x}, \rho)$, that is, the volume of $S_r^n(\mathbf{x}, \rho)$, we take the sum of the number of strings at

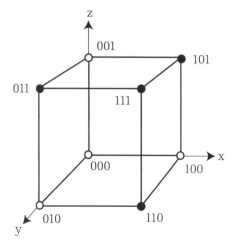

FIGURE 4.5.1

distance k from $\mathbf{x}$ for $k = 0, 1, \ldots, \rho$. Since there are $\binom{n}{k}(r-1)^k$ strings that have distance k from a given string $\mathbf{x}$ of length n, we get the following result.

Theorem 4.5.1 *The volume of the sphere $S_r^n(\mathbf{x}, \rho)$ is*

$$V_r^n(\rho) = \sum_{k=0}^{\rho} \binom{n}{k}(r-1)^k \qquad \square$$

Note that since the volume of a sphere $S_r^n(\mathbf{x}, \rho)$ does not depend on the center $\mathbf{x}$, we can use the notation $V_r^n(\rho)$. The volume of a sphere $S_2^n(\mathbf{x}, \rho)$ in $\mathbb{Z}_2^n$ has the simpler form

$$V_2^n(\rho) = \sum_{k=0}^{\rho} \binom{n}{k}$$

which is just the sum of the first ρ binomial coefficients.

Example 4.5.3 The volume of the sphere $S_2^4(1111, 2)$ is

$$V_2^4(2) = 1 + \binom{4}{1} + \binom{4}{2} = 1 + 4 + 6 = 11$$

Thus, there are 11 binary strings of length 4 whose distance from 1111 is at most 2. $\qquad \square$

Error Correction and the Packing Radius of a Code

Unfortunately, for $n > 3$, it is not possible to draw realistic pictures similar to Figure 4.5.1. However, we can get some representation of the situation for larger values of n by looking at Figure 4.5.2. (Of course, this figure is not metrically accurate—it is only intended as a guide to intuition.)

The solid dots in this figure represent the codewords of an (n, M, d)-code C, and the open dots represent all other words of length n. There is a sphere centered at each codeword of C. The common radius of the spheres is determined by the following simple rule: make the radius as large as possible as long as the spheres remain disjoint (that is, have no words in common). This radius has a name.

Definition Let C be an r-ary (n, M, d)-code. The **packing radius** of C, denoted by $pr(C)$, is the largest possible radius for a set of disjoint spheres,

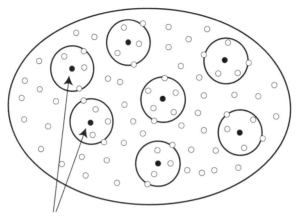

The minimum distance
is between these two
codewords

FIGURE 4.5.2

one centered at each codeword. We refer to the spheres $S_r^n(\mathbf{c}, pr(C))$, centered at each codeword $\mathbf{c}$, as the **packing spheres** for C. □

The value of the packing radius depends on the minimum distance d of the code. If d is even, say $d = 2t + 2$, then we increase the radius of the spheres until just before two spheres become tangent, as shown on the left side of Figure 4.5.3. In this case, the packing radius is $t = \frac{d-2}{2} = \left\lfloor \frac{d-1}{2} \right\rfloor$. If d is odd, say $d = 2t + 1$, then we increase the radius until just *before* two spheres overlap, as shown on the right side of Figure 4.5.3. In this case, the packing radius is $t = \frac{d-1}{2} = \left\lfloor \frac{d-1}{2} \right\rfloor$.

Putting these two cases together gives the following.

Theorem 4.5.2 *The packing radius of an* (n, M, d)-*code is* $pr(C) = \left\lfloor \frac{d-1}{2} \right\rfloor$. □

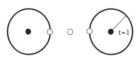

$d(C) = 4 = 2(1) + 2$
so t = 1 and C is
single-error-correcting

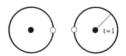

$d(C) = 3 = 2(1) + 1$
so t = 1 and C is
single-error-correcting

FIGURE 4.5.3

We can now describe the error-correcting properties of a code in spherical terms.

Corollary 4.5.3   *A code C is exactly v-error-correcting if and only if* $pr(C) = v$.   □

Perfect Codes

The following concept plays a major role in coding theory.

Definition   An r-ary (n, M, d)-code $C = \{\mathbf{c}_1, \ldots, \mathbf{c}_M\}$, with code alphabet A, is said to be **perfect** if any of the following equivalent conditions holds.

1. Every string in A^n is contained in some packing sphere, in symbols

$$A^n \subseteq \bigcup_{i=1}^{M} S_r^n(\mathbf{c}_i, pr(C))$$

2. The union of the packing spheres is precisely A^n, in symbols

$$A^n = \bigcup_{i=1}^{M} S_r^n(\mathbf{c}_i, pr(C))$$

3. The sum of the volumes of the packing spheres is equal to the number of strings in A^n, in symbols

$$M \cdot V_r^n \left(\left\lfloor \tfrac{d-1}{2} \right\rfloor \right) = r^n$$

or

$$M \cdot \sum_{k=0}^{\left\lfloor \frac{d-1}{2} \right\rfloor} \binom{n}{k} (r-1)^k = r^n \qquad\qquad \square$$

We will leave it to the reader to show that all of these conditions are equivalent. The last condition (in part 3) is known as the **sphere-packing condition**. It can be used to determine the size M of a perfect code with given minimum distance d.

Figure 4.5.4 illustrates a perfect code. It is perfect because the packing spheres perfectly cover all of the strings in A^n, with no overlap and leaving nothing out. In other words, the packing spheres form a partition of A^n.

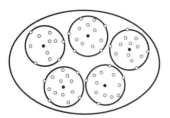

This code is perfect   **FIGURE 4.5.4**

Example 4.5.4 The binary Golay code $\mathcal{G}_{23}$ has parameters $(23, 2^{12}, 7)$. Substituting these numbers into the left-hand side of the sphere-packing condition gives

$$2^{12} \cdot \sum_{k=0}^{3} \binom{23}{k} = 2^{12}(1 + 23 + 253 + 1771) = 2^{12} \cdot 2048 = 2^{12} 2^{11} = 2^{23}$$

and so the sphere-packing condition holds for this code. Thus, the Golay $(23, 2^{12}, 7)$-code is perfect. □

It is important to emphasize that, just because numbers n, M, and d exist that satisfy the sphere-packing condition does not mean that there is a (perfect) code with these parameters. There may, in fact, be no such codes at all, as we will see in a moment.

The important problem of determining all perfect codes has not yet been solved. Of course, one approach to the problem is to try to find all solutions to the sphere-packing condition and then try to find all codes with these parameters. This monumental task has been only partially accomplished and we will now describe some of these solutions. Keep in mind that, just because we find one code that has a given set of parameters doesn't mean that there might not be other codes with these same parameters.

Solution 1 The parameters $(n, r^n, 1)$ satisfy the sphere-packing condition. However, the only r-ary code that fits these parameters is $\mathbb{Z}_r^n$ itself. In this case, the packing radius is 0, each packing sphere consists simply of the single codeword at the center and, while the code is perfect, it is not of much use, since no error correction is possible.

Solution 2 The parameters $(n, 1, 2n + 1)$ satisfy the sphere-packing condition. The set $C = \{\mathbf{0}\}$ consisting of the zero codeword alone is an

$(n,1,?)$-code, with undefined minimum distance, since there is only one codeword. However, if we set the minimum distance to $2n + 1$, then the parameters of the code C satisfy the sphere-packing condition. Again, this code is not of much use, having only one codeword.

Solution 3 The parameters $(2m + 1, 2, 2m + 1)$ satisfy the sphere-packing condition, when $r = 2$. Since the binary repetition codes $\text{Rep}_2(2m + 1)$ of odd length have these parameters, they are perfect codes. (The binary repetition codes are essentially the only binary codes with these parameters.)

Solution 4 The parameters

$$\left(\frac{r^h - 1}{r - 1}, r^{n-h}, 3 \right)$$

where r is a prime power and $h \geq 2$, satisfy the sphere-packing condition. As we have mentioned, the Hamming codes have these parameters and are therefore perfect. However, the Hamming codes are not the only family of codes with these parameters. (See the exercises in Section 6.1.)

A computer search was conducted by the well-known coding theorist J. H. van Lint in 1967. This search showed that the only solutions to the sphere-packing condition for

$$n \leq 1000, d \leq 1000, r \leq 100$$

besides those given by the families above are

Solution 5 $(23, 2^{12}, 7)$

Solution 6 $(90, 2^{78}, 5)$

Solution 7 $(11, 3^6, 5)$.

It happens that there is no code with parameters corresponding to solution 6. (We will ask you to prove this, for linear codes, in the Exercises of Section 5.5.) Solutions 5 and 7 correspond to the Golay codes.

Thus, we see that perfect codes are not only perfect, but are also rare. As further justification of this statement, it has been proven that the only perfect binary v-error-correcting codes, for $v \geq 2$, are the repetition codes (which are not very useful) and the Golay code with parameters $(23, 2^{12}, 7)$. This makes the Golay code very special indeed.

Exercises

1. Compute the following volumes.
   a) $V_2^5(1)$ b) $V_3^5(2)$ c) $V_4^4(3)$ d) $V_2^9(4)$ e) $V_5^{10}(10)$

2. Prove Corollary 4.5.3.

3. Show that a perfect code must have odd minimum distance.

4. Verify that the conditions given in the definition of perfect code are equivalent.

5. Show that the sphere-packing condition is satisfied by the parameters $(n, M, d) = (n, r^n, 1)$.

6. Show that the sphere-packing condition is satisfied by the parameters $(n, M, d) = (n, 1, 2n + 1)$.

7. Show that the sphere-packing condition is satisfied by the parameters $(n, M, d) = (2m + 1, 2, 2m + 1)$, $r = 2$.

8. Show that the sphere-packing condition is satisfied by the parameters $(n, M, d) = (\frac{r^h - 1}{r - 1}, r^{n-h}, 3)$, for $h \geq 2$.

9. Why are the binary repetition codes essentially the only binary codes with the parameters $(2m + 1, 2, 2m + 1)$?

10. Do your own computer search for solutions to the sphere-packing condition.

4.6   Making New Codes from Old Codes

There are many useful ways to create new codes from existing codes. Let us discuss a few of the more important ways.

Extending a Code

The process of adding one or more additional positions to all of the code-words in a code, thereby increasing the length of the code, is referred to as **extending** the code. The most common way to extend a code is by adding an overall parity check. Let us consider the binary case. The **weight** $w(\mathbf{c})$ of a binary word $\mathbf{c}$ is the number of 1s in $\mathbf{c}$. (For example, $w(1001010) = 3$.)

If C is a binary (n, M, d)-code, then we construct a new code as follows. To each codeword $\mathbf{c}$ in C, we add an additional bit in such a way that the resulting codeword has even weight. Thus, if $\mathbf{c}$ has an odd weight, we add a 1, and if $\mathbf{c}$ has an even weight, we add a 0. Denoting the resulting codeword by $\bar{\mathbf{c}}$, we have

$$\text{if } c = u_1 u_2 \ldots u_n \text{ then } \bar{c} = \begin{cases} u_1 u_2 \ldots u_n 0 & \text{if } w(c) \text{ is even} \\ u_1 u_2 \ldots u_n 1 & \text{if } w(c) \text{ is odd} \end{cases}$$

This is referred to as *adding an even parity check* to the code C. The resulting code $\overline{C}$ has length $n + 1$ and size M.

The minimum distance of $\overline{C}$ may be either $d(C)$ or $d(C) + 1$, depending on the parity of $d(C)$. In particular, since all of the words in $\overline{C}$ have even weight, the distance between any two codewords in $\overline{C}$ is even (exercise). Hence, the minimum distance of $\overline{C}$ is also even. It follows that if $d(C)$ is even then $d(\overline{C}) = d(C)$ and if $d(C)$ is odd then $d(\overline{C}) = d(C) + 1$.

In either case, we have

$$\left\lfloor \frac{d(C) - 1}{2} \right\rfloor = \left\lfloor \frac{d(\overline{C}) - 1}{2} \right\rfloor$$

and so (alas) the error-*correcting* capabilities of the code do not increase by adding an even parity check.

More generally, if the code alphabet of C is any finite field, say $\mathbb{Z}_p$ where p is a prime, then we may adjoin an additional element to each codeword in C so that the sum of the elements of a codeword is 0. In symbols, if $\mathbf{c} = c_1 \cdots c_n$ then $\bar{\mathbf{c}} = c_1 \cdots c_n c_{n+1}$, where

$$\sum_{i=1}^{n+1} c_i = 0$$

and so

$$c_{n+1} = -\sum_{i=1}^{n} c_i$$

As an example, if $\mathbf{c} = 12314$ is a codeword in a code over $\mathbb{Z}_5$, then since $1 + 2 + 3 + 1 + 4 = 1$ in $\mathbb{Z}_5$, adding an overall parity check gives the codeword $\overline{\mathbf{c}} = 123144$.

Example 4.6.1 Adding an overall parity check to the binary code $C = \{00, 01, 10, 11\}$ gives the extended code $\overline{C} = \{000, 011, 101, 110\}$. Notice that C has minimum distance 1, but $\overline{C}$ has minimum distance 2. □

Shortening a Code — The Cross-Sections of a Code

Shortening a code refers to the process of keeping only those codewords in a code that have a given symbol in a given position (for instance, a 0 in the first position), and then deleting that position. If C is an (n, M, d)-code, then a shortened code has length $n - 1$ and minimum distance at least d. In fact, shortening a code can result in a substantial increase in the minimum distance and therefore in the error-correcting capabilities of the code, for it can eliminate codewords that are "poorly behaved" as far as distance is concerned. (We ask you to find an example of this in the exercises.) Of course, shortening a code does result in a code with smaller size, which is not so desirable.

The shortened code formed by taking codewords with an s in the ith position is referred to as the **cross-section** $x_i = s$. We will have many occasions to use cross-sections in the sequel.

Example 4.6.2 The code $C = \{0000, 0110, 0011, 1010, 1110\}$ has minimum distance 1. The cross-section $x_1 = 0$ is obtained by taking all codewords that have a 0 in the first position and removing that position, giving $\{000, 110, 011\}$, which has minimum distance 2. □

The $u(u+v)$-Construction

For binary codes, the **u(u + v)-construction** proceeds as follows. Suppose that C_1 is a binary (n, M_1, d_1)-code and C_2 is a binary (n, M_2, d_2)-code.

(Note that both codes must have the same length.) Let

$$C_1 \oplus C_2 = \{\mathbf{c}(\mathbf{c} + \mathbf{d}) \mid \mathbf{c} \in C_1, \mathbf{d} \in C_2\}$$

where the parentheses indicate juxtaposition and the plus sign indicates bitwise addition modulo 2 (that is, $0 + 0 = 0, 0 + 1 = 1 + 0 = 1, 1 + 1 = 0$). For instance, if $\mathbf{c} = 1100$ and $\mathbf{d} = 0110$ then $\mathbf{c} + \mathbf{d} = 1100 + 0110 = 1010$ and so $\mathbf{c}(\mathbf{c} + \mathbf{d}) = 11001010$.

We leave it as an exercise to show that $C_1 \oplus C_2$ has length $2n$, size $M_1 M_2$ and minimum distance equal to $\min\{2d_1, d_2\}$. The $\mathbf{u}(\mathbf{u} + \mathbf{v})$-construction is an important tool in constructing new codes and can be used to construct the Reed-Muller family of codes (see the exercises).

Example 4.6.3 Let $C_1 = \{00, 10\}$ and let $C_2 = \{00, 11\}$. Then

$$C_1 \oplus C_2 = \{0000, 0011, 1010, 1001\}$$

which is a code of length 4, size 4 and minimum distance 2. $\qquad \square$

Equivalence of Codes

Almost every branch of mathematics has at least one notion of equivalence (not to be confused with equality). To state a few instances, in set theory, two sets are equivalent if they have the same size. In logic, two statements are equivalent if they always have the same truth value. In plane geometry, two figures in the plane are equivalent if one can be changed into the other by a rigid motion. In matrix theory, two matrices A and B are equivalent if there exits two invertible matrices P and Q such that $B = PAQ$. In elementary algebra, two equations are equivalent if they have the same solution set. In topology, two topological spaces are equivalent if there is a bijective continuous function between them. In differential geometry, two manifolds are equivalent if there is a bijective differentiable function between them.

Let us now consider equivalence of codes.

Definition Let C be an r-ary (n, M)-code over the alphabet A. Consider the following two-step procedure for transforming C.

1. Permute the code symbols in each codeword in C, using a permutation σ. That is, replace each codeword $\mathbf{c} = c_1 c_2 \cdots c_n \in C$ by the word $c_{\sigma(1)} c_{\sigma(2)} \cdots c_{\sigma(n)}$. Denote the resulting code by D.

2. For each position i, permute the code symbols in the ith position of the codewords in D, using a permutation π_i (possibly different for each position i). That is, replace each codeword $\mathbf{d} = d_1 d_2 \cdots d_n$ with $\pi_1(d_1)\pi_2(d_2)\cdots\pi_n(d_n)$. Call the resulting code E.

A code E is said to be **equivalent** to a code C if E can be obtained from C using the procedure above. □

Note that, since each of the steps in the procedure above is reversible, it follows that if E is equivalent to C, then C is equivalent to E and so we may simply say that two codes are equivalent.

Although the concept of equivalence is extremely important, we will not be using it directly to show that two codes are equivalent. However, we will use it to state certain uniqueness results. For instance, it has been proven that any binary code with the same parameters as the Golay code $\mathcal{G}_{24}$ is equivalent to $\mathcal{G}_{24}$. Thus, the Golay code $\mathcal{G}_{24}$ is, in this sense, unique.

The following intuitive results are quite useful. We will leave the proofs for the exercises.

Lemma 4.6.1   *If the code alphabet A contains 0, then any code over A is equivalent to a code that contains the zero codeword $\mathbf{0} = 0\cdots 0$.* □

Theorem 4.6.2   *Equivalent codes have the same parameters (length, size, and minimum distance).* □

Exercises

1. Add an even parity check to the binary $(5,4,3)$-code

$$C = \{00000, 11100, 00111, 11011\}$$

   What are the parameters of the extended code?

2. Describe the process of adding an *odd* parity check to a binary code. Let C be a binary code and let $\overline{C}$ be the result of adding an even parity check. What happens if you add an even parity check to $\overline{C}$? What happens if you add an odd parity check to $\overline{C}$?

3. If all of the codewords in a binary code C have an even weight, show that all distances between codewords are even. Let C be a code and let $\overline{C}$ be the code obtained from C by adding an even parity check.

Prove that if $d(C)$ is even, then $d(\overline{C}) = d(C)$ and if $d(C)$ is odd then $d(\overline{C}) = d(C) + 1$.

4. What do you get when you take a cross section of a repetition code?

5. Find the cross-section $x_2 = 1$ of the code

$$C = \{0000, 0001, 0010, 0011, 0100, 0101, 0110, 0111, 1110, 1111\}.$$

What are the parameters of this code?

6. Can taking a cross-section of a code result in getting two identical words?

7. Let C be an r-ary (n, M, d)-code over the alphabet $\mathbb{Z}_r$. Form the cross-sections C_i defined by $x_1 = i$, for each $i = 0, 1, \ldots, r-1$. Suppose that C_i is an $(n-1, M_i, d_i)$-code. Show that $\sum M_i = M$ and that $d_i \geq d$ for all i.

8. Let C be an r-ary (n, M, d)-code over the alphabet $\mathbb{Z}_r$. Show that, as long as $d < n$, then for some position i, there is a cross-section that has minimum distance d. What can happen if $d = n$?

9. Construct a binary (7,6,2)-code with a cross section that is a (6,4,3)-code.

10. Suppose that C is an (n, M, d)-code. Show that C is a cross-section of a larger code with parameters $(n + 1, M + 2, 1)$. Thus, the minimum distance of a cross-section can be considerably larger than the minimum distance of the original code. Which are the "poorly behaved" codewords in this case?

11. Construct $C_1 \oplus C_2$, where $C_1 = \{00000000, 11000000\}$ and $C_2 = \{00000000, 11111000, 00011111\}$. What are the parameters of C_1, C_2, and $C_1 \oplus C_2$?

12. Construct $C_1 \oplus C_2$, where $C_1 = \{000, 001, 111\}$ and $C_2 = \{100, 011, 001\}$. What are the parameters of C_1, C_2, and $C_1 \oplus C_2$?

13. Let C be a binary (n, M, d)-code. Let C^c denote the set of complements of all codewords in C.

    (a) Show that $d(\mathbf{c}, \mathbf{d}^c) = n - d(\mathbf{c}, \mathbf{d})$.

    (b) Show that the minimum distance of the code $C \cup C^c$ is

    $$d(C \cup C^c) = \min\{d, n - d_{max}\}$$

    where d_{max} is the maximum distance between codewords in C.

14. If C_1 is an r-ary (n_1, M_1, d_1)-code and C_2 is an r-ary (n_2, M_2, d_2)-code over the same alphabet, then the **direct sum** $C_1 \odot C_2$ is the code

$$C_1 \odot C_2 = \{\mathbf{cd} \mid \mathbf{c} \in C_1, \mathbf{d} \in C_2\}$$

that consists of all words formed by juxtaposing a codeword in C_1 with a codeword in C_2. What are the parameters of the direct sum?

15. Referring to the direct sum construction of the previous exercise, compute $C_1 \odot C_2$ for the codes in Exercise 11. Compare the parameters of $C_1 \oplus C_2$ and $C_1 \odot C_2$.

16. Suppose that C_1 is a binary (n, M_1, d_1)-code and C_2 is a binary (n, M_2, d_2)-code. Show that the $u(u + v)$-construction $C_1 \oplus C_2$ has length $2n$, size $M_1 M_2$, and minimum distance equal to $\min\{2d_1, d_2\}$. *Hint for the latter part*: consider two distinct codewords in $C_1 \oplus C_2$, say $\mathbf{u}_1 = \mathbf{c}_1(\mathbf{c}_1 + \mathbf{d}_1)$ and $\mathbf{u}_2 = \mathbf{c}_2(\mathbf{c}_2 + \mathbf{d}_2)$. If $\mathbf{d}_1 = \mathbf{d}_2$, then show that $d(\mathbf{u}_1, \mathbf{u}_2) \geq 2d_1$. If $\mathbf{d}_1 \neq \mathbf{d}_2$, then consider under what circumstances the addition (modulo 2) of $\mathbf{c}_i$ to $\mathbf{d}_i$ (for $i = 1$ and 2) could wipe out a bit difference between $\mathbf{d}_1$ and $\mathbf{d}_2$.

17. Let C_2 be the (4,8,2)-code consisting of all eight binary words of length 4 that have even weight.

(a) Construct $C_3 = C_2 \oplus \mathrm{Rep}_2(4)$. What are its parameters?

(b) What are the parameters of $C_4 = C_3 \oplus \mathrm{Rep}_2(8)$?

(c) What are the parameters of $C_5 = C_4 \oplus \mathrm{Rep}_2(16)$?

(d) Show that we can construct a binary $(2^m, 2^{m+1}, 2^{m-1})$-code in this fashion. These codes are the first order Reed-Muller codes.

18. Prove Lemma 4.6.1.

19. Prove Theorem 4.6.2.

20. If C is a code over $\mathbb{Z}_p$ and $\overline{C}$ is the code obtained by adding an overall parity check, what is the relationship between the minimum distances of C and $\overline{C}$? Find examples of ternary codes that exhibit each of the possibilities in your answer.

4.7   The Main Coding Theory Problem

Two of the most desireable goals in designing good r-ary (n, M, d)-codes are *high efficiency*, in the sense of having a relatively large number of codewords for a given length and *high minimum distance*, for good error-correcting capabilities. Unfortunately, these goals are at odds with each other, and this is why designing optimal codes is a difficult process. (The other important goal in designing a good code is ease of encoding and decoding.)

The Rate of a Code

To get a meaningful measure of the efficiency of an r-ary (n, M)-code, consider that each of the M codewords can encode a single source symbol. However, we may be comparing apples and oranges here unless we express the source symbols in the same r-ary units as the codewords. To be more specific, each codeword is an r-ary string of length n and so we should also express the source symbols as r-ary strings. Solving $r^k = M$ for k tells us that we require r-ary strings of length $k = \log_r M$. (This number may not be an integer, but that is not a problem since the discussion is purely abstract and the number k is used only in comparison between codes.) Thus, transmission takes place at the rate of $\log_r M$ r-ary source units for every n r-ary code units. This leads to the following definition.

Definition   The **transmission rate** of an r-ary (n, M)-code C is

$$\mathcal{R}(C) = \frac{\log_r M}{n} \qquad \qquad \square$$

Perhaps the best measure of the error-correcting capabilities of an (n, M, d)-code C is not the absolute number $\left\lfloor \frac{d-1}{2} \right\rfloor$ of errors it can correct, but rather the *relative* number of errors it can correct per codeword position. After all, correcting 100 errors per codeword, for example, may sound like a lot, but sounds much less impressive when the codeword length is 100 million!

Definition   The **error correction rate** of an (n, M, d)-code C is

$$\delta(C) = \frac{\left\lfloor \frac{d-1}{2} \right\rfloor}{n} \qquad \qquad \square$$

Let us consider some examples of these measurements.

Example 4.7.1 Consider the binary repetition codes $Rep_2(n)$, which have size 2 and therefore transmission rate

$$\mathcal{R}(Rep_2(n)) = \frac{\log_2 2}{n} = \frac{1}{n}$$

For $n = 2u + 1$, the repetition code is u-error-correcting and can therefore correct up to u errors in each n-bit codeword, for an error correction rate of

$$\delta(C) = \frac{u}{n} = \frac{u}{2u + 1}$$

Notice that the numbers $\frac{u}{2u+1}$ form an increasing sequence that approaches $\frac{1}{2}$. Thus, the longer the codeword length, the better the error correction, not just in absolute terms but also as a percentage of the codeword length. On the other hand, we are stuck with encoding only two source symbols, no matter how long the codeword lengths, and so the transmission rate goes to 0. Thus, repetition codes have excellent error-correcting potential (almost $\frac{1}{2}$ the number of bits sent) but only at the cost of *very* low efficiency. The moral here is that simply repeating the symbols transmitted is not a smart way to proceed. □

Example 4.7.2 At one extreme, we could completely sacrifice efficiency and use a code of size 1, which has perfect error-correction but transmission rate 0. At the other extreme, we could completely sacrifice error correction and simply use every r-ary word of length n as a codeword. Then we get a very high (indeed, the maximum possible) transmission rate

$$\mathcal{R} = \frac{\log_r r^n}{n} = 1$$

but absolutely no error correction. □

Example 4.7.3 The Hamming code $\mathcal{H}_r(h)$ has size $M = r^{n-h}$ and therefore rate

$$\mathcal{R}(\mathcal{H}_r(h)) = \frac{n - h}{n} = 1 - \frac{h}{n}$$

This is a good rate for large n and, in fact, approaches the maximum value 1. However, all Hamming codes are but single-error-correcting and so the error correction rate is

$$\delta(\mathcal{H}_r(h)) = \frac{1}{n}$$

which tends to 0 as n gets large. □

The previous examples should make it clear that a compromise between transmission rate and quality of error correction is in order. Unfortunately, coding theorists are still looking for a really good compromise.

The Famous Numbers $A_r(n, d)$

For given values of n and d, it is customary to let $A_r(n, d)$ denote the largest possible size M (and, consequently, transmission rate) for which there exists an r-ary (n, M, d)-code. Thus,

$$A_r(n, d) = \max\{M \mid \text{there exists an } r\text{-ary}(n, M, d)\text{-code}\}$$

Any (n, M, d)-code C that has maximum size, that is, for which $M = A_r(n, d)$ is called an **optimal code**. The numbers $A_r(n, d)$ play a central role in coding theory, and much effort has been expended in attempting to determine their values. In fact, determining the values of $A_r(n, d)$ has come to be known as the *main coding theory problem*.

Unfortunately, very little is currently known about the numbers $A_r(n, d)$. Our goal here is to determine some values of $A_r(n, d)$, for small values of n and d and to establish some general facts about these numbers. Note that, in order to show that $A_r(n, d) = K$, for some number K, it is enough to show that $A_r(n, d) \leq K$, and then find a specific r-ary (n, K)-code C for which $d(C) \geq d$, which shows that $A_r(n, d) \geq A_r(n, d(C)) \geq K$.

The following two results demonstrate how values of $A_2(n, d)$ can be computed for small n and d. We will use Lemma 4.6.1, which says that, given an (n, M, d)-code C, there is an equivalent (n, M, d)-code C' that contains the zero codeword. (If the code alphabet does not contain the symbol 0, we can simply replace one of the symbols by 0, without effecting the parameters of the code.) Hence, for the purposes of what follows, we may assume that our codes contain the zero codeword.

Theorem 4.7.1 $A_2(4, 3) = 2$ □

Proof   Let C be a (4,M,3)-code. By the remarks preceding this theorem, we may assume that C contains the zero codeword $\mathbf{0}$ = 0000. Now, since $d(C)$ = 3, any other codeword $\mathbf{c}$ in C must satisfy $d(\mathbf{c}, \mathbf{0}) \geq 3$, and so it must have at least three 1s. This leaves five possibilities for additional codewords in C, namely

$$1110, 1101, 1011, 0111 \quad 1111$$

But no pair of these has distance 3 apart, and so only one can be included in C. Hence, C can have at most two codewords, implying that $A_2(4, 3) \leq 2$. Furthermore, since C = {0000, 1110} is a (4,2,3)-code, we have $A_2(4, 3) \geq 2$, and so $A_2(4, 3)$ = 2. ∎

The proof of the following theorem shows how useful cross-sections can be in determining small values of $A_r(n, d)$.

Theorem 4.7.2 $A_2(5, 3)$ = 4   □

Proof   Let C be a (5,M,3)-code and consider the cross-section C_0 defined by x_1 = 0. We know that C_0 has minimum distance $d_0 \geq 3$ and since $A_2(4, 3)$ = 2 and $A_2(4, 4)$ = 2, it follows that C_0 has size $M_0 \leq 2$. Similarly, the cross-section C_1 defined by x_1 = 1 has size $M_1 \leq 2$. Thus, M = $M_0 + M_1 \leq 4$ and $A_2(5, 3) \leq 4$. On the other hand, the code C = {00000, 11100, 00111, 11011} has minimum distance $d(C)$ = 3, and so $A_2(5, 3) \geq 4$, proving the result. ∎

As you might imagine, the approach used in the proof of Theorem 4.7.2 will not go very far in determining values of $A_2(n, d)$, and much more sophisticated methods are needed. Unfortunately, very few actual values of $A_2(n, d)$ are known. Table 4.7.1 summarizes most of our current knowledge. In some cases where we do not know precise values, a range of values is indicated. For instance, the entry 72–79 means that $72 \leq A_2(10, 3) \leq 79$. The source for this table is Sloane (1982).

Table 4.7.1 contains only odd values of d. The following theorem, which we will prove later in this section, shows how to obtain values of $A_2(n, d)$, for d even, from this table.

Theorem 4.7.3   *If $d > 0$ is even, then $A_2(n, d)$ = $A_2(n - 1, d - 1)$.*   □

As an example, $A_2(9, 4)$ = $A_2(8, 3)$ = 20.

TABLE 4.7.1   Values of $A_2(n, d)$

n	d = 3	d = 5	d = 7
5	4	2	—
6	8	2	—
7	16	2	2
8	20	4	2
9	40	6	2
10	72-79	12	2
11	144-158	24	4
12	256	32	4
13	512	64	8
14	1024	128	16
15	2048	256	32
16	2560-3276	256-340	36-37

Let us now turn to the establishment of some general results about the numbers $A_r(n, d)$. Proof of the following simple result is left as an exercise.

Theorem 4.7.4

1. $A_r(n, d) \leq r^n$ for all $1 \leq d \leq n$

2. $A_r(n, 1) = r^n$

3. $A_r(n, n) = r$ □

The next result is often very useful. We leave its proof as an exercise.

Theorem 4.7.5   *For any $n \geq 2$,*

$$A_r(n, d) \leq rA_r(n - 1, d) \qquad \square$$

In order to prove our next result, we need some definitions.

Definition   The **weight** $w(\mathbf{x})$ of a string $\mathbf{x}$ in $\mathbb{Z}_r^n$ is defined to be the number of nonzero positions in $\mathbf{x}$.   □

Definition   The **intersection** $\mathbf{x} \cap \mathbf{y}$ of two binary strings of length n is the binary string that has a 1 in those places for which both $\mathbf{x}$ and $\mathbf{y}$ have 1s and has 0s elsewhere.   □

Note that the intersection $\mathbf{x} \cap \mathbf{y}$ is formed by taking the modulo 2 product of corresponding bits in $\mathbf{x}$ and $\mathbf{y}$.

Example 4.7.4

1. The string 011201021 in $\mathbb{Z}_3^7$ has weight 6.

2. If $\mathbf{x} = 110010$ and $\mathbf{y} = 101100$ then $\mathbf{x} \cap \mathbf{y} = 100000$.   □

Using the concept of intersection of strings, we can express the distance between binary strings in terms of weight.

Lemma 4.7.6 *If* $\mathbf{x}$ *and* $\mathbf{y}$ *are binary strings of length n, then*

$$d(\mathbf{x}, \mathbf{y}) = w(\mathbf{x}) + w(\mathbf{y}) - 2w(\mathbf{x} \cap \mathbf{y}) \qquad\qquad □$$

Proof Suppose that, among the n places in $\mathbf{x}$ and $\mathbf{y}$, there are a_{11} places in which both $\mathbf{x}$ and $\mathbf{y}$ have 1s, there are a_{10} places in which $\mathbf{x}$ has a 1 and $\mathbf{y}$ has a 0, and there are a_{01} places in which $\mathbf{x}$ has a 0 and $\mathbf{y}$ has a 1. Then,

$$d(\mathbf{x}, \mathbf{y}) = a_{10} + a_{01} = (a_{11} + a_{10}) + (a_{11} + a_{01}) - 2a_{11}$$
$$= w(\mathbf{x} +) + w(\mathbf{y}) - 2w(\mathbf{x} \cap \mathbf{y})$$

which proves the lemma.   ■

Now we can prove the following result, which leads directly to a proof of Theorem 4.7.3.

Theorem 4.7.7 *Let d be an odd positive integer. Then a binary* (n, M, d)-*code exists if and only if a binary* $(n + 1, M, d + 1)$-*code exists.*   □

Proof If a binary $(n + 1, M, d + 1)$-code C exists, then it is easy to construct a binary (n, M, d)-code. We simply take two codewords $\mathbf{c}$ and $\mathbf{d}$ in C with minimum distance $d(\mathbf{c}, \mathbf{d}) = d + 1$, choose a position in which they differ and delete this position from every codeword in C. Since the shortened codewords corresponding to $\mathbf{c}$ and $\mathbf{d}$ are now at a distance d from each other, and no two codewords have distance less than d from each other, the resulting code is an (n, M, d)-code.

As to the converse, suppose we have a (n, M, d)-code C, where d is odd. The code $\overline{C}$ created by adding an even parity check to C has length $n + 1$ and size M. Furthermore, since $w(\bar{\mathbf{c}})$ is even for all codewords in $\overline{C}$, Lemma 4.7.6 implies that

$$d(\bar{\mathbf{c}}, \bar{\mathbf{d}}) = w(\bar{\mathbf{c}}) + w(\bar{\mathbf{d}}) - 2w(\bar{\mathbf{c}} \cap \bar{\mathbf{d}})$$

is even for all codewords $\mathbf{c}$ and $\mathbf{d}$. Hence, the minimum distance $d(\overline{C})$ is even. But clearly $d \leq d(\overline{C}) \leq d + 1$, and since d is assumed to be odd, we must have $d(\overline{C}) = d + 1$. Thus, $\overline{C}$ is an $(n + 1, M, d + 1)$-code.   ■

We leave a formal proof of Theorem 4.7.3 from Theorem 4.7.7 to the reader.

Exercises

1. Compute a) $w(0110101)$, b) $w(120120)$, c) $w(1111)$.

2. Compute a) $101101 \cap 111111$, b) $1000 \cap 0111$, c) $101010 \cap 010101$

3. Prove that $d(\mathbf{x}, \mathbf{y}) = w(\mathbf{x} + \mathbf{y})$ for all $\mathbf{x}, \mathbf{y} \in \mathbb{Z}_2^n$, where $+$ is addition modulo 2.

4. Find the transmission rate and the error-correction rate of the Golay codes.

5. Find the transmission rate and the error-correction rate of the Reed-Muller codes.

6. Verify that $A_2(6, 5) = 2$.

7. Verify that $A_2(7, 5) = 2$.

8. Show that $A_2(8, 5) = 4$.

9. Show that $A_2(6, 3) = 8$.

10. Prove that $A_r(n, d) \leq r^n$ for all $1 \leq d \leq n$.

11. Prove that $A_r(n, 1) = r^n$.

12. Prove that $A_r(n, n) = r$.

13. Is it true that we can construct a code of minimum distance d by taking *any* code of minimum distance $d + 1$ and adding an appropriate word or words to it?

14. Show that if $e \leq d$ then $A_r(n, e) \geq A_r(n, d)$.

15. Prove that if $d > 0$ is even, then $A_2(n, d) = A_2(n - 1, d - 1)$.

16. Prove that

$$A_r(n, d) \leq r A_r(n - 1, d)$$

for any $n \geq 2$. Hint: let C be an (n, M, d)-code, where $M = A_r(n, d)$ and let us assume for concreteness that the code alphabet is $\mathbb{Z}_r = \{0, 1, \ldots, r - 1\}$. Among the r distinct cross-sections C_k defined by $x_1 = k$, for $k = 0, \ldots r - 1$, one of them has size at least M/r.

17. Show that if there exists a binary (n, M, d)-code with d even, then there also exists a binary (n, M, d)-code in which all codewords have even weight. *Hint*: use Theorem 4.7.7.

18. Show that $A_r(3, 2) = r^2$. Hint: let C be an optimal code and consider the cross-sections $x_1 = k$, where $k = 0, 1, \ldots, r - 1$. What are the parameters of these cross-sections?

19. Let C be an (n, M, d)-code

    (a) If C is not maximal, is it always possible to add codewords to C until the resulting code is maximal?

    (b) If C is not optimal, is it always possible to add codewords to C until the resulting code is optimal?

    (c) Give an example of a code that is maximal but not optimal.

4.8   Sphere-Covering and Sphere-Packing Bounds

Let us now turn our attention to some upper and lower bounds on the numbers $A_r(n, d)$ that arise from considering spheres in $\mathbb{Z}_r^n$.

Let $C = \{c_1, c_2, \ldots, c_M\}$ be an optimal r-ary (n, M, d)-code over $\mathbb{Z}_r$. Thus $M = A_r(n, d)$. Because C has maximal size, there can be no string in $\mathbb{Z}_r^n$ whose distance from *every* codeword in C is at least d. For if there were such a string, we could simply add it to C, and thereby obtain an $(n, M + 1, d)$-code.

Thus, every string in $\mathbb{Z}_r^n$ has distance at most $d-1$ from some codeword in C and so every string in $\mathbb{Z}_r^n$ is contained in at least one of the spheres $S_r^n(c_i, d - 1)$. For this reason, we say that the spheres $S_r^n(c_i, d - 1)$ **cover** the set $\mathbb{Z}_r^n$, in symbols

$$\mathbb{Z}_r^n \subseteq \bigcup_{i=1}^{M} S_r^n(c_i, d - 1) \tag{4.8.1}$$

Since $|\mathbb{Z}_r^n| = r^n$, taking sizes in (4.8.1) gives

$$r^n \leq |\bigcup_{i=1}^{M} S_r^n(c_i, d - 1)|$$

$$\le \sum_{i=1}^{M} |S_r^n(\mathbf{c}_i, d-1)|$$

$$= \sum_{i=1}^{M} V_r^n(d-1) = V_r^n(d-1) \cdot M$$

Dividing both sides of this by $V_r^n(d-1)$ gives

$$\frac{r^n}{V_r^n(d-1)} \le M$$

Since $M = A_r(n, d)$, we arrive at the following result, called the sphere-covering bound for $A_r(n, d)$.

Theorem 4.8.1 *(The sphere-covering bound for $\mathbf{A_r(n, d)}$) If $V_r^n(\rho)$ denotes the volume of a sphere of radius ρ in $\mathbb{Z}_r^n$, then*

$$\frac{r^n}{V_r^n(d-1)} \le A_r(n, d) \qquad \qquad \square$$

The sphere covering bound is a lower bound for $A_r(n, d)$. We can derive an upper bound for $A_r(n, d)$ by similar methods. In particular, let $C = \{\mathbf{c}_1, \mathbf{c}_2, \dots, \mathbf{c}_M\}$ be an optimal (n, M, d)-code, and set $e = \lfloor \frac{d-1}{2} \rfloor$. Since the packing spheres $S_r^n(\mathbf{c}_i, e)$ are disjoint, we have

$$\sum_{i=1}^{M} |S_r^n(\mathbf{c}_i, e)| \le |\mathbb{Z}_r^n|$$

and since $|S_r^n(\mathbf{c}_i, e)| = V_r^n(e)$, this is equivalent to

$$V_r^n(e) \cdot M \le r^n$$

Dividing by $V_r^n(e)$ gives

$$M \le \frac{r^n}{V_r^n(e)}$$

Since C is optimal, $M = A_r(n, d)$ and so

$$A_r(n, d) \le \frac{r^n}{V_r^n(e)} = \frac{r^n}{V_r^n(\lfloor \frac{d-1}{2} \rfloor)}$$

This inequality is known as the sphere-packing bound for $A_r(n, d)$.

Theorem 4.8.2 (The sphere-packing bound for $A_r(n, d)$.) *If $V_r^n(\rho)$ denotes the volume of a sphere of radius ρ in $\mathbb{Z}_r^n$, then*

$$A_r(n, d) \leq \frac{r^n}{V_r^n\left(\left\lfloor \frac{d-1}{2} \right\rfloor\right)} \qquad \square$$

Combining the sphere-covering and sphere-packing bounds for $A_r(n, d)$ gives

$$\frac{r^n}{V_r^n(d - 1)} \leq A_r(n, d) \leq \frac{r^n}{V_r^n\left(\left\lfloor \frac{d-1}{2} \right\rfloor\right)}$$

Table 4.8.1(a-b) shows some computer generated values of these bounds for $A_2(n, d)$, along with actual known values of $A_2(n, d)$. We have taken advantage of the fact that $A_2(n, d)$ is an integer to round up the lower bound and round down the upper bound. It is interesting to notice that the sphere-packing upper bound tends to be much closer to the actual value than the sphere-covering lower bound.

TABLE 4.8.1(a)   Bounds for $A_2(n, 3)$

n	lower	actual	upper
5	2	4	5
6	3	8	9
7	5	16	16
8	7	20	28
9	12	40	51
10	19	72-79	93
11	31	144-158	170
12	52	256	315
13	155	512	585
14	155	1024	1092
15	271	2048	2048
16	479	2560-3276	3855

TABLE 4.8.1(b)   Bounds for $A_2(n, 5)$

n	lower	actual	upper
5	2	2	2
6	2	2	2
7	2	2	4
8	2	4	6
9	3	6	11
10	3	12	18
11	4	24	30
12	6	32	51
13	8	64	89
14	12	128	154
15	17	256	270
16	27	256-340	478

Exercises

1. What is the relationship between the sphere-packing bound and the sphere-packing condition?

2. We have seen in Theorem 4.7.2 that $A_2(5, 3) = 4$. What does the sphere-packing bound give? How does this relate to existence of codes for solutions to the sphere-packing condition?

3. Compute the sphere-packing and sphere-covering bounds for $A_2(7, 3)$.

4. Compute the sphere-packing and sphere-covering bounds for $A_2(10, 5)$.

5. Compute the sphere-packing and sphere-covering bounds for $A_2(13, 7)$. Compare with the values from Table 4.8.1.

6. Compute the sphere-packing and sphere-covering bounds for $A_2(14, 7)$. Compare with the values from Table 4.8.1.

7. Compute the sphere-packing and sphere-covering bounds for $A_2(15, 7)$. Compare with the values from Table 4.8.1.

8. Is there a binary (8,29,3)-code? Explain.

9. Show that $A_r(r + 1, 3) \leq r^{r-1}$.

10. Show that $A_r(r + 1, 5) \leq \frac{2r^{r-2}}{r-1}$.

4.9   The Singleton and Plotkin bounds

The sphere-packing bound is not the only useful upper bound on the values of $A_r(n, d)$. In this section, we consider two additional bounds.

Theorem 4.9.1 (The Singleton bound)

$$A_r(n, d) \leq r^{n-d+1} \qquad \square$$

Proof   Let C be an (n, M, d)-code. If we remove the last $d - 1$ positions from each codeword in C, the resulting shortened codewords must all be distinct. For if any two were identical, the original codewords could have differed only among their last $d - 1$ places, and hence would have a distance at most $d - 1$. Thus, there are as many shortened codewords as original codewords. But the number of shortened codewords is at most $r^{n-(d-1)} = r^{n-d+1}$. Hence, $A_r(n, d) = M \leq r^{n-d+1}$.   ∎

Example 4.9.1   According to the Singleton bound,

$$A_r(4, 3) \leq r^2$$

On the other hand, the sphere-packing bound is

$$A_r(4, 3) \leq \frac{r^4}{4r - 3}$$

Thus, for $r \geq 4$, the Singleton bound is much better than the sphere-packing bound. $\square$

In order to derive our final upper bound, we require the following lemma, whose proof is left as an exercise.

Lemma 4.9.2   *Let M be a positive integer. As k ranges over the integers from 0 to M, the maximum value of $f(k) = k(M - k)$ is $M^2/4$ when M is even and $(M^2 - 1)/4$ when M is odd.* $\square$

Now let $C = \{c_1, c_2, \ldots, c_M\}$ be a binary (n, M, d)-code, and consider the sum

$$S = \sum_{i < j} d(c_i, c_j)$$

This is the sum of the distances between all pairs of codewords in C.

On the one hand, since $d(\mathbf{c}_i, \mathbf{c}_j) \geq d(C) = d$, for all codewords $\mathbf{c}_i$ and $\mathbf{c}_j$, and since there are $\binom{M}{2}$ pairs of codewords in C, we have

$$S = \sum_{i<j} d(c_i, c_j) \geq d\binom{M}{2} \tag{4.9.1}$$

On the other hand, let us compute the sum S by looking at each position. Suppose the codewords in C have the form

$$c_1 = \mathbf{c}_{11}\mathbf{c}_{12} \cdots \mathbf{c}_{1n}$$
$$c_2 = \mathbf{c}_{21}\mathbf{c}_{22} \cdots \mathbf{c}_{2n}$$
$$\vdots$$
$$c_M = \mathbf{c}_{M1}\mathbf{c}_{M2} \cdots \mathbf{c}_{Mn}$$

Consider the first positions $c_{11}, c_{21}, \ldots, c_{M1}$ from each codeword. If k of these bits are equal to 1, and $M - k$ are equal to 0, then these bits will contribute exactly $k(M-k)$ to the sum S. Now, according to Lemma 4.9.2, if M is even, then $k(M - k) \leq M^2/4$ and if M is odd, then $k(M - k) \leq (M^2 - 1)/4$.

Of course, there is nothing special here about the first position, and we can make the same deductions for any of the n positions. Hence, the total contribution of all positions to the sum S is at most $n(M^2/4)$ for M even and $n((M^2 - 1)/4)$ for M odd. Thus,

$$S \leq \begin{cases} \frac{nM^2}{4} & \text{for } M \text{ even} \\ \frac{n(M^2-1)}{4} & \text{for } M \text{ odd} \end{cases}$$

Putting this together with (4.9.1), we see that

$$\binom{M}{2}d \leq \begin{cases} \frac{nM^2}{4} & \text{for } M \text{ even} \\ \frac{n(M^2-1)}{4} & \text{for } M \text{ odd} \end{cases}$$

After some algebraic simplification, we get, for $n < 2d$,

$$M \leq \begin{cases} \frac{2d}{2d-n} & \text{for } M \text{ even} \\ \frac{2d}{2d-1}n & \text{for } M \text{ odd} \end{cases} \qquad \frac{2d}{2d-n} \tag{4.9.2}$$

Bearing in mind that M is an integer, these can be combined into a single bound. (We leave the details of this as an exercise.)

TABLE 4.9.1

n	Actual value of $A_2(n, 7)$	Plotkin bound on $A_2(n, 7)$	Sphere-packing bound on $A_2(n, 7)$
7	2	2	2
8	2	2	2
9	2	2	3
10	2	2	5
11	4	4	8
12	4	4	13
13	8	8	21
14	16	16	34
15	32	32	56

Theorem 4.9.3 *If $n < 2d$, then*

$$A_2(n, d) \leq 2 \left\lfloor \frac{d}{2d - n} \right\rfloor$$ $\square$

Finally, this can be improved and extended somewhat by separating the cases where d is even and d is odd. We leave the details of this as an exercise as well.

Theorem 4.9.4 (The Plotkin Bound)

1. *If d is even, then, for $n < 2d$,*

$$A_2(n, d) \leq 2 \left\lfloor \frac{d}{2d - n} \right\rfloor$$

and for $n = 2d$,

$$A_2(2d, d) = 4d$$

(b) *If d is odd, then, for $n < 2d + 1$,*

$$A_2(n, d) \leq 2 \left\lfloor \frac{d + 1}{2d + 1 - n} \right\rfloor$$

and for $n = 2d + 1$,

$$A_2(2d + 1, d) \leq 4d + 4$$ $\square$

Table 4.9.1 shows some values of the Plotkin bound, along with the sphere-packing bound. Although the Plotkin bound holds only for a rather restricted range of values for n, it seems superior to the sphere-packing bound in that range. In fact, (as the table suggests) it is thought by some coding theorists that the Plotkin upper bound actually gives the true value of $A_2(n, d)$ for $n \leq 2d + 1$, but this has not been proven.

Exercises

1. Show that $A_{10}(10, 3) \leq 100,000,000$. (See Example 4.4.4.)

2. Compare the Singleton, Plotkin, and sphere-packing upper bounds for $A_2(7, 5)$.

3. Compare the Singleton, Plotkin, and sphere-packing upper bounds for $A_2(8, 5)$.

4. Compare the Singleton, Plotkin, and sphere-packing upper bounds for $A_2(9, 5)$.

5. Compare the Singleton, Plotkin, and sphere-packing upper bounds for $A_2(15, 9)$.

6. Write a computer program to compare Singleton, Plotkin, and sphere-packing upper bounds for various parameters.

7. Prove Lemma 4.9.2.

8. Verify the inequalities in (4.9.2).

9. Show that (4.9.2) leads to the Plotkin bound.

10. The Plotkin bound can be improved and extended slightly by separating the cases where d is even and d is odd.

    (a) Show that, if d is even, then $A_2(2d, d) = 4d$. Hint: use Theorem 4.7.5.

    (b) Show that, if d is odd and $2d > n - 1$, then

    $$A_2(n, d) \leq 2 \left\lfloor \frac{d + 1}{2d + 1 - n} \right\rfloor$$

    (c) Show that if d is odd, then $A_2(2d + 1, d) \leq 4d + 4$.

4.10   Information Theory Revisited – the Noisy Coding Theorem

It is sometimes said that there are two main results in Information Theory. One is the Noiseless Coding Theorem, which we discussed in Chapter 3, and the other is the so-called *Noisy Coding Theorem*, which we will now discuss. Both of these theorems were first presented by Claude Shannon in 1948. Since the Noisy Coding Theorem is far more complicated than its noiseless counterpart, we must be content to state without proof a simple version of the theorem.

Recall that the transmission rate of an r-ary (n, M)-code C is defined by

$$\mathcal{R}(C) = \frac{\log_r M}{n}$$

and that this number is a measure of the efficiency of transmission of source data.

Roughly speaking, the Noisy Coding Theorem says that, if we choose any transmission rate below a certain number, called the capacity of the channel, there exists a code that can transmit at that rate, and yet maintain an error probability $\mathcal{P}(\text{decoding error})$ below some predefined limit. The price we pay for this efficient encoding is that the code size n may be extremely large. Furthermore, the known proofs of this theorem are not constructive. In other words, they tell us only that such a code must exist, but do not show us how to actually find these codes.

For a binary symmetric channel with crossover probability p, the capacity is

$$\mathcal{C}(p) = 1 + p \lg p + (1 - p)\lg(1 - p)$$

where lg is the logarithm base 2. Let us formally state the Noisy Coding Theorem for binary symmetric channels.

Theorem (The Noisy Coding Theorem for Binary Symmetric Channels) *Consider a binary symmetric channel with crossover probability p and capacity $\mathcal{C}(p)$. If $R < \mathcal{C}(p)$, then for any $\epsilon > 0$, there exists, for n sufficiently large, an (n, M)-code C whose transmission rate is at least R, and for which*

$$\mathcal{P}(\textit{decoding error}) < \epsilon \qquad\qquad \square$$

For instance, when $p = 0.01$, we have $C(0.01) = 0.919$, and so we may transmit at a rate of almost 92%, while at the same time keeping the probability of error as low as desired, that is, provided we are willing to make n and M as large as required.

Finally, we note that the Noiseless Coding Theorem holds for more general types of channels but, in general, it is very difficult to determine the capacity of an arbitrary channel.

5 Linear Codes

5.1 The Vector Space $\mathbb{Z}_p^n$

In this section, we discuss some basic notions of linear algebra from the point of view of coding theory. It is not our intention here to be complete, but only to cover those notions that are needed in this book. Also, all proofs are omitted. (It is hoped that the reader will supply proofs modeled after those from a standard linear algebra text.) Readers who have studied linear algebra should find most of these notions to be familiar, but we suggest reading this section carefully in any case.

In this chapter, we study a special class of codes known as linear codes. Linear codes have several advantages over other (nonlinear) codes. In general, they are easier to describe than other codes, nearest neighbor decoding is easier to implement, and the coding and decoding of source messages is easier. For these reasons, linear codes are the most widely studied types of codes.

In addition, the code alphabet of a linear code is a field. This offers the advantage that code symbols can be added and multiplied.

It was mentioned in Chapter 1 that there is an essentially unique field F_q of size q for every prime power $q = p^m$. However, for $m > 1$, the fields F_q are more difficult to work with than the fields of prime size (when

149

$m = 1$). In particular, if p is a prime, then F_p is just the integers modulo p

$$F_p = \mathbb{Z}_p = \{0, 1, 2, \ldots, p - 1\}$$

However, if n is not a prime, then $\mathbb{Z}_n$ is not a field. Hence, F_q is not $\mathbb{Z}_n$. For instance, the field F_4 is best represented in the abstract form $\{0, 1, \alpha, \alpha^2\}$ where $\alpha^3 = 1$ and $\alpha + \alpha^2 = 1$.

Since fields of nonprime size are somewhat more complicated, we will restrict attention to the fields $\mathbb{Z}_p$. This costs us little since, for instance, while the Hamming codes are defined for all prime powers, the binary codes are the most important; the Golay codes are binary or ternary; the Reed-Muller codes are binary; and decimal codes are derived from codes over $\mathbb{Z}_{11}$.

We can define two operations on the set $\mathbb{Z}_p^n$ of strings of length n over $\mathbb{Z}_p$. Addition modulo p is extended to strings simply by adding corresponding elements. In particular, if $\mathbf{u} = u_1 u_2 \cdots u_n$ and $\mathbf{v} = v_1 v_2 \cdots v_n$ are strings in $\mathbb{Z}_p^n$, then

$$\mathbf{u} + \mathbf{v} = u_1 u_2 \cdots u_n + v_1 v_2 \cdots v_n = (u_1 + v_1)(u_2 + v_2) \cdots (u_n + v_n)$$

where the parentheses indicate juxtaposition. We can also multiply a string by a single code symbol. In particular, if $\alpha \in \mathbb{Z}_p$ and $\mathbf{u} = u_1 u_2 \cdots u_n \in \mathbb{Z}_p^n$, then

$$\alpha \mathbf{u} = \alpha \cdot (u_1 u_2 \cdots u_n) = (\alpha u_1)(\alpha u_2) \cdots (\alpha u_n)$$

In this context, the element α is called a scalar and the operation is known as scalar multiplication.

Example 5.1.1 In $\mathbb{Z}_3^4$,

$$1201 + 1212 = (1 + 1)(2 + 2)(0 + 1)(1 + 2) = 2110$$

and

$$2 \cdot (1212) = (2 \cdot 1)(2 \cdot 2)(2 \cdot 1)(2 \cdot 2) = 2121 \qquad \square$$

The set $\mathbb{Z}_p^n$, together with the operations of addition and scalar multiplication defined above, forms an algebraic structure known as a vector space. Let us give a formal definition.

Definition Let $\mathcal{F}$ be a field (such as $\mathbb{Z}_p$ for p prime). A **vector space over** $\mathcal{F}$ is a nonempty set $\mathcal{V}$, together with two operations defined as follows. One operation, called **addition** and denoted by $+$, is a binary operation on $\mathcal{V}$. The other operation, called **scalar multiplication** and denoted by juxtaposition, is a function from $\mathcal{F} \times \mathcal{V}$ to $\mathcal{V}$ (thus, we denote the scalar product of $\alpha \in \mathcal{F}$ and $\mathbf{x} \in \mathcal{V}$ by $\alpha\mathbf{x}$). Furthermore, these two operations satisfy the following conditions:

Associative property: For all $\mathbf{x}$, $\mathbf{y}$, and $\mathbf{z}$ in $\mathcal{V}$,

$$\mathbf{x} + (\mathbf{y} + \mathbf{z}) = (\mathbf{x} + \mathbf{y}) + \mathbf{z}$$

Commutative property: For all $\mathbf{x}$ and $\mathbf{y}$ in $\mathcal{V}$,

$$\mathbf{x} + \mathbf{y} = \mathbf{y} + \mathbf{x}$$

Property of the zero vector: There exists an element $\mathbf{0} \in \mathcal{V}$, called the **zero vector**, such that, for all $\mathbf{x}$ in $\mathcal{V}$,

$$\mathbf{0} + \mathbf{x} = \mathbf{x} + \mathbf{0} = \mathbf{x}$$

Property of negatives: For any $\mathbf{x} \in \mathcal{V}$, there exists another element of $\mathcal{V}$, denoted by $-\mathbf{x}$ and called the **negative** of $\mathbf{x}$, for which

$$\mathbf{x} + (-\mathbf{x}) = (-\mathbf{x}) + \mathbf{x} = \mathbf{0}$$

Properties of scalar multiplication: For all $\mathbf{x}$ and $\mathbf{y}$ in $\mathcal{V}$, and all α and β in $\mathcal{F}$,

$$\alpha(\mathbf{x} + \mathbf{y}) = \alpha\mathbf{x} + \alpha\mathbf{y}$$
$$(\alpha + \beta)\mathbf{x} = \alpha\mathbf{x} + \beta\mathbf{x}$$
$$(\alpha\beta)\mathbf{x} = \alpha(\beta\mathbf{x})$$
$$1\mathbf{x} = \mathbf{x} \qquad \square$$

If $\mathcal{V}$ is a vector space over a field $\mathcal{F}$, we refer to the elements of $\mathcal{V}$ as **vectors** and the elements of $\mathcal{F}$ as **scalars**. The field $\mathcal{F}$ is called the **base field** of $\mathcal{V}$.

Theorem 5.1.1 *If p is a prime number, the set $\mathbb{Z}_p^n$ of all strings of length n over $\mathbb{Z}_p$, together with the operations of addition and scalar multiplication defined earlier, is a vector space over $\mathbb{Z}_p$.* $\qquad \square$

Since each element of $\mathbb{Z}_2$ is its own negative, subtraction is the same as addition; that is, $\alpha - \beta = \alpha + \beta$ for $\alpha, \beta \in \mathbb{Z}_2$. This property carries over to binary strings as well. In particular,

$$\mathbf{x} - \mathbf{y} = \mathbf{x} + \mathbf{y}$$

for all $\mathbf{x}, \mathbf{y} \in \mathbb{Z}_2^n$. Note that this holds in $\mathbb{Z}_p^n$ if and only if $p = 2$. Also, it is useful to note that, when the base field is $\mathbb{Z}_2 = \{0, 1\}$, scalar multiplication in $\mathbb{Z}_2^n$ is essentially trivial, since

$$0\mathbf{x} = \mathbf{0} \text{ and } 1\mathbf{x} = \mathbf{x}$$

Subspaces of $\mathbb{Z}_p^n$

Since we are assuming that all p-ary codes have code alphabet $\mathbb{Z}_p$, a p-ary code C of length n is just a nonempty subset of the vector space $\mathbb{Z}_p^n$. It is a natural step to consider codes that are vector spaces in their own right. If you have studied vector spaces before, then you are no-doubt familiar with the following concept.

Definition   A nonempty subset S of $\mathbb{Z}_p^n$ is called a **subspace** of $\mathbb{Z}_p^n$ if the set S, together with the operations of addition and scalar multiplication inherited from $\mathbb{Z}_p^n$, is itself a vector space. $\square$

In order to tell whether a subset S of $\mathbb{Z}_p^n$ is a subspace using the definition, it would be necessary to check each of the properties in the definition. However, there is a much simpler way.

Theorem 5.1.2   *A nonempty subset S of $\mathbb{Z}_p^n$ is a subspace of $\mathbb{Z}_p^n$ if and only if it satisfies the following two properties*

1. **(closure under addition)**

$$\mathbf{x}, \mathbf{y} \in S \text{ implies } \mathbf{x} + \mathbf{y} \in S$$

2. **(closure under scalar multiplication)**

$$\alpha \in \mathbb{Z}_p, \mathbf{x} \in S \text{ implies } \alpha\mathbf{x} \in S \qquad\qquad \square$$

In the language of codewords, a code C of length n over $\mathbb{Z}_p$ is a subspace of $\mathbb{Z}_p^n$ if the sum of two codewords is another codeword and if any scalar multiple of a codeword is another codeword.

Note that, since a subspace S of $\mathbb{Z}_p^n$ is nonempty by definition, it contains a string $\mathbf{x}$. Hence, since S is closed under scalar multiplication, it must also contain the zero string $0\mathbf{x} = \mathbf{0}$.

Note also that, when the base field is $\mathbb{Z}_2$, all nonempty subsets of $\mathbb{Z}_2^n$ are closed under scalar multiplication (but not necessarily under addition).

Example 5.1.2 To show that the subset $S = \{0000, 0100, 0010, 0110\}$ of $\mathbb{Z}_2^4$ is closed under addition, note that the sum of the zero string and any other string in S is clearly in S. In addition, since $\mathbf{x} + \mathbf{y} = \mathbf{y} + \mathbf{x}$, only one of these sums need be checked. The remaining possibilities are as follows

$$0100 + 0010 = 0110 \in S$$
$$0100 + 0110 = 0010 \in S$$
$$0010 + 0110 = 0100 \in S$$

Hence, S is closed under addition. Since the base field is $\mathbb{Z}_2$, closure under scalar multiplication is automatic. Thus, S is a subspace of $\mathbb{Z}_2^n$.   □

Example 5.1.3 The subset $T = \{0000, 0001, 1000, 0110\}$ of $\mathbb{Z}_2^4$ is not closed under addition because, for instance, $0001 \in T$ and $1000 \in T$ but $0001 + 1000 = 1001 \notin T$. Thus, T is not a subspace of $\mathbb{Z}_2^4$.   □

Example 5.1.4 The set $S = \{000, 100, 200, 001, 002, 101, 102, 201, 202\}$ is a subspace of $\mathbb{Z}_3^3$. It is easy to check that S consists of all strings in $\mathbb{Z}_3^3$ of the form $\alpha 0\beta$, where α and β range over the set of scalars $\mathbb{Z}_3 = \{0,1,2\}$. In symbols,

$$S = \{\alpha 0\beta \mid \alpha, \beta \in \mathbb{Z}_3\}$$

Now, if $\mathbf{x}$ and $\mathbf{y}$ are in S, then $\mathbf{x} = \alpha_1 0\beta_1$ and $\mathbf{y} = \alpha_2 0\beta_2$, whence

$$\mathbf{x} + \mathbf{y} = \alpha_1 0\beta_1 + \alpha_2 0\beta_2 = (\alpha_1 + \beta_1)0(\alpha_2 + \beta_2) \in S$$

Similarly,

$$\alpha\mathbf{x} = \alpha(\alpha_1 0\beta_1) = (\alpha\alpha_1)0(\alpha\beta_1) \in S$$

for any $\alpha \in \mathbb{Z}_3$. Hence, S is closed under addition and scalar multiplication, and so it is a subspace of $\mathbb{Z}_3^3$.   □

Example 5.1.5 The set $S = \{\mathbf{0}\}$, consisting of just the zero string is a subspace of $\mathbb{Z}_p^n$. Also, the set $\mathbb{Z}_p^n$ is a subspace of itself. Verification of these facts is left as an exercise.   □

Bases

One of the most useful features of subspaces of $\mathbb{Z}_p^n$ is that it is possible to describe a subspace without having to list all of its elements. The following definitions will help clarify this.

Definition   Let $\mathbf{x}_1$, $\mathbf{x}_2$, ..., $\mathbf{x}_k$ be strings in $\mathbb{Z}_p^n$, and let $\alpha_1, \alpha_2, \ldots, \alpha_k$ be scalars (i.e., elements of $\mathbb{Z}_p$). The string $\alpha_1 \mathbf{x}_1 + \alpha_2 \mathbf{x}_2 + \cdots + \alpha_k \mathbf{x}_k$ is called a **linear combination** of the strings $\mathbf{x}_1, \mathbf{x}_2, \ldots, \mathbf{x}_k$, with **coefficients** $\alpha_1, \alpha_2, \ldots, \alpha_k$.

If all of the coefficients $\alpha_1, \alpha_2, \ldots, \alpha_k$ are equal to 0, then we say that the linear combination $\alpha_1 \mathbf{x}_1 + \alpha_2 \mathbf{x}_2 + \cdots + \alpha_k \mathbf{x}_k$ is **trivial**. Otherwise, it is **nontrivial**.   □

Example 5.1.6   In $\mathbb{Z}_5^3$, the string $\mathbf{x} = 2(010) + 121 + 3(140) = 411$ is a (nontrivial) linear combination of 010, 121 and 140, with coefficients 2, 1 and 3, respectively.   □

The elements of the subspace

$$S_1 = \{000, 100, 200, 001, 002, 101, 102, 201, 202\}$$

of $\mathbb{Z}_3^3$, encountered in Example 5.1.4, can be described in terms of linear combinations as follows

$$S_1 = \{\alpha(100) + \beta(001) \mid \alpha, \beta \in \mathbb{Z}_3\}$$

Because of this, we say that the set $G_1 = \{100, 001\}$ generates the subspace S_1.

Definition   Let S be a subspace of $\mathbb{Z}_p^n$. A subset $G \subseteq S$ is called a **generating set** for S if *every* string in S can be expressed as a linear combination of the strings in G. More formally, a subset $G = \{\mathbf{g}_1, \ldots, \mathbf{g}_k\}$ of S is a generating set for S if, for any $x \in S$, there exists scalars $\alpha_1, \ldots, \alpha_k \in \mathbb{Z}_p$ for which

$$\mathbf{x} = \alpha_1 \mathbf{g}_1 + \cdots + \alpha_k \mathbf{g}_k$$

To indicate that $G = \{\mathbf{g}_1, \ldots, \mathbf{g}_k\}$ is a generating set for S, we write $S = \langle G \rangle$ or $S = \langle \mathbf{g}_1, \ldots, \mathbf{g}_k \rangle$.   □

Example 5.1.7   The subset $G = \{0100, 0010\}$ is a generating set for the subspace $S = \{0000, 0100, 0010, 0110\}$ of $\mathbb{Z}_2^4$, since any string in S can be written as a linear combination of strings in G. In particular, for the strings not in G itself, we have

$$0000 = 0100 + 0100$$

and

$$0110 = 0100 + 0010$$

(Any string in G is clearly a linear combination of strings in G.) Hence,

$$G = \langle 0100, 0010 \rangle$$

Notice that $H = \{0100, 0110\}$ is also a generating set for S, but that $K = \{0000, 0100\}$ is not a generating set, since there is no way to express the string 0010 as a linear combination of 0000 and 0100. $\square$

As the next theorem states, any nonempty set of strings in $\mathbb{Z}_p^n$ is a generating set for some subspace of $\mathbb{Z}_p^n$.

Theorem 5.1.3 _Let G be a nonempty set of strings in $\mathbb{Z}_p^n$. The set of all linear combinations of strings in G is a subspace of $\mathbb{Z}_p^n$, called the **subspace generated by G**, or the **subspace spanned by G**, and denoted by $\langle G \rangle$._ $\square$

It is not difficult to see that $G_1 = \{100, 001\}$ and $G_2 = \{100, 001, 101\}$ are both generating sets for the subspace $S = \{000, 100, 001, 101\}$ of $\mathbb{Z}_2^3$. However, G_1 is a proper subset of G_2, and so, in some sense, it provides a more efficient description of S than does G_2. This leads to the following definition.

Definition Let S be a subspace of $\mathbb{Z}_p^n$. A generating set B of S is said to be a **basis** for S if it is a _minimal_ generating set, in the sense that no proper subset of B also generates S. $\square$

Example 5.1.8 The set $B = \{100, 001\}$ is a basis for the subspace $S = \{000, 100, 001, 101\}$ of $\mathbb{Z}_2^n$. This follows from the fact that B is a generating set for S and no proper subset of B is a generating set for S. (The nontrivial proper subsets of B are $\{100\}$ and $\{001\}$, and neither of these sets generate all of S.) $\square$

Example 5.1.9 The set $B = \{e_1, e_2, e_3, e_4\} = \{1000, 0100, 0010, 0001\}$ is a basis for $\mathbb{Z}_2^4$. This follows from the fact that any string $\mathbf{x} = u_1 u_2 u_3 u_4$ in $\mathbb{Z}_2^4$ can be written in the form

$$\mathbf{x} = u_1(1000) + u_2(0100) + u_3(0010) + u_4(0001)$$

Hence, $\mathcal{B}$ is a generating set for $\mathbb{Z}_2^4$. On the other hand, it is easy to see that no proper subset of $\mathcal{B}$ can generate all of $\mathbb{Z}_2^4$. Hence, $\mathcal{B}$ is a basis for $\mathbb{Z}_2^4$.

More generally, if $\mathbf{e}_i$ denotes the string in $\mathbb{Z}_p^n$ with zeros in every position except the ith, where it has a 1, then $\mathcal{B} = \{\mathbf{e}_1, \mathbf{e}_2, \ldots, \mathbf{e}_n\}$ is a basis for $\mathbb{Z}_p^n$. It is known as the **standard basis** for $\mathbb{Z}_p^n$.   □

One of the best ways to describe a subspace S of $\mathbb{Z}_p^n$ is to give a basis for it. Loosely speaking, a basis consists of "just enough" strings to describe all of the strings in S, through linear combinations. However, it is important to keep in mind that, while any nonempty set of strings in $\mathbb{Z}_p^n$ is a generating set for some subspace of $\mathbb{Z}_p^n$, not all sets of strings qualify to be bases. To explore this issue further, we require a definition.

Definition   The strings $\mathbf{x}_1, \mathbf{x}_2, \ldots, \mathbf{x}_k$ in $\mathbb{Z}_p^n$ are said to be **linearly independent over $\mathbb{Z}_p$** if no nontrivial linear combination of these strings is equal to the zero string. In symbols, the strings $\mathbf{x}_1, \mathbf{x}_2, \ldots, \mathbf{x}_k$ are linearly independent if the equation

$$\alpha_1 \mathbf{x}_1 + \alpha_2 \mathbf{x}_2 + \cdots + \alpha_k \mathbf{x}_k = \mathbf{0}$$

holds only when $\alpha_1 = 0, \alpha_2 = 0, \ldots, \alpha_k = 0$.

If $\mathbf{x}_1, \mathbf{x}_2, \ldots, \mathbf{x}_k$ are not linearly independent over $\mathbb{Z}_p$, they are **linearly dependent over $\mathbb{Z}_p$**. Thus, $\mathbf{x}_1, \mathbf{x}_2, \ldots, \mathbf{x}_k$ are linearly dependent if the equation

$$\alpha_1 \mathbf{x}_1 + \alpha_2 \mathbf{x}_2 + \cdots + \alpha_k \mathbf{x}_k = 0$$

holds for *some* set of coefficients, at least one of which is not zero.   □

It is important to emphasize that when we say that a set of strings is linearly independent or dependent, we *must* specify the base field. For instance, the set $\{12,21\}$ is linearly independent over $\mathbb{Z}_5$ but it is linearly dependent over $\mathbb{Z}_3$! (Check this for yourself.) Having said this, it is customary, when the context is perfectly clear, to simply say that a set is linearly independent or linearly dependent.

If the elements of the set $S = \{\mathbf{x}_1, \mathbf{x}_2, \ldots, \mathbf{x}_k\}$ are linearly independent, it is also customary to say that the set S is linearly independent.

In order to determine whether strings $\mathbf{x}_1, \mathbf{x}_2, \ldots, \mathbf{x}_k$ of length n are linearly independent or linearly dependent, one usually starts by writing the equation

$$\alpha_1 \mathbf{x}_1 + \alpha_2 \mathbf{x}_2 + \cdots + \alpha_k \mathbf{x}_k = \mathbf{0}$$

and solving for the coefficients $\alpha_1, \alpha_2, \ldots, \alpha_k$. This generally leads to a system of n equations in the k unknowns $\alpha_1, \alpha_2, \ldots, \alpha_k$. If the only possible solution to this system is $\alpha_1 = 0$, $\alpha_2 = 0$, $\ldots$, $\alpha_k = 0$, then the strings are linearly independent. However, if there is a nonzero solution to the system, then the strings are linearly dependent. The following example illustrates the technique.

Example 5.1.10 Consider the strings 1000, 1010, and 0110 in $\mathbb{Z}_2^4$. Suppose that

$$\alpha_1(1000) + \alpha_2(1010) + \alpha_3(0110) = 0000 \qquad (5.1.1)$$

This equation can be simplified to

$$\alpha_1 000 + \alpha_2 0\alpha_2 0 + 0\alpha_3\alpha_3 0 = 0000$$

or

$$(\alpha_1 + \alpha_2)\alpha_3(\alpha_2 + \alpha_3)0 = 0000$$

But two strings are equal if and only if corresponding elements are equal. Hence, this is equivalent to the system of equations in $\mathbb{Z}_2$

$$\alpha_1 + \alpha_2 = 0$$
$$\alpha_3 = 0$$
$$\alpha_2 + \alpha_3 = 0$$

The second of these equations tells us that $\alpha_3 = 0$. Combining this with the third equation shows that $\alpha_2 = 0$. Then the first equation shows that $\alpha_1 = 0$. Thus, the only way (5.1.1) can hold is if $\alpha_1 = \alpha_2 = \alpha_3 = 0$, and so the strings are linearly independent.

On the other hand, consider the strings 1000, 1010, and 0010. The equation

$$\alpha_1(1000) + \alpha_2(1010) + \alpha_3(0010) = 0000 \qquad (5.1.2)$$

This is equivalent to

$$\alpha_1 000 + \alpha_2 0\alpha_2 0 + 00\alpha_3 0 = 0000$$

or

$$(\alpha_1 + \alpha_2)0(\alpha_2 + \alpha_3)0 = 0000$$

which holds if and only if α_1, α_2, and α_3 satisfy the system of equations

$$\alpha_1 + \alpha_2 = 0$$
$$\alpha_2 + \alpha_3 = 0$$

But, this system is satisfied by $\alpha_1 = 1$, $\alpha_2 = 1$, $\alpha_3 = 1$. (Recall that, in $\mathbb{Z}_2^4$, $1 + 1 = 0$.) Since these numbers are not all zero, the strings are linearly dependent. Substituting these values into (5.1.2) gives

$$1000 + 1010 + 0010 = 0000 \qquad \qquad \square$$

Let us make a couple of remarks about solving systems of linear equations. You probably have experience in solving such systems, using Gaussian elimination (alias row reduction) for systems whose coefficients are real numbers. Fortunately, the properties of the real numbers that are used to justify Gaussian elimination apply when the coefficients come from any field. Thus, for instance, to solve the system of equations

$$2x + y = 1$$
$$x - y = 4$$

over the field $\mathbb{Z}_5$, we first multiply the first equation by the inverse of 2, which is $2^{-1} = 3$ (since $2 \cdot 3 \equiv 1$ modulo 5), to get

$$x + 3y = 3$$
$$x - y = 4$$

Next, we subtract the first equation from the second,

$$x + 3y = 3$$
$$-4y = 1$$

Since $-4 = 1$ in $\mathbb{Z}_5$, the second equation is just $y = 1$. Substituting this onto the first equation and solving gives $x = 3 - 3y = 3 - 3 = 0$. Low and behold, $x = 0$, $y = 1$ is actually a solution to the original system, although it may not look like it at first glance!

The following theorem gives various ways to describe a basis, including the definition.

Theorem 5.1.4 *Let S be a subspace of $\mathbb{Z}_p^n$. A subset $\mathcal{B} = \{\mathbf{x}_1, \mathbf{x}_2, \ldots, \mathbf{x}_k\}$ of S is a basis for S if and only if any one of the following equivalent conditions holds.*

1. *(The definition) $\mathcal{B}$ is a minimal generating set for S.*

2. *$\mathcal{B}$ is linearly independent and a generating set for S.*

3. *$\mathcal{B}$ is a maximal linearly independent set in S, that is, $\mathcal{B}$ is linearly independent and no proper superset of $\mathcal{B}$ is linearly independent.*

4. *For every $\mathbf{x} \in S$, there are unique scalars $\alpha_1, \ldots, \alpha_k$ in $\mathbb{Z}_p$ for which*

$$\mathbf{x} = \alpha_1 \mathbf{x}_1 + \alpha_2 \mathbf{x}_2 + \cdots + \alpha_k \mathbf{x}_k$$

(note the word unique). □

Theorem 5.1.4 has some very important and useful consequences, two of which are described in the next theorem.

Theorem 5.1.5 *Let S be a subspace of $\mathbb{Z}_p^n$.*

1. *Any generating set G for S contains a basis for S, that is, some subset of G is a basis for S.*

2. *Any linearly independent subset A of S can be extended to a basis for S, that is, A is contained in a basis for S.* □

In trying to show that a given set of strings is (or is not) a basis for a vector space, any of the equivalent conditions in Theorem 5.1.4 may be used. For instance, if $\mathcal{B}$ is a linearly independent set of strings in $\mathbb{Z}_p^n$ and if $\mathcal{V} = \langle \mathcal{B} \rangle$ is the subspace of $\mathbb{Z}_p^n$ generated by $\mathcal{B}$, then $\mathcal{B}$ satisfies condition 2) of Theorem 5.1.4 and is thus a basis for $\mathcal{V}$. In words, any *linearly independent* set of strings is a basis for the subspace generated by those strings.

Example 5.1.11 Since the set $\mathcal{B} = \{1000, 1010, 0110\}$ is linearly independent (Example 5.1.10) and since $\mathcal{B}$ generates the subspace

$$S = \{0000, 1000, 1010, 0110, 0010, 1110, 1100\}$$

of $\mathbb{Z}_2^n$, it follows that $\mathcal{B}$ is a basis for S.

On the other hand, since $D = \{1000, 0110, 0111, 1110\}$ is not linearly independent ($1000 + 0110 + 1110 = 0000$), it is not a basis for any subspace of $\mathbb{Z}_2^n$. However, removing the string 1110 gives

$$E = \{1000, 0110, 0111\}$$

which is linearly independent and generates $\langle D \rangle$ (why?). Hence, E is a basis for $\langle D \rangle$. □

Example 5.1.12 To determine whether the set $\mathcal{B} = \{111, 123, 011\}$ is a basis for $\mathbb{Z}_5^3$, we may use condition 4) of Theorem 5.1.4. Let $\mathbf{x} = x_1 x_2 x_3$ be any string in $\mathbb{Z}_5^3$ and set

$$\alpha_1 111 + \alpha_2 123 + \alpha_3 011 = x_1 x_2 x_3$$

We must show that there is a *unique* solution $\alpha_1, \alpha_2, \alpha_3$ to this equation. This equation can be written

$$(\alpha_1 + \alpha_2)(\alpha_1 + 2\alpha_2 + \alpha_3)(\alpha_1 + 3\alpha_2 + \alpha_3) = x_1 x_2 x_3$$

which is equivalent to the system

$$\alpha_1 + \alpha_2 = x_1 \tag{5.1.3}$$
$$\alpha_1 + 2\alpha_2 + \alpha_3 = x_2 \tag{5.1.4}$$
$$\alpha_1 + 3\alpha_2 + \alpha_3 = x_3 \tag{5.1.5}$$

Subtracting the first equation from the others gives (in $\mathbb{Z}_5$)

$$\alpha_1 + \alpha_2 = x_1 \tag{5.1.6}$$
$$\alpha_2 + \alpha_3 = 4x_1 + x_2 \tag{5.1.7}$$
$$2\alpha_2 + \alpha_3 = 4x_1 + x_3 \tag{5.1.8}$$

Subtracting twice the second equation from the third gives

$$\alpha_1 + \alpha_2 = x_1 \tag{5.1.9}$$
$$\alpha_2 + \alpha_3 = 4x_1 + x_2 \tag{5.1.10}$$
$$4\alpha_3 = x_1 + 3x_2 + x_3 \tag{5.1.11}$$

Solving these equations from the bottom up gives

$$\alpha_3 = 4^{-1}(x_1 + 3x_2 + x_3) = 4x_1 + 2x_2 + 4x_3$$

and

$$\alpha_2 = 4x_1 + x_2 - \alpha_3 = 4x_2 + x_3$$

and finally

$$\alpha_1 = x_1 - \alpha_2 = x_1 + x_2 + 4x_3$$

Thus, there is a unique solution to this system and $\mathcal{B}$ is indeed a basis for $\mathbb{Z}_5^3$. □

A subspace S of $\mathbb{Z}_p^n$ may have more than one basis. However, it is a remarkable fact that *all bases of a given subspace have the same size*. Thus, for instance, since G = {100, 001} is a basis of S = {000, 100, 001, 101} (see Example 5.1.8), all bases of S will have size 2. This allows us to make the following definition.

Theorem 5.1.6 *Let S be a subspace of $\mathbb{Z}_p^n$. Then all bases for S have the same size. This size is called the* **dimension** *of S and is denoted by $dim(S)$.* □

Example 5.1.13

1. The subspace S of Example 5.1.8 has dimension 2.

2. Since the standard basis $\mathcal{B}$ = $\{\mathbf{e}_1, \mathbf{e}_2, \ldots, \mathbf{e}_n\}$ has size n, we have $dim(\mathbb{Z}_p^n)$ = n. Note that the dimension of $\mathbb{Z}_p^n$ is the same for all values of p, but the size p^n of $\mathbb{Z}_p^n$ depends on both n and p. □

Generally speaking, to determine whether or not a subset $\mathcal{B}$ of a subspace S is a basis for S, we need to check two conditions—$\mathcal{B}$ must be linearly independent and it must generate S. The next theorem says that if $\mathcal{B}$ has the right size, then we need only check one of the two conditions.

Theorem 5.1.7 *Let S be a subspace of $\mathbb{Z}_p^n$ of dimension k. A subset $\mathcal{B}$ of S is a basis for S if and only if either of the following two equivalent conditions holds.*

1. *$\mathcal{B}$ has size k and it is linearly independent.*

2. *$\mathcal{B}$ has size k and generates S.* □

This theorem would certainly have come in handy in Example 5.1.13. In particular, since $dim(\mathbb{Z}_5^3)$ = 3 and since $|\mathcal{B}|$ = 3, we need only check that $\mathcal{B}$ is linearly independent. In particular, we need only solve the system (5.1.4) when x_1 = x_2 = x_3 = 0, which is considerably easier!

Exercises

1. Show that $\alpha\mathbf{0}$ = $\mathbf{0}$ for any $\alpha \in \mathbb{Z}_p$. Show also that if $\alpha\mathbf{x}$ = $\mathbf{0}$ for some $\alpha \in \mathbb{Z}_p$ and $\mathbf{x} \in \mathbb{Z}_p^n$, then either α = 0 or $\mathbf{x}$ = $\mathbf{0}$.

2. Show that if a subset S of $\mathbb{Z}_p^n$ contains the zero string, then S must be linearly dependent.

3. Show that the set $S = \{\mathbf{0}\}$, consisting of just the zero string is a subspace of $\mathbb{Z}_p^n$. Show that $\mathbb{Z}_p^n$ is a subspace of itself.

4. Prove that every vector in $\mathbb{Z}_p^n$ has a negative.

5. If we define multiplication of strings in $\mathbb{Z}_p^n$ by

$$(u_1 \cdots u_n)(v_1 \cdots v_n) = (u_1 v_1) \cdots (u_n v_n)$$

show that the product of two nonzero strings may be the zero string.

6. Determine whether the following subsets are subspaces.

   (a) $S = \{000, 100, 200, 010, 020, 110, 120, 210, 220\} \subseteq \mathbb{Z}_3^3$.

   (b) $S = \{123, 456, 579\} \subseteq \mathbb{Z}_{11}^3$.

   (c) $S = \{000, 123, 241, 314, 432\} \subseteq \mathbb{Z}_5^3$.

7. Is the set E of all binary strings in $\mathbb{Z}_2^n$ with even weight a subspace of $\mathbb{Z}_2^n$? What about the set O of all binary strings of odd weight? What if the strings are not binary?

8. Is the set $S = \{\mathbf{x} \in \mathbb{Z}_2^n \mid \text{first element of } \mathbf{x} \text{ is } 0\}$ a subspace of $\mathbb{Z}_2^n$?

9. Is the set $S = \{\mathbf{x} \in \mathbb{Z}_2^n \mid \text{first element of } \mathbf{x} \text{ is } 1\}$ a subspace of $\mathbb{Z}_2^n$?

10. Is the set $S = \{\mathbf{x} = x_1 \cdots x_n \in \mathbb{Z}_p^n \mid x_1 = x_n\}$ a subspace of $\mathbb{Z}_p^n$?

11. Is the set $S = \{\mathbf{x} = x_1 \cdots x_n \in \mathbb{Z}_p^n \mid x_1 + x_1 = 0\}$ a subspace of $\mathbb{Z}_p^n$? Does it matter whether or not $p = 2$?

12. Write out all of the elements of the set $\langle 111, 101, 011 \rangle \subseteq \mathbb{Z}_2^3$.

13. Write out all of the elements of the set $\langle 1402 \rangle \subseteq \mathbb{Z}_5^4$.

14. Solve the following system over $\mathbb{Z}_2$

$$\alpha_1 + \alpha_2 + \alpha_3 = 1$$
$$\alpha_2 - \alpha_3 = 0$$
$$\alpha_1 + \alpha_2 = 1$$

15. Solve the following system over $\mathbb{Z}_2$

$$\alpha_1 + \alpha_2 + \alpha_3 = 1$$
$$\alpha_2 - \alpha_3 + \alpha_4 = 0$$
$$\alpha_1 + \alpha_2 = 1$$

16. Solve the following system over $\mathbb{Z}_3$

$$\alpha_1 + 2\alpha_2 + \alpha_3 = 2$$

$$\alpha_2 - 2\alpha_3 = 1$$
$$2\alpha_1 + \alpha_2 = 1$$

17. Solve the following system over $\mathbb{Z}_5$

$$\alpha_1 + \alpha_2 + \alpha_3 = 1$$
$$2\alpha_1 + \alpha_2 + 3\alpha_3 = 3$$
$$3\alpha_1 + 2\alpha_2 - \alpha_3 = 4$$

18. Determine which of the following subsets are linearly independent and which are bases. For those that are linearly dependent, find a basis for $\langle S \rangle$ contained in S.

    (a) $S = \{101, 110\} \subseteq \mathbb{Z}_2^3$

    (b) $S = \{121, 212\} \subseteq \mathbb{Z}_3^3$

    (c) $S = \{413, 134\} \subseteq \mathbb{Z}_5^3$

    (d) $S = \{110, 101, 010\} \subseteq \mathbb{Z}_2^3$

    (e) $S = \{1001, 1111, 1010, 1011, 0011\} \subseteq \mathbb{Z}_2^4$

    (f) $S = \{120, 112, 012\} \subseteq \mathbb{Z}_3^3$

    (g) $S = \{120, 112, 011\} \subseteq \mathbb{Z}_3^3$

    (h) $S = \{1111, 1231, 0123, 1001\} \subseteq \mathbb{Z}_5^4$

19. Prove that a subspace of $\mathbb{Z}_p^n$ of dimension k has size p^k.

20. Prove Theorem 5.1.1.

21. Prove Theorem 5.1.2.

22. Prove Theorem 5.1.3.

23. Prove Theorem 5.1.4.

24. Prove Theorem 5.1.5.

25. Prove Theorem 5.1.6.

26. Prove Theorem 5.1.7.

5.2 Linear Codes

We can now define the most important and most studied type of code.

Definition   A code $C \subseteq \mathbb{Z}_p^n$ that is also a subspace of $\mathbb{Z}_p^n$ is called a **linear code**. If C has dimension k and minimum distance $d(C) = d$, then C is an **[n, k, d]-code**. When we do not care to emphasize the minimum distance d, we use the notation **[n, k]-code**. The numbers n, k, and d are called the **parameters** of the linear code.   □

Note that a linear code C, being a subspace of $\mathbb{Z}_p^n$, must contain the zero codeword **0**.

Note also the use of square brackets in the definition above. This notation applies *only* to linear codes, whereas the notation (n, M, d)-code applies to all codes. For a linear code C, we may use both notations. The dimension k and size M of a p-ary linear code are related by the formula $M = p^k$. To see this, let $\mathcal{B} = \{\mathbf{x}_1, \ldots, \mathbf{x}_k\}$ be a basis for C, then any codeword $\mathbf{x}$ can be written in the form

$$\mathbf{x} = \alpha_1 \mathbf{x}_1 + \alpha_2 \mathbf{x}_2 + \cdots + \alpha_k \mathbf{x}_k$$

for a unique set of scalars. Hence, the association $x \leftrightarrow (\alpha_1, \ldots, \alpha_k)$ is a one-to-one correspondence from C to the set D of all ordered k-tuples with components in $\mathbb{Z}_p$. Hence,

$$M = |C| = |D| = p^k$$

We can summarize as follows.

Theorem 5.2.1   *A p-ary linear $[n, k, d]$-code has size p^k and is thus an (n, p^k, d)-code.*   □

We need not justify the importance of linear codes any further than saying that the Hamming codes, the Golay codes, and the Reed-Muller codes that we will study in the next chapter are all linear codes. However, let us consider some additional examples of linear codes. We will use these simple codes to illustrate various concepts in the sequel.

Example 5.2.1   The binary code $C_1 = \{0000, 1011, 0110, 1101\}$ is a subspace of $\mathbb{Z}_2^4$ and so it is a linear code. Since $\mathcal{B} = \{1011, 0110\}$ is a basis for C_1, it follows that $\dim(C_1) = 2$. Hence, C_1 is a $[4, 2]$-code.

Since the strings 0121 and 2210 are linearly independent over $\mathbb{Z}_3$, they form a basis for a ternary $[4, 2]$-linear code

$$C_2 = \langle 0121, 2210 \rangle = \{\alpha(0121) + \beta(2210) \mid \alpha, \beta \in \mathbb{Z}_3\}$$

The set $\mathcal{B}$ = {1000011, 0100101, 0010110, 0001111} is linearly independent over $\mathbb{Z}_2$ and so it forms a basis for a binary linear [7, 4]-code C_3 = $\langle \mathcal{B} \rangle$. In fact, we will see that C_3 is one of the Hamming codes.   □

Example 5.2.2   Consider the binary linear code

$$C = \langle 11001, 01101, 10100 \rangle$$

Since the codewords 11001, 01101, and 10100 are not linearly independent over $\mathbb{Z}_2$, they generate a linear code of dimension less than 3. In fact, since $\mathcal{B}$ = {11001, 01101} is a maximal linearly independent subset of {11001, 01101, 10100}, the code C has basis $\mathcal{B}$ and therefore dimension 2 and size 2^2 = 4. Hence, C is a binary [5, 2]-code. In fact, since all linear codes contain the zero codeword, we see immediately that

$$C = \{00000, 11001, 01101, 10100\}$$ □

The Minimum Weight of a Linear Code

In order to determine the minimum distance for an arbitrary code C of size M, we need to check each of the $\binom{M}{2}$ distances $d(\mathbf{c}, \mathbf{d})$ between codewords. The next theorem shows that for linear codes, we can greatly simplify this task. First we need a definition, and a simple lemma. Recall that the **weight** $w(c)$ of a word is the number of places in $\mathbf{c}$ with nonzero entries.

Definition   The **weight** of a code C, denoted by $w(C)$, is the minimum weight of all *nonzero* codewords in C.   □

Lemma 5.2.2   *If C is a linear code, then*

$$d(\mathbf{c}, \mathbf{d}) = w(\mathbf{c} - \mathbf{d})$$

for all codewords $\mathbf{c}$ and $\mathbf{d}$ in C.   □

Proof   This follows from the fact that the codeword $\mathbf{c} - \mathbf{d}$ has nonzero entries precisely in those places where $\mathbf{c}$ and $\mathbf{d}$ differ, that is, in $d(\mathbf{c}, \mathbf{d})$ places. Hence, $w(\mathbf{c} - \mathbf{d}) = d(\mathbf{c}, \mathbf{d})$.   ■

Theorem 5.2.3   *If C is a linear code, then $d(C) = w(C)$.*   □

Proof   Let $\mathbf{c}$ and $\mathbf{d}$ be codewords at minimum distance $d(\mathbf{c}, \mathbf{d}) = d(C)$ apart. Then,

$$d(C) = d(\mathbf{c}, \mathbf{d}) = w(\mathbf{c} - \mathbf{d}) \geq w(C)$$

On the other hand, since $w(C)$ is the minimum weight in C, there must exist a codeword $\mathbf{e}$ in C for which $w(\mathbf{e}) = w(C)$. Hence

$$w(C) = w(\mathbf{e}) = w(\mathbf{e} - 0) = d(\mathbf{e}, \mathbf{0}) \geq d(C)$$

Combining the two inequaltities gives $d(C) = w(C)$, as desired.   ■

Example 5.2.3   The linear code $C_1 = \{0000, 1011, 0110, 1101\}$ of Example 5.2.1 has minimum weight 2. Hence $d(C) = 2$.

To determine the minimum weight of the code C_2 of Example 5.2.1, we need to list all 9 codewords

$$C_2 = \langle 0121, 2210 \rangle = \{\alpha(0121) + \beta(2210) \mid \alpha, \beta \in \mathbb{Z}_3\}$$
$$= \{0000, 0121, 0212, 2210, 2001, 2122, 1120, 1211, 1002\}$$

Notice that the minimum weight of C_2 (and hence the minimum distance) is 2 even though the weight of each basis codeword is 3. In general, there does not seem to be a simple way to deduce the minimum weight of a code directly from the minimum weights of the codewords in a basis for the code.

We will leave it as an exercise to show that the minimum weight of the Hamming code C_3 of Example 5.2.1 is 3.   □

It is important to emphasize that Theorem 5.2.3 holds only for linear codes. This theorem implies that we can detemine the minimum distance of a linear code by checking the M codeword weights, rather than the $\binom{M}{2}$ codeword distances. This is a significant savings in time, since $\binom{M}{2}$ is on the order of M^2. (For example, a code of size $M = 10{,}000$ has $\binom{10000}{2} = 49{,}995{,}000$ codeword pairs!)

The Generator Matrix of a Linear Code

Another advantage of linear codes is that a p-ary linear $[n, k]$-code C can be described simply by giving a basis for C, which consists of k linearly independent codewords in C, rather than having to list all of the p^k individual codewords in the code. For a binary code of dimension 30, for instance, a basis has size 30 but the code size is $2^{30} = 1,073,741,824$.

It is customary to put the codewords of a basis for a linear code C into a matrix, as described in the following definition.

Definition   Let C be a linear code, with basis $\mathcal{B} = \{\mathbf{b}_1, \mathbf{b}_2, \ldots, \mathbf{b}_k\}$ If

$$\mathbf{b}_1 = b_{11}b_{12}\cdots b_{1n}$$
$$\mathbf{b}_2 = b_{21}b_{22}\cdots b_{2n}$$
$$\vdots$$
$$\mathbf{b}_k = b_{k1}b_{k2}\cdots b_{kn}$$

then the $k \times n$)-matrix

$$G = \begin{bmatrix} b_{11} & b_{12} & \cdots & b_{1n} \\ b_{21} & b_{22} & \cdots & b_{2n} \\ & & \vdots & \\ b_{k1} & b_{k2} & \cdots & b_{kn} \end{bmatrix}$$

whose rows are the codewords in $\mathcal{B}$, is called a **generator matrix** for C.   □

Example 5.2.4   The code C_1 of Example 5.2.1 has basis $\mathcal{B} = \{1011, 0110\}$ and so a *generator matrix* for C_1 is

$$G_1 = \begin{bmatrix} 1 & 0 & 1 & 1 \\ 0 & 1 & 1 & 0 \end{bmatrix}$$

The code C_2 of Example 5.2.1 has generator matrix

$$G_2 = \begin{bmatrix} 0 & 1 & 2 & 1 \\ 2 & 2 & 1 & 0 \end{bmatrix}$$

The Hamming code C_3 of Example 5.2.1 has generator matrix

$$G_3 = \begin{bmatrix} 1 & 0 & 0 & 0 & 0 & 1 & 1 \\ 0 & 1 & 0 & 0 & 1 & 0 & 1 \\ 0 & 0 & 1 & 0 & 1 & 1 & 0 \\ 0 & 0 & 0 & 1 & 1 & 1 & 1 \end{bmatrix}$$ □

The next theorem follows from the results of Section 5.1.

Theorem 5.2.4   *Let G be a matrix with elements in $\mathbb{Z}_p$. Then G is a generator matrix for some linear code C over $\mathbb{Z}_p$ if and only if the rows of G, thought of as strings over $\mathbb{Z}_p$, are linearly independent.* $\square$

Exercises

1. Is the r-ary repetition code $\text{Rep}_r(n)$ linear? If so, give a basis, state the dimension, and find the minimum weight. If not, show why not.

2. Is the code E_n consisting of all codewords in $\mathbb{Z}_2^n$ with even weight a linear code? If so, give a basis, state the dimension, and find the minimum weight. If not, show why not.

3. Let $C = \{x_1 \cdots x_n \in \mathbb{Z}_p^n \mid x_1 = 0\}$. Is C a linear code? If so, give a basis, state the dimension and find the minimum weight. If not, show why not.

4. Let $C = \{x_1 \cdots x_n \in \mathbb{Z}_p^n \mid x_1 = 1\}$. Is C a linear code? If so, give a basis, state the dimension, and find the minimum weight. If not, show why not.

5. Prove that, if $\mathbf{c}$ and $\mathbf{d}$ are binary strings, then $w(\mathbf{c} + \mathbf{d})$ is even if and only if $w(\mathbf{c})$ and $w(\mathbf{d})$ have the same parity. Is this true for nonbinary strings?

6. Show that if C is a binary linear $[n, k, d]$-code, then the code $\overline{C}$ obtained by adding an even parity check to C is also linear. What are the parameters $[\overline{n}, \overline{k}, \overline{d}]$ of $\overline{C}$?

7. Prove that for a binary linear code C, either all of the codewords have even weight or else exactly half of the codewords have even weight. Is a similar statement true for ternary linear codes?

8. Find three distinct bases for the binary linear code $C = \langle 111, 101, 010 \rangle$.

9. Find a basis for the binary linear code $C = \langle 11111, 11101, 00110, 00100 \rangle$.

10. Write out all of the codewords for the binary code with generator matrix

$$G = \begin{bmatrix} 0 & 1 & 1 \\ 1 & 1 & 0 \end{bmatrix}$$

and find the parameters of the code.

11. Write out all of the codewords for the binary code with generator matrix

$$G = \begin{bmatrix} 1 & 0 & 0 & 1 & 0 & 0 & 1 \\ 0 & 1 & 0 & 1 & 0 & 1 & 1 \\ 0 & 0 & 1 & 0 & 1 & 1 & 1 \end{bmatrix}$$

and find the parameters of the code.

12. Write out all of the codewords for the ternary code with generator matrix

$$G = \begin{bmatrix} 0 & 1 & 2 \\ 2 & 1 & 0 \end{bmatrix}$$

find the parameters of the code.

13. Show that the minimum weight of the Hamming code C_3 of Example 5.2.1 is 3.

14. Write out all of the codewords for the code $C = \langle 0123, 0314, 0432 \rangle$ over $\mathbb{Z}_5$ and find the parameters of the code. Hint: check for linear independence first!

15. Is it possible to construct a binary linear (5,13)-code? Explain.

16. Write out all of the codewords for the ternary code with generator matrix

$$\begin{bmatrix} 1 & 0 & 1 & 1 \\ 0 & 1 & 1 & 2 \end{bmatrix}$$

and find the the parameters of the code. Show that C is perfect.

17. If C and D are binary linear codes of the same length, show that the $u(u + v)$-construction $C \oplus D$ is also linear.

5.3 Correcting Errors in a Linear Code

Nearest neighbor decoding involves finding a codeword closest to the received word. If a code has no particular structure, it may be necessary to employ the "brute force" method of computing the distance from the received word to each codeword. This may be impractical, if not impossible, for large codes. Fortunately, there are much better methods for decoding with linear codes, and we will devote much of the rest of this chapter to discussing them.

Let $C = \{\mathbf{c}_1, \mathbf{c}_2, \ldots, \mathbf{c}_M\}$ be a p-ary $[n, k, d]$-code, with size $M = p^k$, and let $\mathbf{c}_1 = 0$. We construct a table containing all strings in $\mathbb{Z}_p^n$ as follows. The first row of the table contains the codewords themselves, with $c_1 = \mathbf{0}$ in the first position,

$\mathbf{0}$	$\mathbf{c}_2$	$\mathbf{c}_3$	$\cdots$	$\mathbf{c}_M$

Next, we choose a string of smallest weight that has not yet appeared in the table, call it $\mathbf{f}_2$, and place it below $\mathbf{0}$. Then we add $\mathbf{f}_2$ to each of the codewords in the top row of the table, and place the results in the second row,

$\mathbf{0}$	$\mathbf{c}_2$	$\mathbf{c}_3$	$\cdots$	$\mathbf{c}_M$
$\mathbf{f}_2$	$\mathbf{f}_2 + \mathbf{c}_2$	$\mathbf{f}_2 + \mathbf{c}_3$	$\cdots$	$\mathbf{f}_2 + \mathbf{c}_M$

Again we choose a string of smallest weight that has not yet appeared in the table, place it in the first column, and fill out that row by adding this string to each codeword,

$\mathbf{0}$	$\mathbf{c}_2$	$\mathbf{c}_3$	$\cdots$	$\mathbf{c}_M$
$\mathbf{f}_2$	$\mathbf{f}_2 + \mathbf{c}_2$	$\mathbf{f}_2 + \mathbf{c}_3$	$\cdots$	$\mathbf{f}_2 + \mathbf{c}_M$
$\mathbf{f}_3$	$\mathbf{f}_3 + \mathbf{c}_2$	$\mathbf{f}_3 + \mathbf{c}_3$	$\cdots$	$\mathbf{f}_3 + \mathbf{c}_M$

Continuing in this way until all strings in $\mathbb{Z}_p^n$ are included, we arrive at a table of the form

$\mathbf{0}$	$\mathbf{c}_2$	$\mathbf{c}_3$	$\cdots$	$\mathbf{c}_M$
$\mathbf{f}_2$	$\mathbf{f}_2 + \mathbf{c}_2$	$\mathbf{f}_2 + \mathbf{c}_3$	$\cdots$	$\mathbf{f}_2 + \mathbf{c}_M$
$\mathbf{f}_3$	$\mathbf{f}_3 + \mathbf{c}_2$	$\mathbf{f}_3 + \mathbf{c}_3$	$\cdots$	$\mathbf{f}_3 + \mathbf{c}_M$
$\vdots$	$\vdots$	$\vdots$	$\vdots$	$\vdots$
$\mathbf{f}_q$	$\mathbf{f}_q + \mathbf{c}_2$	$\mathbf{f}_q + \mathbf{c}_3$	$\cdots$	$\mathbf{f}_q + \mathbf{c}_M$

where each string in the first column has minimum weight among all strings that follow it in the table. Such a table is referred to as a **standard array** for the linear code C. Each row of the array is called a **coset** of C, and the string in the first position in a given row is the **coset leader** for that coset. Hence, the coset leader has minimum weight among all strings in its coset (and in subsequent cosets). Note that any string in a standard array is the sum of the codeword at the top of its column and the coset leader at the far left of its row.

The following basic facts about standard arrays will be used repeatedly.

Theorem 5.3.1 *Let C be a linear $[n, k]$-code with standard array A.*

1. *Every string in $\mathbb{Z}_p^n$ appears exactly once in A.*

2. *The number of rows of A is p^{n-k}.*

3. *Two strings $\mathbf{x}$ and $\mathbf{y}$ in $\mathbb{Z}_p^n$ lie in the same coset (row) of A if and only if their difference $\mathbf{x} - \mathbf{y}$ is a codeword.* $\square$

Proof First, we observe that the difference $\mathbf{f}_i - \mathbf{f}_j$ of two distinct coset leaders in A cannot be a codeword, for if $\mathbf{f}_i - \mathbf{f}_j = c \in C$, then $\mathbf{f}_i = \mathbf{f}_j + \mathbf{c}$. Assuming (as we may) that $i > j$, this says that the coset leader $\mathbf{f}_i$ appears in the earlier coset whose coset leader is $\mathbf{f}_j$. But this is not possible by the rules for choosing a coset leader.

For part 1), it is clear from the way standard arrays are constructed that every string in $\mathbb{Z}_p^n$ appears in A. (If a string is not yet in A, then we are not done constructing A.) To show that no string appears twice in A, we show that if $\mathbf{f}_i + \mathbf{c}_u = \mathbf{f}_j + \mathbf{c}_v$, then $\mathbf{f}_i = \mathbf{f}_j$ and $\mathbf{c}_u = \mathbf{c}_v$. For if $\mathbf{f}_i + \mathbf{c}_u = \mathbf{f}_j + \mathbf{c}_v$, then $\mathbf{f}_i - \mathbf{f}_j = \mathbf{c}_v - \mathbf{c}_u \in C$, which is not possible by the previous remarks unless $i = j$. But if $i = j$, then $\mathbf{f}_i = \mathbf{f}_j$ and therefore $\mathbf{c}_u = \mathbf{c}_v$.

We leave proof of part 2) as an exercise. As for part 3), if $\mathbf{x}$ and $\mathbf{y}$ are in the same coset, then we may write $\mathbf{x} = \mathbf{f}_i + \mathbf{c}_u$ and $\mathbf{y} = \mathbf{f}_i + \mathbf{c}_v$ and so

$$\mathbf{x} - \mathbf{y} = (\mathbf{f}_i + \mathbf{c}_u) - (\mathbf{f}_i + \mathbf{c}_v) = \mathbf{c}_u - \mathbf{c}_v \in C$$

Conversely, if $(\mathbf{f}_i + \mathbf{c}_u) - (\mathbf{f}_j + \mathbf{c}_v) = \mathbf{c} \in C$, then

$$\mathbf{f}_i - \mathbf{f}_j = \mathbf{c} - \mathbf{c}_u + \mathbf{c}_v \in C$$

which is impossible unless $i = j$, whence $\mathbf{f}_i + \mathbf{c}_u$ and $\mathbf{f}_j + \mathbf{c}_v$ lie in the same coset. ■

Part 1) of the previous theorem can be summarized by saying that the cosets of C are disjoint, nonempty sets whose union is all of $\mathbb{Z}_p^n$. In the language of set theory, the cosets of C form a partition of $\mathbb{Z}_p^n$.

Example 5.3.1 A standard array for the binary $[4, 2]$-code $C_1 = \{0000, 1011, 0110, 1101\}$ of Example 5.2.1 is

0000	1011	0110	1101
1000	0011	1110	0101
0100	1111	0010	1001
0001	1010	0111	1100

Since $p^{n-k} = 2^{4-2} = 4$, we know that the above array is complete.   □

Standard arrays are not unique. For instance, in the array of the previous example, we could have chosen 0010 to be the coset leader for the third row of the array. This would have led to a different standard array for the code C_1.

We now come to the purpose of standard arrays, which is to implement nearest neighbor decoding.

Theorem 5.3.2 *Let C be a p-ary $[n, k]$-code, with standard array A. For any string $\mathbf{x}$ in $\mathbb{Z}_p^n$, the codeword $\mathbf{c}$ that lies at the top of the column containing $\mathbf{x}$ is a nearest neighbor codeword to $\mathbf{x}$.*   □

Proof   Consider a standard array

$\mathbf{0}$	$\mathbf{c}_2$	$\mathbf{c}_3$	$\cdots$	$\mathbf{c}_M$
$\mathbf{f}_2$	$\mathbf{f}_2 + \mathbf{c}_2$	$\mathbf{f}_2 + \mathbf{c}_3$	$\cdots$	$\mathbf{f}_2 + \mathbf{c}_M$
$\mathbf{f}_3$	$\mathbf{f}_3 + \mathbf{c}_2$	$\mathbf{f}_3 + \mathbf{c}_3$	$\cdots$	$\mathbf{f}_3 + \mathbf{c}_M$
$\vdots$	$\vdots$	$\vdots$	$\vdots$	$\vdots$
$\mathbf{f}_q$	$\mathbf{f}_q + \mathbf{c}_2$	$\mathbf{f}_q + \mathbf{c}_3$	$\cdots$	$\mathbf{f}_q + \mathbf{c}_M$

Suppose that the received string is $\mathbf{x} = \mathbf{f}_i + \mathbf{c}_j$. We want to show that $\mathbf{c}_j$ has minimum distance to x among all codewords, that is,

$$d(\mathbf{x}, \mathbf{c}_j) = \min_{\mathbf{c} \in C}\{d(\mathbf{x}, \mathbf{c})\}$$

Since $d(\mathbf{x}, \mathbf{c}) = w(\mathbf{x} - \mathbf{c})$, we have

$$\min_{\mathbf{c} \in C}\{d(\mathbf{x}, \mathbf{c})\} = \min_{\mathbf{c} \in C}\{w(\mathbf{x} - \mathbf{c})\} = \min_{\mathbf{c} \in C}\{w(\mathbf{f}_i + \mathbf{c}_j - \mathbf{c})\}$$

But since $\mathbf{c}_j - \mathbf{c} \neq \mathbf{c}_j - \mathbf{d}$ for $\mathbf{c} \neq \mathbf{d}$, we conclude that as $\mathbf{c}$ ranges over all codewords in C, so does $\mathbf{c}_j - \mathbf{c}$, and so the last expression above equals

$$\min_{\mathbf{c} \in C}\{w(\mathbf{f}_i + \mathbf{c})\}$$

which is the minimum weight among all strings in the coset containing $\mathbf{x}$. But, according to the way that standard arrays are constructed, this minimum weight is $w(\mathbf{f}_i)$. Hence,

$$\min_{\mathbf{c} \in C}\{d(\mathbf{x}, \mathbf{c})\} = \min_{\mathbf{c} \in C}\{w(\mathbf{f}_i + \mathbf{c})\} = w(\mathbf{f}_i) = w(\mathbf{x} - \mathbf{c}_j) = d(\mathbf{x}, \mathbf{c}_j) \quad \blacksquare$$

Notice that the difference $\mathbf{x} - \mathbf{c}$ between the received string $\mathbf{x}$ and the nearest neighbor interpretation $\mathbf{c}$ at the top of the column containing $\mathbf{x}$, is the coset leader for the coset containing $\mathbf{x}$. This difference (and hence this coset leader) is called the **error string**.

Example 5.3.2   Referring to the code C_1 and its standard array in Example 5.3.1, if the string 0010 is received, this array gives the nearest neighbor interpretation 0110, with error string 0100.   □

Note that, when using a standard array for nearest neighbor decoding, a received string is decoded correctly *if and only if* the error string is one of the coset leaders $\mathbf{f}_1 = \mathbf{0}, \mathbf{f}_2, \mathbf{f}_3, \ldots, \mathbf{f}_q$.

Moreover, nearest neighbor ties are always decided when using a standard array, since all received words are decoded. Thus, standard array decoding is complete decoding. In fact, nearest neighbor ties are manifested in a standard array by the appearance of a string in a given coset with the same weight as the coset leader for that coset. For example, consider the binary code $C = \{000, 110\} = \langle 110 \rangle$. A standard array for C is

000	110
100	010
001	111
101	011

Notice that both strings in row 2 have the same weight. There are two codewords with minimum distance 1 from the word 010 and so we have a nearest neighbor tie. However, this standard array decodes 010 as 110. If we had started the second row of the array with 010, the array would decode 010 as 000. We leave it as an exercise to prove that a linear code C admits ties if and only if any standard array for C has this property.

Recall that if C is a linear $[n, k, d]$-code, then it is v-error-correcting, where $v = \lfloor \frac{d-1}{2} \rfloor$. Put another way, any errors that result in an error string of weight v or less are corrected. It follows that the coset leaders of any standard array for C must include all strings of weight v or less. Of course, there may be other coset leaders as well. However, if C is also a perfect code, then no other errors are corrected and so the coset leaders of A are precisely the strings of weight v or less. (As an exercise, you may wish to derive the sphere-packing condition from this fact.)

The Probability of Correct Decoding

We can easily obtain an exact expression (as opposed to an upper or lower bound, see Section 4.4) for the probability of correct decoding using a linear $[n, k]$-code C, with standard array

0	$\mathbf{c}_2$	$\mathbf{c}_3$	$\cdots$	$\mathbf{c}_M$
$\mathbf{f}_2$	$\mathbf{f}_2 + \mathbf{c}_2$	$\mathbf{f}_2 + \mathbf{c}_3$	$\cdots$	$\mathbf{f}_2 + \mathbf{c}_M$
$\mathbf{f}_3$	$\mathbf{f}_3 + \mathbf{c}_2$	$\mathbf{f}_3 + \mathbf{c}_3$	$\cdots$	$\mathbf{f}_3 + \mathbf{c}_M$
$\vdots$	$\vdots$	$\vdots$	$\vdots$	$\vdots$
$\mathbf{f}_q$	$\mathbf{f}_q + \mathbf{c}_2$	$\mathbf{f}_q + \mathbf{c}_3$	$\cdots$	$\mathbf{f}_q + \mathbf{c}_M$

Assuming a binary symmetric channel with crossover probability p, if a coset leader $\mathbf{f}_i$ has weight $w(\mathbf{f}_i)$, then the probability that a codeword picks up errors in precisely the positions where $\mathbf{f}_i$ has 1s is

$$\mathcal{P} \text{ (error string is } \mathbf{f}_i) = p^{w(\mathbf{f}_i)}(1 - p)^{n - w(\mathbf{f}_i)}$$

Summing this over all coset leaders, we get the following theorem.

Theorem 5.3.3 *Let C be a binary $[n, k]$-code, and suppose that the coset leaders in a standard array for C are $\mathbf{f}_1, \mathbf{f}_2, \mathbf{f}_3, \ldots, \mathbf{f}_q$. Then, assuming a binary symmetric channel with crossover probability p, the probability that a received string will be decoded correctly by the standard array is*

$$\mathcal{P} \text{ (correct decoding)} = \sum_{i=1}^{q} p^{w(\mathbf{f}_i)}(1 - p)^{n - w(\mathbf{f}_i)}$$

Put another way, if we let w_i be the number of coset leaders of weight i, then this probability is

$$\mathcal{P} \text{ (correct decoding)} = \sum_{i=0}^{n} w_i p^i (1 - p)^{n - i}$$

$\square$

Example 5.3.3 Consider the binary $[4, 2]$-code $C_1 = \{0000, 1011, 0110, 1101\}$ of Example 5.2.1, with standard array (from Example 5.3.1)

0000	1011	0110	1101
1000	0011	1110	0101
0100	1111	0010	1001
0001	1010	0111	1100

Since $w_0 = 1$, $w_1 = 3$, $w_2 = w_3 = w_4 = 0$, we have

$$\mathcal{P} \text{ (correct decoding)} = (1 - p)^4 + 3p(1 - p)^3 = (1 - p)^3(1 + 2p)$$

For example, if the crossover probability is $p = 0.01$, then

$$\mathcal{P} \text{ (correct decoding)} = (1 - 0.01)^3(1 + 0.02) \approx 0.9897$$

Hence, $\mathcal{P}(\text{decode error}) = 1 - \mathcal{P}(\text{correct decoding}) \approx 0.0103$.

It is instructive to compare this with the alternative of no encoding for error correction. Since $|C_1| = 4$, we can think of the source as consisting of the 4 symbols 00, 01, 10, and 11 and send these symbols as is, without encoding. In this event, the probability that a source symbol gets through without error is $(1 - p)^2 = 0.9801$, which is less than $\mathcal{P}(\text{correct decoding})$ using the code C_1. $\square$

In general, the problem of determining the number w_i of coset leaders of weight i is quite difficult, and in fact these numbers are not known for many important classes of codes. However, in the case of perfect codes, we can easily determine these numbers. We have already remarked that the coset leaders in a standard array for a perfect $[n, k, d]$-code are precisely the strings of weight v or less, where $v = \lfloor \frac{d-1}{2} \rfloor$. Since there are $\binom{n}{i}$ strings in $\mathbb{Z}_2^n$ of weight i, we have $w_i = \binom{n}{i}$ for $0 \le i \le v$, and $w_i = 0$ for $i > v$.

Theorem 5.3.4 *Let C be a perfect binary $[n, k, d]$-code. The probability of correct decoding using any standard array is given by*

$$\mathcal{P} \text{ (correct decoding)} = \sum_{i=0}^{\lfloor \frac{d-1}{2} \rfloor} \binom{n}{i} p^i (1 - p)^{n-i}$$

□

Example 5.3.4 The binary Hamming code C_3 of Example 5.2.1 is a perfect [7, 4, 3]-code. Hence, according to Theorem 5.3.4,

$$\mathcal{P} \text{ (correct decoding)} = (1-p)^7 + 7p(1-p)^6 = (1-p)^6(1+6p)$$

Let us compare this with the probability of receiving correct information without using error correction. Since $|C_3| = 2^3 = 8$, this code is capable of encoding eight source symbols. If we let these be the eight binary strings of length 3, the probability of receiving a source symbol correctly is

$$\mathcal{P} \text{ (correct receipt)} = (1-p)^3$$

The following table compares $\mathcal{P}$(correct decoding) with $\mathcal{P}$(correct receipt), for various symbol error probabilities.

p	$\mathcal{P}$(correct decoding)	$\mathcal{P}$(correct receipt)
0.1	0.8503	0.7290
0.05	0.9556	0.8574
0.01	0.9980	0.9703
0.005	0.9995	0.9851

The table clearly shows that using the code C significantly improves the probability of obtaining correct information, but at the expense of having to send 7 bits per source symbol rather than 3. □

Burst Error Correction

You will recall that, in defining a communications channel (Section 4.1), we included the requirements that symbol errors be independent of time, that is, that $\mathcal{P}(a_j$ received $\mid a_i$ sent) does not change with time and that the outcome of a given symbol transmission does not depend on previous symbol transmissions. While these assumptions make life a lot simpler, they are not always realistic. For instance, one could reasonably argue that, if a bit error occurs in the transmission of a certain bit, that error may very well have been the result of a phenomenon (electrical disturbance, for instance) that lasts longer than the time it takes to send a single bit, and so adjacent bits are likely to be affected as well. Similarly, a defect on the surface of a compact disk, for instance, is likely to be larger than the amount of space it takes to store a single bit and so errors will tend to occur in bunches, rather than independently.

This leads us to the concept of a *burst error*. It is our intention to define this concept and discuss a few simple results related to burst errors. However, codes that are good at correcting independent errors may do a poor job at burst error correction and vice-versa and so we will not pursue burst error correction beyond this short discussion (and an exercise or two).

Definition A **burst** in $\mathbb{Z}_p^n$ of length b is a string in $\mathbb{Z}_p^n$ whose nonzero coordinates are confined to b consecutive positions, the first and last of which must be nonzero. □

For example, the string 0001100100 in $\mathbb{Z}_2^{10}$ is a burst of length 5. Note that not all of the coordinates between the first and last 1s need be nonzero.

Note that if a linear code is to correct any burst of length b or less, then no such burst can be a codeword (why?). The next theorem gives an upper bound on the dimension of a linear code that contains no bursts of length b or less.

Theorem 5.3.5 *Let C be a linear $[n, k]$-code over $\mathbb{Z}_p$. If C contains no bursts of length b or less, then $k \le n - b$.* □

Proof Consider the set S of all strings in $\mathbb{Z}_p^n$ with 0s in the last $n - b$ positions. (The first b positions may contain any values, including 0.) Since the difference between any two strings in the same row of a standard array for C is a codeword, if any two distinct strings in S lie in the same coset of a standard array, then their difference would be a nonzero burst of length at most b lying in C, which is not possible since C is assumed not to contain any such bursts. Hence, the number of cosets of C, which is p^{n-k}, must be greater than or equal to the size of S, which is p^b. In other words, $p^{n-k} \ge p^b$, whence $n - k \ge b$. ■

We have seen that the more errors we expect a code to detect or correct, the smaller must be the size of the code. The situation for burst error detection is settled by the following result.

Theorem 5.3.6 *If a linear $[n, k]$-code C can detect all burst errors of length b or less, then $k \leq n - b$. Furthermore, there is a linear $[n, n - b]$-code that will detect all burst errors of length b or less.* $\square$

Proof   We leave it as an exercise to show that C can detect all burst errors of length b or less if and only if C contains no such bursts. Hence, Theorem 5.3.5 implies that $k \leq n - b$. To prove the second statement, we construct a generator matrix G of size $(n - b) \times n$ for a linear $[n, n - b]$-code C that can detect all bursts of length b or less.

The matrix G is constructed in two parts. The leftmost part consists of the row matrices $\mathbf{e}_1, \ldots, \mathbf{e}_b$ of length b, repeated as many times as is necessary to fill in $n - b$ rows. The rightmost part consists of the identity matrix of size $(n - b) \times (n - b)$. For example, if $n = 8$ and $b = 3$, then

$$
G_1 = \left[\begin{array}{ccc|ccccc}
1 & 0 & 0 & 1 & 0 & 0 & 0 & 0 \\
0 & 1 & 0 & 0 & 1 & 0 & 0 & 0 \\
0 & 0 & 1 & 0 & 0 & 1 & 0 & 0 \\
1 & 0 & 0 & 0 & 0 & 0 & 1 & 0 \\
0 & 1 & 0 & 0 & 0 & 0 & 0 & 1
\end{array} \right]
$$

(the line is just for clarity). Notice that the rightmost $n - b = 5$ columns of G form an identity matrix and the leftmost $b = 3$ columns, reading the rows, are e_1, e_2, e_3, e_1, e_2.

By the first statement of this theorem, we need only show that no linear combination of the rows of G is a burst of length b or less. Assume to the contrary that $\mathbf{x} = \alpha_1 \mathbf{r}_1 + \cdots + \alpha_{n-b} \mathbf{r}_{n-b}$ is a burst of length b or less. Notice that if $\alpha_i \neq 0$, then $\mathbf{x}$ has α_i in the $(i + b)$th position. It follows that the indices i for which α_i are nonzero must lie within b consecutive values. However, for any b or fewer consecutive rows of G, the leftmost portion is just the matrices $\mathbf{e}_1, \ldots, \mathbf{e}_b$ appearing in some order. For instance, in rows 2, 3, and 4, the leftmost portion of G_1 is $\mathbf{e}_2, \mathbf{e}_3, \mathbf{e}_1$. The key point is that any b or fewer consecutive rows of G have no nonzero positions in common (either on the left or the right) and so, if α_i is nonzero, then $\mathbf{x}$ has nonzero entries in those positions in which $\mathbf{r}_i$ has nonzero entries. But the nonzero entries of $\mathbf{r}_i$ are not confined to b or fewer consecutive positions. It follows that all of the coefficients must be 0, implying that $\mathbf{x} = \mathbf{0}$. This contradiction implies that no linear combination of the rows of G is a burst of length b or less.   ∎

Now let us consider burst error correction.

Theorem 5.3.7 *If a linear* $[n, k]$*-code* C *can correct all burst errors of length* b *or less, using a standard array, then* $k \leq n - 2b$. □

Proof Our goal is to apply Theorem 5.3.5 (with b replaced by $2b$). If $2 \leq \ell \leq 2b$, then any burst $\mathbf{x}$ of length ℓ can be written as the difference $\mathbf{x}_1 - \mathbf{x}_2$ of two distinct bursts of length at most b. For instance, if $n = 11$ and $b = 3$, then the burst $\mathbf{x} = 00011101100$ of length 6 can be written as the difference of a burst of length 3 and a burst of length 2 as follows

$$00011101100 = 00011100000 + 00000001100$$

Since C can correct $\mathbf{x}_1$ and $\mathbf{x}_2$, these strings must be coset leaders and therefore cannot lie in the same coset of any standard array for C. This implies that their difference, which is $\mathbf{x}$, is not a codeword. Since a burst of length 1 cannot be a codeword, we may apply Theorem 5.3.5 (with b replaced by $2b$), to get $k \leq n - 2b$. ■

We have observed in the previous proof that if a code C can correct any burst error of length b or less, then no two such bursts can lie in the same coset of a standard array for C. Thus, by counting the total number of bursts of length b or less, we get a lower bound on the number of cosets of C, and hence an upper bound on the dimension of C, quite different from that given in Theorem 5.3.7. We leave proof of the following result as an exercise.

Theorem 5.3.8 *If a linear* $[n, k]$*-code* C *over* $\mathbb{Z}_p$ *can correct all burst errors of length* b *or less, using a standard array, then*

$$k \leq n - b + 1 - \log_p[(n - b + 1)(p - 1) + 1]$$ □

In the exercises, we ask you to compare the quality of Theorems 5.3.7 and 5.3.8.

Exercises

1. Show that the number of rows in a standard array for a linear $[n, k]$-code over $\mathbb{Z}_p$ is p^{n-k}.

2. Consider the binary code $C = \{000, 100, 101, 111\}$. Try to construct a standard array for C. What went wrong and why?

3. Construct a standard array for the binary linear code with generator matrix

$$\begin{bmatrix} 0 & 1 & 1 \\ 1 & 1 & 0 \end{bmatrix}$$

Decode the words a) 111, b) 100. Find the probability of correct decoding, assuming a binary symmetric channel with crossover probability p. Give an example of a single error that is decoded incorrectly.

4. Construct a standard array for the binary linear code with generator matrix

$$\begin{bmatrix} 1 & 0 & 1 & 1 & 0 \\ 0 & 1 & 0 & 1 & 1 \end{bmatrix}$$

Decode the received words a) 11111, b) 00000, c) 01000. Find an example of a codeword and received word such that two bit errors have occurred with correct decoding and another example of two bit errors with incorrect decoding. Find the probability of correct decoding, assuming a binary symmetric channel with crossover probability p.

5. Construct a standard array for the ternary code C_2 of Example 5.2.1.

6. Let C be a linear $[n, k, d]$-code. Prove that C admits ties (that is, some word x has at least two nearest neighbor codewords) if and only if there is a row in any standard array for C that has a non coset-leader with the same weight as the coset-leader.

7. Formulate Theorem 5.3.3 for a r-ary code C (r a prime) and a channel, with channel alphabet $\mathbb{Z}_r$, wherein the probability that a code symbol is changed to any other particular code symbol is p.

8. How big is the standard array for the Hamming code C_3 of Example 5.2.1? Compute a set of coset leaders for this array.

9. Write a computer program to compute standard arrays for binary codes. Use it to compute a standard array for the Hamming code C_3 of Example 5.2.1.

10. Compute the probability of a decoding error for the binary Hamming code $\mathcal{H}_2(h)$, assuming a binary symmetric channel. What are the parameters (length, dimension, size, and minimum distance) of this code and what is this probability when $p = 0.001$ and $h = 3, 4$, or 5?

11. Compute the probability of a decoding error under a binary symmetric channel with crossover probability 0.001 using the perfect binary (23,4097,7)-Golay code $\mathcal{G}_{23}$. Compare this with the results of the previous exercise for the Hamming codes.

12. Let C be a binary linear code. Let C^c be the set of all complements of codewords in C.

    (a) Show that if $1 \in C$ then $C = C^c$.

    (b) If C is a linear code, is C^c also a linear code?

    (c) Show that if C is a linear code, then so is $C \cup C^c$.

13. Suppose that C is a p-ary linear $[n, k]$-code that is capable of correcting any set of v or fewer errors, using a standard array. Thus, any error string of weight v or less must be a coset leader. Compute the number of strings of weight v or less and use that number to formulate an upper bound on the size of C. Does this look familiar?

14. Let C be a linear code. Show that C can detect all burst errors of length b or less if and only if C contains no such bursts.

15. Design a linear code of length 11 that can detect any burst of length 4 or less. Give a generator matrix for your code.

16. Can a linear $[10, 5]$-code correct all burst errors of length 3? Explain.

17. Prove Theorem 5.3.8. Hint: if N is the total number of bursts of length b or less, then since each burst must be in a different coset of a standard array for C, but none lie in C itself, we deduce that $p^{n-k} \geq N + 1$. In order to compute N, let the bursts of length b or less be divided into two types—those whose first nonzero entry lies somewhere within the first $n - b + 1$ positions and those whose first nonzero entry lies somewhere within the last $b - 1$ positions (that is, position numbers $n - b + 2$ through n). Show that there are $(n - b + 1)(p - 1)p^{b-1}$ of the former type and $p^{b-1} - 1$ of the latter. Hence, $N + 1 = [(n-b+1)(p-1)+1]p^{b-1}$. Complete the proof.

18. How would you compare the quality of the bounds in Theorem 5.3.7 and Theorem 5.3.8?

5.4   The Dual of a Linear Code

We have seen several ways of constructing new codes from old ones. Let us now describe another method—perhaps the most important one for linear codes. We begin with a definition.

Definition   Let $\mathbf{x} = x_1 x_2 \cdots x_n$ and $\mathbf{y} = y_1 y_2 \cdots y_n$ be strings in $\mathbb{Z}_p^n$, p a prime. The **inner product** of $\mathbf{x}$ and $\mathbf{y}$, denoted by $\mathbf{x} \cdot \mathbf{y}$, is the element of $\mathbb{Z}_p$ defined by

$$\mathbf{x} \cdot \mathbf{y} = x_1 y_1 + x_2 y_2 + \cdots + x_n y_n$$

where the sum and product are taken in $\mathbb{Z}_p$ (that is, modulo p).   □

Example 5.4.1   The inner product of $\mathbf{x}_1 = 10012$ and $\mathbf{x}_2 = 12122$ in $\mathbb{Z}_3^5$ is

$$\mathbf{x}_1 \cdot \mathbf{x}_2 = (10012) \cdot (12122) = 1 \cdot 1 + 0 \cdot 2 + 0 \cdot 1 + 1 \cdot 2 + 2 \cdot 2 = 1$$

The inner product of $\mathbf{x}_3 = 1010$ and $\mathbf{x}_4 = 1111$ in $\mathbb{Z}_2^4$ is

$$\mathbf{x}_3 \cdot \mathbf{x}_4 = (1010) \cdot (1111) = 1 \cdot 1 + 0 \cdot 1 + 1 \cdot 1 + 0 \cdot 1 = 0$$ □

Notice that the inner product of two strings in $\mathbb{Z}_p^n$ is an element of $\mathbb{Z}_p$, and not another string in $\mathbb{Z}_p^n$. The inner product satisfies some important properties, which are given in the next theorem.

Theorem 5.4.1   *For all strings* $\mathbf{x}$, $\mathbf{y}$, $\mathbf{z}$ *in* $\mathbb{Z}_p^n$ *and all* α *in* $\mathbb{Z}_p$,

1. $\mathbf{x} \cdot \mathbf{y} = \mathbf{y} \cdot \mathbf{x}$
2. $(\mathbf{x} + \mathbf{y}) \cdot \mathbf{z} = \mathbf{x} \cdot \mathbf{z} + \mathbf{y} \cdot \mathbf{z}$
3. $(\alpha \mathbf{x}) \cdot \mathbf{y} = \mathbf{x} \cdot (\alpha \mathbf{y}) = \alpha(\mathbf{x} \cdot \mathbf{y})$ □

Proof   We will prove part 3), and leave the others as exercises. If $\mathbf{x} = x_1 x_2 \cdots x_n$ and $\mathbf{y} = y_1 y_2 \cdots y_n$, then

$$(\alpha \mathbf{x}) \cdot \mathbf{y} = ((\alpha x_1)(\alpha x_2) \cdots (\alpha x_n)) \cdot (y_1 y_2 \cdots y_n)$$
$$= \alpha x_1 y_1 + \alpha x_2 y_2 + \cdots + \alpha x_n y_n$$
$$= \alpha(x_1 y_1 + x_2 y_2 + \cdots + x_n y_n) = \alpha(\mathbf{x} \cdot \mathbf{y})$$

The proof that $\mathbf{x} \cdot (\alpha \mathbf{y}) = \alpha(\mathbf{x} \cdot \mathbf{y})$ is quite similar.   ■

Definition   Let $\mathbf{x}$ and $\mathbf{y}$ be strings in $\mathbb{Z}_p^n$. If $\mathbf{x} \cdot \mathbf{y} = 0$, we say that $\mathbf{x}$ and $\mathbf{y}$ are **orthogonal**.   □

Thus, the strings $\mathbf{x}_3$ and $\mathbf{x}_4$ in Example 5.4.1 are orthogonal.

For any $a \in \mathbb{Z}_p^n$, we let $\{\mathbf{a}\}^\perp$ denote the set of all strings in $\mathbb{Z}_p^n$ that are orthogonal to $\mathbf{a}$. Thus,

$$\{\mathbf{a}\}^\perp = \{\mathbf{x} \in \mathbb{Z}_p^n \mid \mathbf{a} \cdot \mathbf{x} = 0\}$$

This set is called the **orthogonal complement** of $\mathbf{a}$. (The expression $\{\mathbf{a}\}^\perp$ is read "a purp.")

Theorem 5.4.2 *For any string a in $\mathbb{Z}_p^n$, the set $\{\mathbf{a}\}^\perp$ is a linear code.* $\qquad\square$

Proof If $\mathbf{x}$ and $\mathbf{y}$ are in $\{\mathbf{a}\}^\perp$, then $\mathbf{a} \cdot \mathbf{x} = 0$ and $\mathbf{a} \cdot \mathbf{y} = 0$. Thus, Theorem 5.4.1 implies that

$$\mathbf{a} \cdot (\mathbf{x} + \mathbf{y}) = \mathbf{a} \cdot \mathbf{x} + \mathbf{a} \cdot \mathbf{y} = 0$$

and

$$\mathbf{a} \cdot (\alpha \mathbf{x}) = \alpha(\mathbf{a} \cdot \mathbf{x}) = \alpha 0 = 0$$

for all $\alpha \in \mathbb{Z}_p$. This shows that $\{\mathbf{a}\}^\perp$ is closed under both addition and scalar multiplication, and so it is a subspace of $\mathbb{Z}_p^n$. $\qquad\blacksquare$

Note that if $\mathbf{a} = a_1 a_2 \cdots a_n$ and $\mathbf{x} = x_1 x_2 \cdots x_n$, then

$$\mathbf{a} \cdot \mathbf{x} = a_1 x_1 + a_2 x_2 + \cdots + a_n x_n$$

Hence, $\mathbf{x}$ is in $\{\mathbf{a}\}^\perp$ if and only if the equation

$$a_1 x_1 + a_2 x_2 + \cdots + a_n x_n = 0 \qquad\qquad (5.4.1)$$

holds in $\mathbb{Z}_p$.

Example 5.4.2 Consider the set $\{\mathbf{1}\}^\perp$, where $= 1 \cdots 1 \in \mathbb{Z}_2^n$. In this case, equation (5.4.1) is

$$x_1 + x_2 + \cdots + x_n = 0 \qquad\qquad (5.4.2)$$

But, in $\mathbb{Z}_2$, the sum on the left is equal to 0 if and only if an even number of the x_is are equal to 1. Hence,

$$\{\mathbf{1}\}^\perp = \{\mathbf{x} \in \mathbb{Z}_2^n \mid w(\mathbf{x}) \text{ is even}\}$$

A *binary string* has **even parity** if it has even weight and **odd parity** if it has odd weight. Hence, $\{\mathbf{1}\}^\perp$ is the set of all strings in $\mathbb{Z}_2^n$ with even parity. For this reason, equation (5.4.2) is often referred to as an even parity check equation. $\qquad\square$

Taking a hint from the terminology in the previous example, we call equation (5.4.1) the **parity check equation** for the code $\{\mathbf{a}\}^\perp$.

We now generalize.

Definition   Let $A = \{\mathbf{a}_1, \mathbf{a}_2, \ldots, \mathbf{a}_s\} \subseteq \mathbb{Z}_p^n$ be any code. The set

$$A^\perp = \{\mathbf{a}_1, \mathbf{a}_2, \ldots, \mathbf{a}_s\}^\perp = \{\mathbf{x} \in \mathbb{Z}_p^n \mid \mathbf{x} \cdot \mathbf{a}_i = 0 \text{ for all } i\}$$

is called the **orthogonal complement** of A.   ☐

The proof of the following theorem is very similar to that of Theorem 5.4.2, and is left as an exercise.

Theorem 5.4.3   *The orthogonal complement $A^\perp$ of any code A is a linear code called the* **dual code** *of A.*   ☐

We can describe the dual code $\{\mathbf{a}_1, \mathbf{a}_2, \ldots, \mathbf{a}_s\}^\perp$ as the solutions to certain equations, similar to equation (5.4.1). Before doing so, however, we need to make a notational point. Up to now, we have been using boldface letters to denote strings, for example, $\mathbf{x} = x_1 x_2 \cdots x_n$. We will also use boldface letters to denote row matrices. Thus, for example, we may use $\mathbf{x}$ to denote both the string $x_1 x_2 \cdots x_n$ and the matrix

$$\begin{bmatrix} x_1 & x_2 & \cdots & x_n \end{bmatrix}$$

This should not cause any confusion since the context will make it clear whether we are talking about a string or about a matrix.

Definition   Let $A = \{\mathbf{a}_1, \mathbf{a}_2, \ldots, \mathbf{a}_s\} \subseteq \mathbb{Z}_p^n$ and let

$$\mathbf{a}_1 = a_{11} a_{12} \cdots a_{1n}$$
$$\mathbf{a}_2 = a_{21} a_{22} \cdots a_{2n}$$
$$\vdots$$
$$\mathbf{a}_s = a_{s1} a_{s2} \cdots a_{sn}$$

The equations

$$
\begin{aligned}
a_{11} x_1 &+ a_{12} x_2 + \cdots + a_{1n} x_n = 0 \\
a_{21} x_1 &+ a_{22} x_2 + \cdots + a_{2n} x_n = 0 \\
&\qquad\qquad \vdots \\
a_{s1} x_1 &+ a_{s2} x_2 + \cdots + a_{sn} x_n = 0
\end{aligned}
$$

are called the **parity check equations** for the code $A^{\perp}$. We can also write this in the matrix form

$$\begin{bmatrix} x_1 & x_2 & \cdots & x_n \end{bmatrix} \begin{bmatrix} a_{11} & a_{12} & \cdots & a_{1n} \\ a_{21} & a_{22} & \cdots & a_{2n} \\ & & \vdots & \\ a_{s1} & a_{s2} & \cdots & a_{sn} \end{bmatrix}^t = [00 \cdots 0]$$

(Note the transpose on the second matrix.) Letting $\mathcal{A}$ denote the matrix (a_{ij}), we can write this more concisely as

$$x \mathcal{A}^t = 0$$

The matrix $\mathcal{A}$ is called a **parity check matrix** for the code $A^{\perp}$. $\square$

The following theorem is just a rewording of the relevant definitions.

Theorem 5.4.4   Let A be a code. A string $\mathbf{x} = x_1 x_2 \cdots x_n$ is in the dual code $A^{\perp}$ if and only if its components $x_1, \ldots, x_n$ satisfy the parity check equations for $A^{\perp}$. Expressed in terms of the parity check matrix, we have

$$A^{\perp} = \{\mathbf{x} \in \mathbb{Z}_p^n \mid \mathbf{x} \mathcal{A}^t = \mathbf{0}\} \qquad \square$$

Example 5.4.3   Consider the binary strings $\mathbf{a}_1 = 1110$ and $\mathbf{a}_2 = 1001$. The parity check equations for the binary linear code $C = \{\mathbf{a}_1, \mathbf{a}_2\}^{\perp}$ are

$$\begin{aligned} x_1 & -x_2 & -x_3 & & = 0 \\ x_1 & & & -x_4 & = 0 \end{aligned}$$

This system is easily solved to give $C = \{0000, 0110, 1101, 1011\}$, which is none other than the linear code C_1 of Example 5.2.1. A parity check matrix for this code is thus

$$P_1 = \begin{bmatrix} 1 & 1 & 1 & 0 \\ 1 & 0 & 0 & 1 \end{bmatrix}$$

Let $\mathbf{a}_1 = 0110$ and $\mathbf{a}_2 = 1201$ be ternary strings. The parity check equations for the ternary code $C = \{\mathbf{a}_1, \mathbf{a}_2\}^{\perp}$ are

$$\begin{aligned} x_2 & -x_3 & & = 0 \\ x_1 & -2x_2 & -x_4 & = 0 \end{aligned}$$

whose solutions are easily seen to be the 9 codewords in the code C_2 of Example 5.2.1. A parity check matrix for this code is thus

$$P_2 = \begin{bmatrix} 0 & 1 & 1 & 0 \\ 1 & 2 & 0 & 1 \end{bmatrix}$$

The parity check equations for the binary code $\{0111100, 1011010, 1101001\}^{\perp}$ are

$$
\begin{array}{llllllll}
 & x_2 & +x_3 & +x_4 & +x_5 & & & = 0 \\
x_1 & & +x_3 & +x_4 & & +x_6 & & = 0 \\
x_1 & +x_2 & & +x_4 & & & +x_7 & = 0
\end{array}
$$

We will leave it to you to show that this code is the Hamming code C_3 of Example 5.2.1. A parity check matrix for the Hamming code C_3 is thus

$$
P_3 = \begin{bmatrix} 0 & 1 & 1 & 1 & 1 & 0 & 0 \\ 1 & 0 & 1 & 1 & 0 & 1 & 0 \\ 1 & 1 & 0 & 1 & 0 & 0 & 1 \end{bmatrix}
$$

         □

The concept of a parity check matrix is very important, and so we make the following definition.

Definition   A **parity check matrix** for a linear p-ary $[n, k]$-code C is a matrix P with the property that

$$
C = \{\mathbf{x} \in \mathbb{Z}_p^n \mid \mathbf{x}P^t = \mathbf{0}\}
$$

         □

Note that, unlike a generator matrix, we make no requirement that the rows of P be linearly independent. Of course, parity check matrices in which the rows are linearly independent are smaller and therefore more efficient than other parity check matrices.

The following theorem gives some of the most basic properties of dual codes.

Theorem 5.4.5

1. If C and D are codes and $C \subseteq D$ then $C^{\perp} \supseteq D^{\perp}$. (Note the reversal of inclusion.)

2. Let C be a linear $[n, k]$-code over $\mathbb{Z}_p$, with generator matrix G. If G has rows $\{\mathbf{g}_1, \mathbf{g}_2, \ldots, \mathbf{g}_k\}$ then $C^{\perp} = \{\mathbf{g}_1, \mathbf{g}_2, \ldots, \mathbf{g}_k\}^{\perp}$. In words, $C^{\perp}$ is the set of all strings that are orthogonal to every row of G. In symbols,

$$
C^{\perp} = \{x \in \mathbb{Z}_p^n \mid xG^t = 0\}
$$

Thus, G is also a parity check matrix for $C^{\perp}$.

3. If C is a linear $[n, k]$-code then $C^{\perp}$ is a linear $[n, n-k]$-code. In other words,

$$
\dim(C^{\perp}) = n - \dim(C)
$$

4. *For any linear code C, we have $C^{\perp\perp} = C$.* $\qquad\qquad\qquad$ □

Proof We leave proof of part 1) as an exercise. For part 2), since $\{\mathbf{g}_1, \mathbf{g}_2, \ldots, \mathbf{g}_k\} \subseteq C$, part 1) implies that $C^{\perp} \subseteq \{\mathbf{g}_1, \mathbf{g}_2, \ldots, \mathbf{g}_k\}^{\perp}$. The reverse inclusion follows from the fact that any string that is orthogonal to every basis string $\mathbf{g}_i$ is also orthogonal to every linear combination of the basis strings, and hence to every codeword in C.

For part 3), let $\mathcal{B} = \{\mathbf{b}_1, \mathbf{b}_2, \ldots, \mathbf{b}_k\}$ be a basis for C. Consider the parity check equations

$$
\begin{array}{ccccccc}
b_{11}x_1 & + & b_{12}x_2 & + & \cdots & + & b_{1n}x_n & = 0 \\
b_{21}x_1 & + & b_{22}x_2 & + & \cdots & + & b_{2n}x_n & = 0 \\
& & & & & & & \vdots \\
b_{k1}x_1 & + & b_{k2}x_2 & + & \cdots & + & b_{kn}x_n & = 0
\end{array}
$$

whose solutions are precisely the codewords in $C^{\perp}$. Since $\{\mathbf{b}_1, \ldots, \mathbf{b}_k\}$ is linearly independent, we may appeal to a theorem of linear algebra (which we will not prove here) that says that there is a set of k coordinates $x_{i_1}, \ldots, x_{i_k}$ such that the above system is equivalent to a system in which each x_{i_j} appears with nonzero coefficient in *exactly* one equation and the coefficient of x_{i_j} in that equation is 1. For example, if the coordinates happen to be $x_1, \ldots, x_k$, the equivalent system has the form

$$
\begin{array}{ccccccc}
x_1 & & & + & d_{1,k+1}x_{k+1} & + & \cdots & + & d_{1,n}x_n & = 0 \\
& x_2 & & + & d_{2,k+1}x_{k+1} & + & \cdots & + & d_{2,n}x_n & = 0 \\
& & \ddots & & & & & & & \vdots \\
& & x_k & + & d_{k,k+1}x_{k+1} & + & \cdots & + & d_{k,n}x_n & = 0
\end{array}
$$

where the coefficients d_{ij} lie in $\mathbb{Z}_p$. In a system such as this one, we may assign any values to $x_{k+1}, \ldots, x_n$ and then simply solve for $x_1, \ldots, x_k$. Since there are p choices for each of the $n - k$ variables $x_{k+1}, \ldots, x_n$, there are p^{n-k} solutions and so $C^{\perp}$ has size p^{n-k} and hence dimension $n - k$.

For part 4), observe first that $C^{\perp\perp} = (C^{\perp})^{\perp}$ is the set of all strings orthogonal to every string in $C^{\perp}$. But, by definition of $C^{\perp}$, all codewords in C are orthogonal to every string in $C^{\perp}$, whence $C \subseteq C^{\perp\perp}$. According to part 3),

$$
\dim(C^{\perp\perp}) = n - \dim(C^{\perp}) = n - (n - \dim(C)) = \dim(C)
$$

It is a theorem of linear algebra (which we ask you to prove as an exercise) that if a code D is a subspace of a code E and if $\dim(D) = \dim(E)$, then $D = E$. It follows that $C = C^{\perp\perp}$.   ■

Example 5.4.4   Consider the code C_1 of Example 5.2.1, with generator matrix

$$G_1 = \begin{bmatrix} 1 & 0 & 1 & 1 \\ 0 & 1 & 1 & 0 \end{bmatrix}$$

According to Theorem 5.4.5, the dual code is given by the solutions to the parity check equations

$$\begin{aligned} x_1 \quad\quad + x_3 \quad + x_4 \quad &= 0 \\ x_2 \quad + x_3 \quad\quad &= 0 \end{aligned}$$

In this system, we may assign any values to x_3 and x_4 and solve for x_1 and x_2. The solutions are $C_1^{\perp} = \{0000, 1110, 1001, 0111\} = \langle 1110, 1001 \rangle$. We leave it to the reader to check that each codeword in this code is orthogonal to each codeword in $C_1 = \langle 1011, 0110 \rangle$.   □

It might be useful to clarify the relationship between generator matrices and parity check matrices, which we do in the next theorem.

Theorem 5.4.6

1. *Any matrix $\mathcal{A}$ is a parity check matrix for some linear code (perhaps the zero code).*

2. *Any matrix with linearly independent rows is both a generator matrix and a parity check matrix.*

3. *A generator matrix for a code C is a parity check matrix for the dual code $C^{\perp}$.*

4. *Any linear code C has a parity check matrix. In particular, a generator matrix for the dual code $C^{\perp}$ is a parity check matrix for C.*   □

Proof   Theorem 5.4.4 tells us that, for any $s \times n$ matrix $\mathcal{A}$ with entries in $\mathbb{Z}_p$, the set

$$\{ \mathbf{x} \in \mathbb{Z}_p^n \mid \mathbf{x}\mathcal{A}^t = \mathbf{0} \}$$

is a linear code, with parity check matrix $\mathcal{A}$. In fact, it is the dual code to the code consisting of the rows of the matrix $\mathcal{A}$. This proves part 1). Part 2) follows from part 1) and the fact that if the rows of a matrix are linearly independent, they form a basis for the code generated by these rows. Part

3) follows directly from part 2) of Theorem 5.4.5. Finally, suppose that C is a linear code and that the dual code $C^{\perp}$ has generator matrix G. Then G is a parity check matrix for the dual of $C^{\perp}$, which is $C^{\perp\perp} = C$. This proves part 4). ∎

We will discuss later in this chapter a way to obtain a parity check matrix for a linear code directly from a generator matrix, without having to explicity compute the dual code and find a basis for that code.

We now have two convenient ways to define a linear code C—by giving a generator matrix or by giving a parity check matrix. For instance, the code C_1 of Example 5.2.1 has generator matrix G_1 and parity check matrix P_1 given by

$$G_1 = \begin{bmatrix} 1 & 0 & 1 & 1 \\ 0 & 1 & 1 & 0 \end{bmatrix} \text{ and } P_1 = \begin{bmatrix} 1 & 1 & 1 & 0 \\ 1 & 0 & 0 & 1 \end{bmatrix}$$

Thus, C_1 is the set of all binary strings of length 4 that are linear combinations of the rows of G_1 and C_1 is the set of all binary strings of length 4 that are orthogonal to the rows of P_1. Either matrix completely defines the code.

Self-Dual Codes

If you studied linear algebra, you are probably familiar with inner product spaces over the real or complex number fields. You also undoubtedly know that a nonzero vector over the reals cannot be orthogonal (perpendicular) to itself. However, in dealing with the finite fields $\mathbb{Z}_p$, things are different. For example, the binary string 1010 is orthogonal to itself, since

$$1010 \cdot 1010 = 1 + 0 + 1 + 0 = 0$$

In fact, it is possible for a code C to have the property that every codeword in C is orthogonal to every other codeword! Here is an example.

Example 5.4.5   Consider the binary linear code

$$C = \langle 1100, 0011 \rangle = \{0000, 1100, 0011, 1111\}$$

Since the basis codewords 1100 and 0011 are orthogonal to themselves and to each other, we have $C \subseteq C^{\perp}$. But $\dim(C) = 2 = \dim(C^{\perp})$ and so $C = C^{\perp}$. □

Definition   A code C for which $C = C^\perp$ is said to be **self-dual**.   □

In the exercises, we ask you to prove that a self-dual code must have even length and that, for every even number n, there is a *binary* self-dual code of length n. It is also true (although we will not prove it here) that a *ternary* self-dual $[n, \frac{n}{2}]$-code exists if and only if n is divisible by 4.

Determining the Minimum Distance from a Parity Check Matrix

As we have said, a linear code can be described either by giving a generator matrix G or a parity check matrix P. Both methods have advantages. For instance, it is easier to generate all codewords in a code C from a generator matrix—just form all linear combinations of the rows. We know that each linear combination is a distinct codeword and that all codewords are obtained in this way. On the other hand, to use a parity check matrix to generate all codewords requires solving a system of linear equations over a finite field. However, it is easier to determine whether or not a given string is a codeword using a parity check matrix.

We also mentioned that there does not seem to be a simple, direct method for determining the minimum weight of a linear code from a generator matrix. However, the following theorem shows that it is easy to do so from a parity check matrix.

Theorem 5.4.7   *Let P be a parity check matrix for a linear $[n, k, d]$-code C. Then the minimum distance d is the smallest integer r for which there are r linearly dependent columns in P. (Here we are thinking of the columns of P as strings.)*   □

Proof   Let the columns of P be denoted by $\mathbf{p}_1, \mathbf{p}_2, \ldots, \mathbf{p}_n$. The crucial step in the proof is to note that, for any row matrix

$$\begin{bmatrix} c_1 & c_2 & \cdots & c_n \end{bmatrix}$$

we have

$$\begin{bmatrix} c_1 & c_2 & \cdots & c_n \end{bmatrix} P^t = c_1 \cdot (\text{1st row of } P^t) + c_2 \cdot (\text{2nd row of } P^t)$$
$$+ \cdots + c_n \cdot (n\text{th row of } P^t)$$
$$= c_1 \cdot \mathbf{p}_1 + c_2 \cdot \mathbf{p}_2 + \cdots + c_n \cdot \mathbf{p}_n$$

which is just a linear combination of the columns of P. Now suppose that r is the minimum number of linearly dependent columns. Then there exists constants $c_1, \ldots, c_n$, of which exactly r are nonzero, for which

$$\begin{bmatrix} c_1 & c_2 & \cdots & c_n \end{bmatrix} P^t = \mathbf{0} \tag{5.4.3}$$

But this says that $\mathbf{c} = c_1 \cdots c_n$ is in C. Since $w(\mathbf{c}) = r$, we get $d = w(C) \leq r$. On the other hand, if $\mathbf{c} = c_1 \cdots c_n$ is a codeword of minimum weight $d = w(C)$, then (5.4.3) holds, and we deduce that the d columns of P corresponding to the d nonzero elements c_i are linearly dependent. Hence, $r \leq d$. This shows that $r = d$. ∎

Example 5.4.6 Consider the [10, 8]-code C over $\mathbb{Z}_{11}$, with parity check matrix

$$P = \begin{bmatrix} 111111111\ 1 \\ 12345678910 \end{bmatrix}$$

To show that any two distinct columns of P are linearly independent, note that any two distinct columns have the form $\mathbf{c}_1 = \begin{bmatrix} 1 \\ a \end{bmatrix}$ and $\mathbf{c}_2 = \begin{bmatrix} 1 \\ b \end{bmatrix}$, where $1 \leq a, b \leq 10$ and $a \neq b$. Using string form, we see that the equation

$$\alpha(1a) + \beta(1b) = 00$$

is equivalent to

$$(\alpha + \beta)(\alpha a + \beta b) = 00$$

which is equivalent to the system of equations

$$\alpha + \beta = 0$$

$$\alpha a + \beta b = 0$$

The first equation gives $\beta = -\alpha$. Substituting this into the second equation gives $\alpha a - \alpha b = 0$, or $\alpha(a - b) = 0$. But since $a \neq b$, this implies that $\alpha = 0$, whence $\beta = -\alpha = 0$. Thus, the only solution to this system is $\alpha = 0$, $\beta = 0$, and so the strings are linearly independent.

On the other hand, the first three columns of P, thought of as strings, are linearly dependent, for we have

$$10(11) + 2(12) = 13$$

in $\mathbb{Z}_{11}$. Therefore, according to Theorem 5.4.7, C has minimum distance 3. □

The Gilbert-Varshamov Bound

Recall that $A_r(n, d)$ denotes the largest possible size M for which there exists an r-ary (n, M, d)- code (linear or nonlinear). Recall also the sphere-covering lower bound on $A_r(n, d)$, is given by

$$\frac{r^n}{V_r^n(d-1)} \leq A_r(n, d)$$

It happens that we can improve upon this bound, in some cases, by considering linear codes and using Theorem 5.4.7. First, let us take another quick look at the sphere-covering bound. In attempting to construct an (n, M, d)-code, we could simply pick any string c_1 for the first codeword. The second codeword c_2 can be any string provided that it does not lie in the sphere $S_r^n(c_1, d-1)$. The third codeword can be any string provided that it does not lie in either of the spheres $S_r^n(c_1, d-1)$ or $S_r^n(c_2, d-1)$. We can continue to pick codewords $c_1, \ldots, c_M$ in this manner until the spheres $S_r^n(c_1, d-1), \ldots, S_r^n(c_M, d-1)$ cover all of $\mathbb{Z}_r^n$. But this cannot happen until the sum of the volumes of these spheres is at least the size of $\mathbb{Z}_r^n$, that is, until

$$M \cdot V_r^n(d-1) \geq r^n$$

Thus, we can always form an (n, M, d)- code with $M \geq r^n/V_r^n(d-1)$. This proves the sphere-covering bound.

For linear codes, rather than picking codewords directly, we pick columns for a parity check matrix P. To construct a linear $[n, k, d]$-code, we pick n columns of length $n - k$, with the requirement that no $d - 1$ of these columns are linearly dependent. The latter amounts to saying that no column is a linear combination of $d - 2$ (or fewer) of the previous columns.

The first column of P can be any nonzero string in $\mathbb{Z}_p^{u-k}$. In general, we must choose the ith column (for $i \geq 2$) of P so that it is not a linear combination of any $d - 2$ (or fewer) of the previously chosen columns. For a particular choice of $j \leq d - 2$ of these $i - 1$ columns, there are $(p - 1)^j$ choices for nonzero coefficients to form linear combinations of these columns. Since there are $\binom{i-1}{j}$ choices of j columns, the number of linear combinations (with nonzero coefficients) of exactly j columns is

$$\binom{i-1}{j}(p-1)^j$$

Summing this over $j = 1, \ldots, d - 2$, we see that the number of linear combinations of $d - 2$ or fewer columns is

$$N_i = \sum_{j=1}^{d-2} \binom{i-1}{j} (p-1)^j$$

Since the total number of strings of length $n - k$ is p^{n-k}, we deduce that, excluding the zero column as well, there are enough strings left to choose the ith column provided that $N_i + 1 < p^{n-k}$. It follows that we may form the entire parity check matrix of n columns if $N_n + 1 < p^{n-k}$. But $N_n + 1 = V_p^{n-1}(d-2)$ and so we have established the following result.

Theorem 5.4.8 (Gilbert-Varshamov bound)   *There exists a p-ary linear* $[n, k, d]$*-code if*

$$p^k < \frac{p^n}{V_p^{n-1}(d-2)}$$

Thus, if p^k is the largest power of p satisfying this inequality, we have $A_p(n, k) \geq p^k$. The inequality displayed above is known as the **Gilbert-Varshamov inequality**.   □

A simple example will show that the Gilbert-Varshamov bound is better than the sphere-covering bound. However, note that since it was derived using linear codes, it applies only to radixes of prime power, whereas the sphere-packing bound applies to all radixes.

Example 5.4.7   The sphere-covering lower bound says that

$$A_2(5, 3) \geq \frac{2^5}{1 + \binom{5}{1} + \binom{5}{2}} = 2$$

On the other hand, the Gilbert-Varshamov bound says that there exists a binary linear $(5, 2^k, 3)$-code if

$$2^k < \frac{2^5}{1 + \binom{4}{1}} = \frac{32}{5}$$

and so we may take $k = 2$, showing that there is a binary linear $(5,4,3)$-code, whence $A_2(5, 3) \geq 4$. Indeed, Table 4.8.1(a) shows that $A_2(5, 3) = 4$ and so, in this case, the Gilbert-Varshamov bound is exact.   □

Exercises

1. Compute the following inner products
   (a) $(10110) \cdot (11000)$ in $\mathbb{Z}_2$
   (b) $(12120) \cdot (20212)$ in $\mathbb{Z}_3$
   (c) $(12345) \cdot (67890)$ in $\mathbb{Z}_{11}$
   (d) $\mathbf{x} \cdot \mathbf{x}^c$ in $\mathbb{Z}_2$

2. Verify that $\mathbf{x} \cdot \mathbf{y} = \mathbf{y} \cdot \mathbf{x}$ and $(\mathbf{x} + \mathbf{y}) \cdot \mathbf{z} = \mathbf{x} \cdot \mathbf{z} + \mathbf{y} \cdot \mathbf{z}$, for all $\mathbf{x}$, $\mathbf{y}$, $\mathbf{z} \in \mathbb{Z}_p^n$.

3. Prove that the orthogonal complement $A^\perp$ of a nonempty subset A of $\mathbb{Z}_p^n$ is a linear code.

4. Solve the second set of parity check equations and confirm that the solutions give the code C_2 of Example 5.2.1.

5. Show that the code $A^\perp$ of Example 5.4.3 is the Hamming code C_3 of Example 5.2.1.

6. Prove that if $C \subseteq D$, then $D^\perp \subseteq C^\perp$.

7. Show that $\{\mathbf{a}_1, \mathbf{a}_2, \ldots, \mathbf{a}_s\}^\perp = \langle \mathbf{a}_1, \mathbf{a}_2, \ldots, \mathbf{a}_s \rangle^\perp$.

8. Find the dual code $C_2^\perp$ of the code C_2 in Example 5.2.1 by solving the parity check equations.

9. Find the dual code of the repetition code $\text{Rep}_p(n)$.

10. Find the dual code of the Hamming code C_3 of Example 5.2.1 by solving the parity check equations.

11. Show that the code C_2 has minimum distance 2 by using Theorem 5.4.7 and the parity check matrix P_2 given in Example 5.4.3.

12. Find the minimum distance of the binary code with parity check matrix
$$P = \begin{bmatrix} 1 & 0 & 0 & 1 \\ 0 & 0 & 1 & 1 \\ 0 & 1 & 0 & 1 \end{bmatrix}$$

13. Find the minimum distance of the binary code with parity check matrix
$$P = \begin{bmatrix} 1 & 0 & 0 & 1 & 1 \\ 0 & 0 & 1 & 0 & 1 \\ 0 & 1 & 1 & 1 & 1 \end{bmatrix}$$

14. Show that the Hamming code C_3 has minimum distance 3 by using Theorem 5.4.7 and the parity check matrix P_3 given in Example 5.4.3.

15. Find a parity check matrix for the repetition code $\text{Rep}_r(n)$.

16. Prove that if D is a subspace of a code C and $\dim(D) = \dim(C)$ then $D = C$.

17. Show that if C is a binary self-dual code then all codewords in C have even weight and $\mathbf{1} \in C$.

18. A linear code C is **self-orthogonal** if $C \subseteq C^\perp$. Prove that a linear code C, with generator matrix G, is self-orthogonal if and only if the rows of G are orthogonal to themselves and to each other.

19. Prove that the length of a self-dual code must be even. What is the dimension of a a self-dual code? Prove that, for any even number n, there is a binary self-dual code C_n of length n. Hint: construct a generator matrix for C_n.

20. A linear code C is **self-orthogonal** if $C \subseteq C^\perp$. Let G be a generator matrix for a p-ary linear code C, where $p = 2$ or 3. Show that C is self-orthogonal if and only if distinct rows of G are orthogonal and each row of G has weight divisible by p.

21. Determine the Gilbert-Varshamov lower bound on $A_2(6, 3)$ and compare it with the actual value.

22. Determine the Gilbert-Varshamov lower bound on $A_2(7, 3)$ and compare it with the actual value.

23. Determine the Gilbert-Varshamov lower bound on $A_2(8, 5)$ and compare it with the actual value.

5.5 Syndrome Decoding

One of the advantages of parity check matrices is that they can be used for efficient implementation of nearest neighbor decoding.

Definition Let P be the parity check matrix for a code $C \subseteq \mathbb{Z}_p^n$. The **syndrome** $S(\mathbf{x})$ of a string $\mathbf{x}$ in $\mathbb{Z}_p^n$ is the product $\mathbf{x}P^t$. □

Example 5.5.1 The code C_1 of Example 5.2.1 has parity check matrix

$$P_1 = \begin{bmatrix} 1 & 1 & 1 & 0 \\ 1 & 0 & 0 & 1 \end{bmatrix}$$

The syndrome of the string $\mathbf{x} = 0111$ is

$$S(\mathbf{x}) = \mathbf{x}P^t = \begin{bmatrix} 0 & 1 & 1 & 1 \end{bmatrix} \begin{bmatrix} 1 & 1 \\ 1 & 0 \\ 1 & 0 \\ 0 & 1 \end{bmatrix} = \begin{bmatrix} 0 & 1 \end{bmatrix} \qquad \square$$

According to the definition, a syndrome $S(\mathbf{x})$ is a row matrix. However, it will be convenient to also consider the syndrome as a string. Thus, with regard to the previous example, we may also write $S(\mathbf{x}) = 01$.

Note also that the parity check equation $\mathbf{x}P^t = 0$ is equivalent to $S(\mathbf{x}) = \mathbf{0}$ and so $\mathbf{x} \in C$ if and only if its syndrome is $\mathbf{0}$. Let us record this simple result in a theorem, along with a few simple properties of the syndrome. Proof of these properties is left as an exercise.

Theorem 5.5.1 *Let P be the parity check matrix for a linear code $C \subseteq \mathbb{Z}_p^n$ and let S be the syndrome function for P. Then, for all $\mathbf{x}, \mathbf{y} \in \mathbb{Z}_p^n$ and $\alpha \in \mathbb{Z}_p$,*

1. *if P has size $m \times n$, then $S(\mathbf{x}) \in \mathbb{Z}_p^m$*

2. $S(\mathbf{x} + \mathbf{y}) = S(\mathbf{x}) + S(\mathbf{y})$

3. $S(\alpha\mathbf{x}) = \alpha S(\mathbf{x})$

4. $\mathbf{x} \in C$ *if and only if $S(\mathbf{x}) = \mathbf{0}$.* $\qquad \square$

The main importance of the syndrome comes from the following theorem.

Theorem 5.5.2 *Let C be a linear code. Two strings $\mathbf{x}$ and $\mathbf{y}$ are in the same coset of any standard array for C if and only if they have the same syndrome.* $\qquad \square$

Proof A standard array for C has the form

$\mathbf{0}$	$\mathbf{c}_2$	$\mathbf{c}_3$	$\cdots$	$\mathbf{c}_M$
$\mathbf{f}_2$	$\mathbf{f}_2 + \mathbf{c}_2$	$\mathbf{f}_2 + \mathbf{c}_3$	$\cdots$	$\mathbf{f}_2 + \mathbf{c}_M$
$\mathbf{f}_3$	$\mathbf{f}_3 + \mathbf{c}_2$	$\mathbf{f}_3 + \mathbf{c}_3$	$\cdots$	$\mathbf{f}_3 + \mathbf{c}_M$
$\vdots$	$\vdots$	$\vdots$	$\vdots$	$\vdots$
$\mathbf{f}_q$	$\mathbf{f}_q + \mathbf{c}_2$	$\mathbf{f}_q + \mathbf{c}_3$	$\cdots$	$\mathbf{f}_q + \mathbf{c}_M$

Let $\mathbf{x} = \mathbf{f}_i + \mathbf{c}_k$ and $\mathbf{y} = \mathbf{f}_j + \mathbf{c}_\ell$. Then

$$S(\mathbf{x}) = S(\mathbf{f}_i + \mathbf{c}_k) = S(\mathbf{f}_i) + S(\mathbf{c}_k) = S(\mathbf{f}_i)$$

and similarly,

$$S(\mathbf{y}) = S(\mathbf{f}_j)$$

Thus, $S(\mathbf{x}) = S(\mathbf{y})$ if and only if $S(\mathbf{f}_i) = S(\mathbf{f}_j)$, that is, $S(\mathbf{f}_i - \mathbf{f}_j) = 0$. But the latter is equivalent to $\mathbf{f}_i - \mathbf{f}_j \in C$, which happens if and only if $i = j$, that is, if and only if $\mathbf{x}$ and $\mathbf{y}$ lie in the same coset of the array. ∎

How do we use Theorem 5.5.2 in nearest neighbor decoding? Recall that, under nearest neighbor decoding, the error string $\mathbf{e}$ in a received word $\mathbf{x}$ is the coset leader of the coset containing $\mathbf{x}$ and that the nearest neighbor codeword is $\mathbf{c} = \mathbf{x} - \mathbf{e}$. But, the syndrome of $\mathbf{x}$ is equal to the syndrome of $\mathbf{e}$ and since the syndromes of the coset leaders are all distinct, we can find $\mathbf{e}$ simply by comparing the syndrome of $\mathbf{x}$ to the syndromes of the coset leaders.

Thus, we need only a list of coset leaders and their syndromes, which we refer to as a **syndrome table** for C

coset leader	syndrome
0	**0**
$\mathbf{f}_2$	$S(\mathbf{f}_2)$
$\mathbf{f}_3$	$S(\mathbf{f}_3)$
$\vdots$	$\vdots$
$\mathbf{f}_q$	$S(\mathbf{f}_q)$

Now, nearest neighbor decoding can be implemented by the following simple algorithm.

1. Compute the syndrome $S(\mathbf{x})$ of the received string.

2. Compare it with the list of syndromes of the coset leaders. If $S(\mathbf{x}) = S(\mathbf{f}_i)$, then $\mathbf{f}_i$ is the error string and $\mathbf{c} = \mathbf{x} - \mathbf{f}_i$ is a nearest neighbor codeword.

What could be simpler?

Example 5.5.2 Consider again the parity check matrix

$$P_1 = \begin{bmatrix} 1 & 1 & 1 & 0 \\ 1 & 0 & 0 & 1 \end{bmatrix}$$

for the code C_1 of Example 5.2.1. A standard array for the code C_1 is

0000	1011	0110	1101
1000	0011	1110	0101
0100	1111	0010	1001
0001	1010	0111	1100

and the syndrome table is

coset leader	syndrome
0000	00
1000	11
0100	10
0001	01

If, for example, the string $\mathbf{x} = 1110$ is received, its syndrome is $S(1110) = [1110]P^t = 11$. Checking the syndrome table, we find that the corresponding coset leader is $\mathbf{f} = 1000$. Hence, a nearest neighbor codeword to $\mathbf{x}$ is $\mathbf{x} - \mathbf{f} = 0110$.   □

One final note. We have seen that a standard array for a p-ary $[n, k]$-code C has p^{n-k} rows. If P is a parity check matrix for C and P has linearly independent rows, then it has size $(n-k) \times n$ and therefore each syndrome $\mathbf{x}P^t$ is an element of $\mathbb{Z}_p^{n-k}$. Since $|\mathbb{Z}_p^{n-k}| = p^{n-k}$, we conclude that the set of syndromes is precisely the entire space $\mathbb{Z}_p^{n-k}$. (This is illustrated in Example 5.5.2.)

Exercises

1. How many rows are there in a syndrome table for a p-ary $[n, k, d]$-code? Can you read this number directly from the size of a generator matrix? Of a parity check matrix?

2. Let P be the parity check matrix for a linear code $C \subseteq \mathbb{Z}_p^n$. Show that (a) $S(\mathbf{x} + \mathbf{y}) = S(\mathbf{x}) + S(\mathbf{y})$ for any strings $\mathbf{x}$ and $\mathbf{y}$ in $\mathbb{Z}_p^n$ and (b) $S(\alpha \mathbf{x}) = \alpha S(\mathbf{x})$, for any $\alpha \in \mathbb{Z}_p$.

3. Use the parity check matrix P_3 in Example 5.4.3 to construct a syndrome table for the Hamming code C_3 of Example 5.2.1. Use the table to decode the strings a) 1101101, b) 1111111, c) 0000001.

4. Construct a syndrome table for the repetition code $\text{Rep}_3(3)$. Use it to decode the strings a) 111, b) 121, c) 120.

5. Use the parity check matrix P_2 in Example 5.4.3 to construct a syndrome table for the ternary code C_2 of Example 5.2.1. Use this table to decode the strings a) 0110, b) 2222, c) 2012.

6. Write a computer program to construct a syndrome table, given a parity check matrix for a code.

7. In Section 4.5, we promised that we would ask you to show that there is no binary linear $(90, 2^{78}, 5)$-code. Please do that now. Hint: Since such a code C, if it exists, must be perfect, the coset leaders are precisely the strings in $\mathbb{Z}_2^{90}$ that have weight at most 2. Suppose that r of the 90 coset leaders of weight 1 have syndromes of odd weight. How many have syndromes of even weight? How many of the coset leaders of weight 2 have syndromes of odd weight? How many coset leaders have syndromes of odd weight? Show that this number is also equal to 2^{11}. Is there a solution for r?

5.6 Equivalent Linear Codes

A linear code C can be described by giving a generator matrix or by giving a parity check matrix. The question naturally arises as to how we might obtain a parity check matrix from a generator matrix, and vice-versa.

In order to give a practical answer to this question, we introduce the concept of equivalent *linear* codes (we have already seen one definition of equivalence that applies to all codes). Suppose that C is a linear code, with generator matrix

$$
G = \begin{bmatrix}
b_{11} & b_{12} & \cdots & b_{1n} \\
b_{21} & b_{22} & \cdots & b_{2n} \\
& & \vdots & \\
b_{k1} & b_{k2} & \cdots & b_{kn}
\end{bmatrix}
$$

Thus, the rows $\mathbf{b}_1, \mathbf{b}_2, \ldots, \mathbf{b}_k$ form a basis for C.

If we interchange any two rows of G, the resulting matrix will still generate C, since interchanging two rows simply amounts to changing the order of the basis strings. Similarly, multiplying a row of G by a nonzero scalar results in another generator matrix for C, as does adding a scalar multiple of one row to another row. In particular, we have the following theorem, whose proof is left as an exercise.

Theorem 5.6.1 *Let G be a generator matrix for a linear code C. If any of the following operations are performed on the rows of G, the resulting matrix is also a generator matrix for C.*

1. *Interchange any two rows of G.*

2. *Multiply any row of G by a nonzero scalar.*

3. *Add a multiple of one row of G to another row of G.*

These three operations are known as **elementary row operations**.   □

In the parlance of linear algebra, this theorem says that any matrix that is *row equivalent* to G is also a generator matrix for the code C.

Note that adding one row of G to another is a special case of an elementary row operation, as is subtracting one row of G from another.

Example 5.6.1 Since the rows of the matrix

$$G = \begin{bmatrix} 1 & 2 & 1 & 2 & 1 \\ 1 & 2 & 1 & 0 & 0 \\ 2 & 0 & 0 & 1 & 2 \end{bmatrix}$$

are linearly independent over $\mathbb{Z}_3$, this matrix is a generator matrix for a ternary $[5, 3]$-code C. Adding 2 times the first row to the second row of G gives

$$G_1 = \begin{bmatrix} 1 & 2 & 1 & 2 & 1 \\ 0 & 0 & 0 & 1 & 2 \\ 2 & 0 & 0 & 1 & 2 \end{bmatrix}$$

which, according to Theorem 5.6.1, is also a generator matrix for C.   □

According to Theorem 5.6.1, elementary row operations do not change the code. However, in general, if we perform similar operations on the *columns* of G, the resulting matrix will no longer generate the code C.

On the other hand, if G generates C, and if we interchange two columns of G, or multiply a column of G by a nonzero constant, the code D generated by the resulting matrix will have the same parameters (length, size, and minimum distance) as C. That is, if C is an $[n, k, d]$-code, then so is D. In a sense then, D is "just as good" as C. This leads us to our next definition.

Definition Let C be a linear code, with generator matrix G. If any of the following operations are performed on the rows or columns of G, the code generated by the resulting matrix is said to be **scalar multiple equivalent** to C.

1. Interchange any two rows of G.

2. Multiply any row of G by a nonzero scalar.

3. Add a multiple of one row of G to another row of G.

4. Interchange any two columns of G.

5. Multiply any column of G by a nonzero scalar.

We refer to these operations as the **five elementary matrix operations**. □

We leave it as an exercise to show that if two linear codes are scalar multiple equivalent, then they are also equivalent in the sense of Chapter 4. (But the converse of this statement is not true.)

In order to discuss the importance of scalar multiple equivalence, we need some preliminary remarks.

Definition A $k \times n$ matrix A is in **left standard form** if it has the form

$$A = [I_k \mid M_{k,n-k}]$$

where I_k is the identity matrix of size $k \times k$, and $M_{k,n-k}$ is a matrix of size $k \times (n-k)$. □

For example, the matrices M and N below are in left standard form, but the matrix P is not.

$$M = \begin{bmatrix} 1 & 0 & 1 & 2 & 3 \\ 0 & 1 & 0 & 1 & 2 \end{bmatrix}, \quad N = \begin{bmatrix} 1 & 0 & 0 & 1 & 1 \\ 0 & 1 & 0 & 0 & 0 \\ 0 & 0 & 1 & 1 & 0 \end{bmatrix}, \quad P = \begin{bmatrix} 1 & 0 & 1 & 1 \\ 0 & 1 & 0 & 0 \\ 0 & 0 & 1 & 1 \end{bmatrix}$$

There are advantages to describing a linear code by means of a generator matrix that is in left standard form. Unfortunately, not all linear codes have such a generator matrix. However, as the next result shows, all linear codes are *scalar multiple equivalent* to codes that do.

Theorem 5.6.2 *Let G be a generator matrix for a linear code C. Then it is possible, by applying the five elementary matrix operations, to bring G to left standard form. Therefore, C is scalar multiple equivalent to a code that has a generator matrix in left standard form.* □

Proof   Once we have proven the first statement of the theorem, the second statement follows from the definition of scalar multiple equivalence. To prove the first statement, we describe a step-by-step method for bringing a generator matrix G to left standard form.   ■

Procedure for Bringing a Generator Matrix to Left Standard Form

Let G be the generator matrix

$$
G = \begin{bmatrix} b_{11} & b_{12} & \cdots & b_{1n} \\ b_{21} & b_{22} & \cdots & b_{2n} \\ & & \vdots & \\ b_{k1} & b_{k2} & \cdots & b_{kn} \end{bmatrix}
$$

We proceed one column at a time.

Column 1:   To prepare the first column, first make certain that the $(1,1)$-th entry in the matrix is a 1, which can be done as follows. Since G is a generator matrix, the rows of G are linearly independent and so no row consists entirely of 0s. Hence, by interchanging columns if necessary, we can arrange it so that the $(1,1)$-th entry is nonzero. If this entry is α then multiply the first row by the inverse α^{-1} in $\mathbb{Z}_p$, which turns the $(1,1)$-th entry into a 1. Next, zero-out the other entries in the first column by subtracting an appropriate multiple of row 1 from each of the other rows. The resulting matrix has the form

$$
G_1 = \begin{bmatrix} 1 & c_{12} & \cdots & c_{1n} \\ 0 & c_{22} & \cdots & c_{2n} \\ & & \vdots & \\ 0 & c_{k2} & \cdots & c_{kn} \end{bmatrix}
$$

Column 2:   Since the rows of G are linearly independent, so are the rows of G_1. It follows as above that we can, by interchanging columns if necessary, insure that c_{22} is nonzero. Multiplying the second row by c_{22}^{-1} will turn this entry into a 1. The second row can now be used to zero-out all other entries in column 2. Note that since the first column has only 0s below the first row, zeroing-out the entry c_{12} will not affect the $(1,1)$-th entry, which will remain a 1. The resulting matrix has the form

$$
G_2 = \begin{bmatrix}
1 & 0 & 0 & \cdots & 0 \\
0 & 1 & d_{23} & \cdots & d_{2n} \\
0 & 0 & d_{33} & \cdots & d_{3n} \\
& & \vdots & & \\
0 & 0 & d_{k3} & \cdots & d_{kn}
\end{bmatrix}
$$

This process may be continued until a matrix in left standard form is obtained. Let us consider an example.

Example 5.6.2 Let us bring the ternary generator matrix

$$
G = \begin{bmatrix}
1 & 2 & 2 & 0 & 0 \\
1 & 0 & 1 & 1 & 1 \\
0 & 1 & 1 & 2 & 1
\end{bmatrix}
$$

into left standard form.

Column 1: Interchange columns 1 and 2.

$$
\begin{bmatrix}
2 & 1 & 2 & 0 & 0 \\
0 & 1 & 1 & 1 & 1 \\
1 & 0 & 1 & 2 & 1
\end{bmatrix}
$$

Multiply row 1 by $2^{-1} = 2$ (in $\mathbb{Z}_3$).

$$
\begin{bmatrix}
1 & 2 & 1 & 0 & 0 \\
0 & 1 & 1 & 1 & 1 \\
1 & 0 & 1 & 2 & 1
\end{bmatrix}
$$

Subtract row 1 from row 3.

$$
\begin{bmatrix}
1 & 2 & 1 & 0 & 0 \\
0 & 1 & 1 & 1 & 1 \\
0 & 1 & 0 & 2 & 1
\end{bmatrix}
$$

Column 2: Subtract 2 times row 2 from row 1. Then subtract row 2 from row 3.

$$
\begin{bmatrix}
1 & 0 & 2 & 2 & 2 \\
0 & 1 & 1 & 1 & 1 \\
0 & 0 & 2 & 1 & 0
\end{bmatrix}
$$

Column 3: Multiply row 3 by $2^{-1} = 2$.

$$\begin{bmatrix} 1 & 0 & 2 & 2 & 2 \\ 0 & 1 & 1 & 1 & 1 \\ 0 & 0 & 1 & 2 & 0 \end{bmatrix}$$

Subtract 2 times row 3 from row 1. Then subtract row 3 from row 2.

$$\begin{bmatrix} 1 & 0 & 0 & 1 & 2 \\ 0 & 1 & 0 & 2 & 1 \\ 0 & 0 & 1 & 2 & 0 \end{bmatrix}$$

This matrix is in left standard form, and we are done. $\square$

One of the advantages of describing a linear code by means of a generator matrix in left standard form is that such a description makes it easy to encode and decode source messages. We will discuss this matter in detail in the next section. Another advantage is that it is easy to construct a parity check matrix from a generator matrix that is in left standard form.

Theorem 5.6.3 *The matrix* $G = [I_k \mid B]$ *is a generator matrix for an* $[n, k]$-*code C if and only if the matrix*

$$P = [-B^t \mid I_{n-k}]$$

is a parity check matrix for C. $\square$

Proof   Let

$$G = \begin{bmatrix} 1 & 0 & \cdots & 0 & b_{11} & \cdots & b_{1,n-k} \\ 0 & 1 & \cdots & 0 & b_{21} & \cdots & b_{2,n-k} \\ & & \vdots & & & \vdots & \\ 0 & 0 & 0 & 1 & b_{k,1} & \cdots & b_{k,n-k} \end{bmatrix}$$

Then P has the form

$$P = \begin{bmatrix} -b_{11} & \cdots & -b_{k1} & 1 & 0 & \cdots & 0 \\ -b_{12} & \cdots & -b_{k2} & 0 & 1 & \cdots & 0 \\ & \vdots & & & & \vdots & \\ -b_{1,n-k} & \cdots & -b_{k,n-k} & 0 & 0 & \cdots & 1 \end{bmatrix}$$

We want to compute the matrix product $\mathbf{g}_i P^t$, where $\mathbf{g}_i$ is the ith row of G. The jth entry of this product is the product of $\mathbf{g}_i$ and the jth column of

P^t, which is the jth row of P (in column format). Hence, this entry is

$$[0\cdots 1\cdots 0b_{i1}\cdots b_{i,n-k}]\begin{bmatrix} -b_{1j} \\ \vdots \\ -b_{kj} \\ 0 \\ \vdots \\ 1 \\ \vdots \\ 0 \end{bmatrix} = -b_{ij} + b_{ij} = 0$$

This shows that the matrix product $\mathbf{g}_i P^t$ is the zero matrix.

Now suppose that G is a generator matrix for C. If D is the code whose parity check matrix is P, then the syndrome of $\mathbf{g}_i$ is $\mathbf{g}_i P^t = \mathbf{0}$ and so $\mathbf{g}_i \in D$. Since this holds for all rows of G, we must have $C \subseteq D$. But since the $n-k$ rows of P are linearly independent, the dual code $D^{\perp}$, which has P as a generator matrix, has dimension $n-k$. It follows that

$$\dim(D) = n - \dim(D^{\perp}) = n - (n-k) = k = \dim(C)$$

Hence $C = D$, which shows that P is a parity check matrix for C, as desired.

Finally, suppose that P is a parity check matrix for C. Then since the syndrome of each row of G is $\mathbf{0}$, each row of G is in C and since the $k = \dim(C)$ rows of G are linearly independent, G is a generator matrix for C. ∎

Example 5.6.3 Harking back to Example 5.2.4, the generating matrix for the code C_1 of Example 5.2.1 was seen to be

$$G_1 = \begin{bmatrix} 1 & 0 & 1 & 1 \\ 0 & 1 & 1 & 0 \end{bmatrix}$$

This matrix is in left standard form and so the corresponding parity check matrix is

$$P_1 = \begin{bmatrix} 1 & 1 & 1 & 0 \\ 1 & 0 & 0 & 1 \end{bmatrix}$$

as noted in Example 5.4.3. Similarly, the Hamming code C_3 of Example 5.2.1 has generating matrix

$$G_3 = \begin{bmatrix} 1 & 0 & 0 & 0 & 0 & 1 & 1 \\ 0 & 1 & 0 & 0 & 1 & 0 & 1 \\ 0 & 0 & 1 & 0 & 1 & 1 & 0 \\ 0 & 0 & 0 & 1 & 1 & 1 & 1 \end{bmatrix}$$

which is also in left standard form. Hence, a parity check matrix for C_3 is

$$P_3 = \begin{bmatrix} 0 & 1 & 1 & 1 & 1 & 0 & 0 \\ 1 & 0 & 1 & 1 & 0 & 1 & 0 \\ 1 & 1 & 0 & 1 & 0 & 0 & 1 \end{bmatrix} \qquad \square$$

Example 5.6.4   Recall from Example 5.6.2 that the generator matrix

$$G = \begin{bmatrix} 0 & 1 & 1 & 1 & 1 \\ 2 & 1 & 2 & 0 & 0 \\ 1 & 0 & 1 & 2 & 1 \end{bmatrix}$$

for a ternary [5, 3]-code can be brought into the left standard form

$$H = \begin{bmatrix} 1 & 0 & 0 & 1 & 2 \\ 0 & 1 & 0 & 2 & 1 \\ 0 & 0 & 1 & 2 & 0 \end{bmatrix}$$

Since this process required the use of column operations, the code D generated by H is not the same as the code C generated by G. However, the codes C and D are scalar multiple equivalent. By Theorem 5.6.3, D has parity check matrix

$$P = \begin{bmatrix} -1 & -2 & -2 & 1 & 0 \\ -2 & -1 & 0 & 0 & 1 \end{bmatrix} = \begin{bmatrix} 2 & 1 & 1 & 1 & 0 \\ 1 & 2 & 0 & 0 & 1 \end{bmatrix}$$

Hence, C is scalar multiple equivalent to a code with parity check matrix P. $\square$

A matrix of the form $A = [M \mid I_k]$ is said to be in **right standard form**. Note that the parity check matrix obtained from Theorem 5.6.3 is in right standard form. Thus, it is easy to go back and forth between generator matrices in left standard form and parity check matrices in right standard form.

Moreover, if P is any parity check matrix for C, applying any of the five elementary matrix operations results in a parity check matrix for a scalar multiple equivalent code. Furthermore, we can bring any parity check matrix *with linearly independent rows* into right standard form with such operations. Therefore, given a parity check matrix for a linear code, we can obtain a scalar multiple equivalent code with parity check matrix in right standard form, and then use Theorem 5.6.3 to obtain a generator matrix in left standard form.

Exercises

1. Suppose that C is an (n, p, n)-code over $\mathbb{Z}_p$. Show that C is scalar multiple equivalent to the repetition code $\text{Rep}_p(n)$.

2. Prove Theorem 5.6.1.

3. Find a generating matrix in left standard form for the ternary code C_2 of Example 5.2.1 and compute the corresponding parity check matrix for this code.

4. Find a parity check matrix for the binary code C with generator matrix

$$\begin{bmatrix} 1 & 1 & 1 \\ 1 & 0 & 1 \end{bmatrix}$$

5. Find a parity check matrix for the binary code C with generator matrix

$$\begin{bmatrix} 1 & 1 & 1 & 0 \\ 1 & 0 & 1 & 1 \\ 0 & 1 & 1 & 1 \end{bmatrix}$$

6. Find a parity check matrix for the code C over $\mathbb{Z}_7$ with generator matrix

$$\begin{bmatrix} 1 & 2 & 3 & 4 & 5 \\ 1 & 0 & 2 & 0 & 3 \\ 2 & 3 & 4 & 0 & 0 \end{bmatrix}$$

7. Let E_n be the binary code of length n that consists of all codewords of even weight. Find a generator matrix for E_n in left standard form and a parity check matrix for E_n.

8. We know that every generator matrix G is also a parity check matrix. Is every parity check matrix a generator matrix?

9. Prove that performing any of the five row or column operations mentioned in this section on a parity check matrix produces a parity check matrix for a scalar multiple equivalent code.

10. Prove that if P is a parity check matrix but the rows of P are not linearly independent, we may remove some rows from P and still have a parity check matrix for the same code.

11. Let C be a binary linear code with generator matrix G. Let $\overline{C}$ be the code obtained from C by adding an even parity check. Describe a generator matrix for $\overline{C}$.

12. Let C be a binary linear code, with parity check matrix P. Let $\overline{C}$ be the code obtained from C by adding an even parity check. Describe a parity check matrix for $\overline{C}$.

13. Apply elementary row operations to the following parity check matrix for a binary code C

$$P = \begin{bmatrix} 0 & 0 & 0 & 1 & 1 & 1 & 1 \\ 0 & 1 & 1 & 0 & 0 & 1 & 1 \\ 1 & 0 & 1 & 0 & 1 & 0 & 1 \end{bmatrix}$$

to obtain another parity check matrix in right standard form. Use this matrix to find a generator matrix for the code C.

14. Show that if two linear codes are scalar multiple equivalent, then they are also equivalent in the sense of Chapter 4.

5.7   Source Encoding with a Linear Code

Up to now, we have spent much of our time discussing efficient methods for implementing nearest neighbor decoding, which is designed for error correction. We have not yet addressed the issue of how to encode source messages and, once errors are corrected, how to "decode" the corresponding code messages back into source data.

Before continuing, we must face a slight conflict in terminology, regarding the word *decoding*. We have been using the term *decode* to stand for the application of a decision rule to replace a received word with a codeword. However, the term decode is also used to refer to the process of replacing a codeword with the original source data that it represents.

(This usage is more in line with the term *encode*, which means to represent a source message by a codeword.) To avoid confusion, let us refer to this as *source decoding* .

Source information may first be encoded for efficient transmission using, for example, a Huffman encoding scheme. Since Huffman encoding schemes produce variable length codewords, the output of an efficiency encoder would, in this case, be a sequence of binary strings of varying lengths. This output can then be grouped into strings of a constant length k, for the purpose of further encoding to correct errors in transmission.

For instance, the source message CODES can be encoded using the Huffman codes in Table 2.1.1 to get

$$00000/1001/10100/010/0011$$

(the slashes are for readability) and this can be broken into 7 source strings of length 3

$$000/001/001/101/000/100/011$$

Thus, in general, for the purpose of error correction, we may think of the source information as consisting of strings in $\mathbb{Z}_p^k$, which we will refer to as source strings. (Even if the original source information is not first encoded for efficiency, we can still identify the source messages with strings in an appropriate $\mathbb{Z}_p^k$.)

Since an $[n, k]$-code C over $\mathbb{Z}_p$ has size $|C| = p^k$, it is capable of encoding all source strings in $\mathbb{Z}_p^k$. In fact, if G is a generator matrix for the $[n, k]$-code C, then we can encode the source string $\mathbf{s} = s_1 s_2 \cdots s_k \in \mathbb{Z}_p^k$ as the codeword $\mathbf{s}G$.

Example 5.7.1 Consider the binary $[5, 3]$-code C with generator matrix

$$G = \begin{bmatrix} 1 & 1 & 1 & 1 & 1 \\ 1 & 0 & 1 & 0 & 1 \\ 0 & 0 & 1 & 1 & 0 \end{bmatrix}$$

Since $\dim(C) = 3$, it is capable of encoding all source strings in $\mathbb{Z}_2^3$. For example, the codeword for the source string $s = 111$ is

$$c = sG = \begin{bmatrix} 1 & 1 & 1 \end{bmatrix} \begin{bmatrix} 1 & 1 & 1 & 1 & 1 \\ 1 & 0 & 1 & 0 & 1 \\ 0 & 0 & 1 & 1 & 0 \end{bmatrix} = \begin{bmatrix} 0 & 1 & 1 & 0 & 0 \end{bmatrix}$$

or, in string notation, $c = 01100$.   ☐

Thus, encoding with a linear code C amounts to simply multiplying the source string by a generator matrix for C. Unfortunately, in general, source decoding is not quite as simple.

Example 5.7.2 Consider the code in the previous example. Suppose that we have received a string x and corrected it for errors to get the codeword $c = 11001$. What was the source string?

To answer this question, we set $\mathbf{s} = s_1 s_2 s_3$ and solve the equation $\mathbf{s}G = \mathbf{c}$, or in matrix notation

$$\begin{bmatrix} s_1 & s_2 & s_3 \end{bmatrix} \begin{bmatrix} 1 & 1 & 1 & 1 & 1 \\ 1 & 0 & 1 & 0 & 1 \\ 0 & 0 & 1 & 1 & 0 \end{bmatrix} = \begin{bmatrix} 1 & 1 & 0 & 0 & 1 \end{bmatrix}$$

Taking the product on the left gives

$$\begin{bmatrix} s_1 + s_2 & s_1 s_1 + s_2 + s_3 & s_1 + s_3 & s_1 + s_2 \end{bmatrix} = \begin{bmatrix} 1 & 1 & 0 & 0 & 1 \end{bmatrix}$$

Since two matrices are equal if and only if their corresponding entries are equal, this leads to the system of equations

$$s_1 + s_2 = 1$$
$$s_1 = 1$$
$$s_1 + s_2 + s_3 = 0$$
$$s_1 + s_3 = 0$$
$$s_1 + s_2 = 1$$

We will leave it to you to show that the only solution to this system is $s_1 = 1$, $s_2 = 0$, $s_3 = 1$, and so the source string is $\mathbf{s} = 101$.   ☐

The previous example shows that, in general, in order to perform source decoding, we must solve a system of linear equations. It is worth pointing out that, since the string $\mathbf{c}$ obtained from the nearest neighbor decision rule is a codeword, and since every codeword corresponds to a unique source string, the system of equations in question must have a unique solution. (In the language of linear operators, the operator G has rank k and hence is an isomorphism from $\mathbb{Z}_p^k$ onto the code C.)

Unfortunately, for a large code, solving the corresponding system of equations can be a formidable task. However, when the generator matrix G has left standard form, the task becomes extremely simple.

Let C be an $[n, k]$-code with generator matrix G in left standard form

$$G = \begin{bmatrix} 1 & 0 & \cdots & 0 & b_{11} & \cdots & b_{1,n-k} \\ 0 & 1 & \cdots & 0 & b_{21} & \cdots & b_{2,n-k} \\ & & \vdots & & & \vdots & \\ 0 & 0 & \cdots & 1 & b_{k,1} & \cdots & b_{k,n-k} \end{bmatrix}$$

Then, because the left portion of G is an identity matrix, if $\mathbf{s} = s_1 s_2 \cdots s_k$ is a source string, its codeword $\mathbf{c} = \mathbf{s}G$ has the form

$$\mathbf{c} = \mathbf{s}G = s_1 s_2 \cdots s_k x_1 x_2 \cdots x_{n-k}$$

where $x_1, x_2, \ldots, x_{n-k}$ are in $\mathbb{Z}_p$. In other words, the source message $\mathbf{s}$ is a prefix of the codeword for $\mathbf{s}$. This makes source decoding almost trivial, for all we have to do is read the first k symbols from the codeword $\mathbf{c}$ to get the source $\mathbf{s}$!

Example 5.7.3 Consider our old friend the Hamming $[7, 4]$-code C_3, with generator matrix

$$G_3 = \begin{bmatrix} 1 & 0 & 0 & 0 & 0 & 1 & 1 \\ 0 & 1 & 0 & 0 & 1 & 0 & 1 \\ 0 & 0 & 1 & 0 & 1 & 1 & 0 \\ 0 & 0 & 0 & 1 & 1 & 1 & 1 \end{bmatrix}$$

which is in left standard form. To encode a source string $\mathbf{s} = s_1 s_2 s_3 s_4$, we multiply $\mathbf{s}$ on the right by G

$$\mathbf{c} = \mathbf{s}G = \begin{bmatrix} s_1 & s_2 & s_3 & s_4 \end{bmatrix} \begin{bmatrix} 1 & 0 & 0 & 0 & 0 & 1 & 1 \\ 0 & 1 & 0 & 0 & 1 & 0 & 1 \\ 0 & 0 & 1 & 0 & 1 & 1 & 0 \\ 0 & 0 & 0 & 1 & 1 & 1 & 1 \end{bmatrix}$$

$$= \begin{bmatrix} s_1 & s_2 & s_3 & s_4 & s_2 + s_3 + s_4 & s_1 + s_3 + s_4 & s_1 + s_2 + s_4 \end{bmatrix}$$

and so $\mathbf{c} = s_1 s_2 s_3 s_4 (s_2 + s_3 + s_4)(s_1 + s_3 + s_4)(s_1 + s_2 + s_4)$. As expected, the first 4 elements of $\mathbf{c}$ are just the elements of the source string $\mathbf{s}$. Now, if we are given a codeword to decode, say $\mathbf{c} = 1001100$, then we see immediately that the source string is $\mathbf{s} = 1001$. $\square$

Source encoding with a generator matrix in left standard form makes some of the concepts involved stand out more clearly. In particular, if C

is a $[n, k]$-code, the source encoding process is

$$\mathbf{s} = s_1 s_2 \cdots s_k \to \mathbf{c} = \mathbf{s}G = s_1 s_2 \cdots s_k x_1 x_2 \cdots x_{n-k}$$

The extra $n - k$ elements $x_1, x_2, \ldots, x_k$ of the codeword $\mathbf{c}$ are present only to correct errors in transmission, for they certainly contribute nothing to the meaning of the source message. In other words, $x_1, x_2, \ldots, x_{n-k}$ is the *redundant data* that we spoke about in the introduction to this book. The number $n-k$ of redundant symbols is often referred to as the **redundancy** of the code. On the other hand, when we encode with a generator matrix that is not in left standard form, the redundant data is "mixed in" with the source information, so that retrieving the latter (i.e., source decoding) requires considerably more computation.

The use of parity check matrices that are in *right* standard form also has some interesting features. Let us illustrate with an example.

Example 5.7.4   Consider the binary $[7, 3]$-code C, with parity check matrix

$$P = \begin{bmatrix} 1 & 1 & 1 & 1 & 0 & 0 & 0 \\ 0 & 1 & 0 & 0 & 1 & 0 & 0 \\ 1 & 0 & 1 & 0 & 0 & 1 & 0 \\ 0 & 1 & 1 & 0 & 0 & 0 & 1 \end{bmatrix}$$

which is in right standard form. A string $\mathbf{x} = x_1 x_2 \cdots x_7$ is in C if and only if $\mathbf{x}P^t = \mathbf{0}$, that is,

$$\begin{bmatrix} x_1 & x_2 & \cdots & x_7 \end{bmatrix} \begin{bmatrix} 1 & 0 & 1 & 0 \\ 1 & 1 & 0 & 1 \\ 1 & 0 & 1 & 1 \\ 1 & 0 & 0 & 0 \\ 0 & 1 & 0 & 0 \\ 0 & 0 & 1 & 0 \\ 0 & 0 & 0 & 1 \end{bmatrix} = \begin{bmatrix} 0 & 0 & 0 & 0 \end{bmatrix}$$

Taking the product on the left, we get

$$\begin{bmatrix} x_1 + x_2 + x_3 + x_4 & x_2 + x_5 & x_1 + x_3 + x_6 & x_2 + x_3 + x_7 \end{bmatrix} = \begin{bmatrix} 0 & 0 & 0 & 0 \end{bmatrix}$$

which is equivalent to the system of parity check equations

$$x_1 + x_2 + x_3 + x_4 = 0$$
$$x_2 + x_5 = 0$$

$$x_1 + x_3 + x_6 = 0$$
$$x_2 + x_3 + x_7 = 0$$

which has a nice form due to the right standard form of the parity check matrix. In fact, since $-1 = 1$ in $\mathbb{Z}_2$, we may write these equations in the form

$$x_4 = x_1 + x_2 + x_3$$
$$x_5 = x_2$$
$$x_6 = x_1 + x_3$$
$$x_7 = x_2 + x_3$$

and so

$$C = \left\{ x_1 x_2 x_3 (x_1 + x_2 + x_3) x_2 (x_1 + x_3)(x_2 + x_3) \mid x_1, x_2, x_3 \in \mathbb{Z}_2 \right\}$$

This description of C is very pleasant, for we can easily generate codewords from it by just picking values for x_1, x_2, and x_3 and substituting, or we can easily determine whether or not a given string is a codeword. $\quad\square$

Exercises

1. If we wish to encode all source strings in $\mathbb{Z}_2^k$ using the binary code $C = \langle 1000, 1011, 1110 \rangle$, what is the maximum value of k? Use a generator matrix for C in left standard form to encode the source string $\mathbf{1}$.

2. If we wish to encode all source strings in $\mathbb{Z}_3^k$ using the ternary code $C = \langle 1020, 1211, 1120 \rangle$, what is the maximum value of k? Use a generator matrix for C in left standard form to encode the source string $\mathbf{1}$.

3. If we wish to encode all source strings in $\mathbb{Z}_2^k$ using the binary code $C = \langle 1000, 1101, 0101 \rangle$, what is the maximum value of k? Use a generator matrix for C in left standard form to encode the source string $\mathbf{1}$.

4. Assume that, for a binary code C, the generator matrix

$$G = \begin{bmatrix} 1 & 1 & 1 & 1 & 1 \\ 1 & 0 & 1 & 0 & 1 \\ 0 & 0 & 1 & 1 & 0 \end{bmatrix}$$

was used for source encoding. Decode the codeword 11001.

5. For the ternary code C_2 of Example 5.2.1, assume that the generator matrix

$$G_2 = \begin{bmatrix} 0 & 1 & 2 & 1 \\ 2 & 2 & 1 & 0 \end{bmatrix}$$

was used for source encoding. Decode the codeword 1211.

6. Assuming that the generator matrix G_3 for the Hamming code C_3 is used for source encoding, decode the codewords a) 1010101, b)1110000, c) 1111111.

7. The Hamming code C_3 has parity check matrix

$$P_3 = \begin{bmatrix} 0 & 1 & 1 & 1 & 1 & 0 & 0 \\ 1 & 0 & 1 & 1 & 0 & 1 & 0 \\ 1 & 1 & 0 & 1 & 0 & 0 & 1 \end{bmatrix}$$

Use this matrix to determine the form of a codeword $\mathbf{c} = c_1 \cdots c_7$ in C_3. Now consider the matrix H obtained from P_3 by adding the first row of P_3 to each of the other rows of P_3. Use this parity check matrix to describe the codewords in C_3. Show directly that these describe the same code.

8. Use the parity check matrix

$$P_2 = \begin{bmatrix} 0 & 1 & 1 & 0 \\ 1 & 2 & 0 & 1 \end{bmatrix}$$

for the ternary code C_2 to describe the codewords in C_2. Reconcile this with the description of C_2 given in Example 5.2.1.

9. A (not necessarily linear) (n, p^k)-code C over $\mathbb{Z}_p$ is called **systematic** if there are k coordinate positions $i_1, \ldots, i_k$ with the property that, by restricting the codewords for these k positions, we get all p^k possible strings of length k over $\mathbb{Z}_p$. The set $\{i_1, \ldots, i_k\}$ is called an **information set** and the codeword symbols in these positions are called **information symbols**.

   (a) Show that the binary code $C = \{0000, 0110, 1001, 1010\}$ is systematic. What is the information set?

   (b) Show that the binary code $D = \{000, 100, 010, 001\}$ is not systematic.

   (c) What is the significance of a code being systematic?

10. If the codes C and D are equivalent, is the probability of a decoding error the same for both codes? Be careful here about the channel!

6

C H A P T E R

Some Special Codes

6.1 The Hamming and Golay Codes

The r-ary Hamming codes $\mathcal{H}_r(h)$ are probably the most famous of all error-correcting codes. These codes were discovered independently by Marcel Golay in 1949 and Richard Hamming in 1950. They are perfect, linear codes that decode in a very elegant manner. We shall construct these codes by constructing their parity check matrices. Hamming codes exist for any prime power radix r and for any integer $h > 0$. However, we will construct these codes only for prime r. (The construction is essentially the same for prime powers r as for prime r.)

According to Theorem 5.4.7, the minimum distance of a linear $[n, k]$-code with parity check matrix P is the smallest integer d for which there exists d linearly dependent columns in P. Hence, the parity check matrix of an $[n, k, 3]$-code has the property that no two of its columns are linearly dependent, that is, no column is a scalar multiple of another column, but some set of three columns is linearly dependent.

For a given code alphabet $\mathbb{Z}_p$ (p a prime), we can easily construct such a parity check matrix P with h rows, and with the maximum possible number of columns. First a bit of terminology. If a string over $\mathbb{Z}_p$ is thought of as a p-ary number, we refer to the leftmost nonzero position as the **most significant position** in the number. For instance, the most significant position in x = 00210 is position 3. Also, the value in the most significant position is called the **most significant value**. (The most significant value of **x** is 2.) We also use these terms to refer to the topmost nonzero position and value in a column of a matrix. Now, for the columns of P, we take the p-ary representations of those positive integers, in increasing order starting with 1, that have most significant value 1. Some examples will make this clear.

For p = 2, the most significant value is always a 1, so we don't need to worry about this and we can simply take the binary representations of the positive integers 1, 2, ... For instance, when h = 3 rows, we get

$$H_2(3) = \begin{bmatrix} 0 & 0 & 0 & 1 & 1 & 1 & 1 \\ 0 & 1 & 1 & 0 & 0 & 1 & 1 \\ 1 & 0 & 1 & 0 & 1 & 0 & 1 \end{bmatrix}$$

whose columns are the binary representation of the numbers 1, 2, 3, 4, 5, 6, and 7. For p = 3, we must skip those integers whose most significant value is 2. If h = 3, for instance, we get

$$H_3(3) = \begin{bmatrix} 0 & 0 & 0 & 0 & 1 & 1 & 1 & 1 & 1 & 1 & 1 & 1 & 1 \\ 0 & 1 & 1 & 1 & 0 & 0 & 0 & 1 & 1 & 1 & 2 & 2 & 2 \\ 1 & 0 & 1 & 2 & 0 & 1 & 2 & 0 & 1 & 2 & 0 & 1 & 2 \end{bmatrix}$$

The first column of $H_3(3)$ is the ternary representation of the number 1. However, the ternary representation of 2 is 002, which must be skipped since its most significant value is 2. The ternary representations of 3, 4, and 5 have most significant value 1 and so form the next three columns of $H_3(3)$. The ternary representation of 6 is 020, which must be skipped, and so on.

The matrices $H_p(h)$ are called **Hamming matrices**. We now establish some facts about the Hamming matrices.

Fact 1   The Hamming matrix $H_p(h)$ has size $h \times n$, where $n = \frac{p^h - 1}{p - 1}$.

Since $H_p(h)$ was constructed using columns of length h, it has h rows. To count the number of columns in $H_p(h)$, observe that the most significant position of a column of $H_p(h)$ can be any number between 1 and h. If the most significant position is m, then the column consists of $m - 1$ zeros, followed by a 1, followed by $h - m$ additional symbols, each of which can be any of the p symbols in $\mathbb{Z}_p$. Hence, there are p^{h-m} columns with most significant position m. Summing over m ranging from 1 to h, the total number of columns of $H_p(h)$ is

$$n = p^0 + p^1 + \cdots + p^{h-1} = \frac{p^h - 1}{p - 1}$$

Fact 2   The rows of $H_p(h)$ are linearly independent over $\mathbb{Z}_p$.

This can be seen by observing that each of the column matrices $\mathbf{e}_1, \ldots, \mathbf{e}_h$ is a column of $H_p(h)$. We leave details of this as an exercise.

Fact 3   No two distinct columns of $H_p(h)$ are linearly dependent.

To see this, consider two distinct columns of $H_p(h)$, written in row notation, say

$$\mathbf{c} = \underbrace{0 \cdots 0}_{u \text{ zeros}} 1 c_1 \cdots c_{h-u-1} \text{ and } \mathbf{d} = \underbrace{0 \cdots 0}_{v \text{ zeros}} 1 d_1 \cdots d_{h-v-1}$$

Clearly, if $u \neq v$, then $\mathbf{c}$ and $\mathbf{d}$ do not have the same number of leading 0s, and so neither can be a scalar multiple of the other. But if $u = v$ and $\mathbf{c} = \alpha \mathbf{d}$, for some $\alpha \in \mathbb{Z}_p$ then comparing the $u + 1$st positions gives $1 = \alpha 1$ and so $\alpha = 1$, implying that $\mathbf{c} = \mathbf{d}$. Since distinct columns of $H_p(h)$ contain distinct numbers, this cannot happen. Hence, no column is a scalar multiple of another column.

Fact 4   The first three columns of $H_p(h)$ are linearly dependent.

The first column of $H_p(h)$ is always $0 \cdots 01$, the second column is always $0 \cdots 010$ and the third column is always $0 \cdots 011$, which is the sum of the first two columns.

The previous facts lead to the following theorem.

Theorem 6.1.1   *The Hamming matrix $H_p(h)$ is a parity check matrix for a linear $[n, k, d]$-code $\mathcal{H}_p(h)$ over $\mathbb{Z}_p$ with parameters*

$$n = \frac{p^h - 1}{p - 1}, k = n - h, d = 3$$

*This code is called the **p-ary Hamming code** with parameter h. Thus, $\mathcal{H}_p(h)$ is an exactly single-error-correcting code.*   □

The binary case is by far the most common, where $\mathcal{H}_2(h)$ is a binary linear $[n, k, d]$-code, with parameters

$$n = 2^h - 1, k = n - h, d = 3$$

The following table shows that the size $M = 2^{n-h}$ of the binary Hamming codes gets large very fast.

h	$n = 2^h - 1$	$M = 2^{2^h-1-h}$
2	3	2
3	7	16
4	15	2048
5	31	67,108,864
6	63	1.44×10^{17}

The Hamming codes are special, partly because they are perfect and easy to decode and partly because they are "unique" in the sense that any linear code C with the same parameters as a Hamming code is scalar multiple equivalent to that Hamming code. Proof of this statement is left as an exercise. We also ask the reader, in the exercises, to show (with some help) that there are nonlinear codes with the same parameters as some of the Hamming codes.

Decoding with a Hamming Code

The form of the Hamming matrices $H_p(h)$ allows for perhaps the most elegant decoding procedure of any code. This is especially true for the binary Hamming codes, which we discuss first.

Theorem 6.1.2 *If a codeword from the binary Hamming code $\mathcal{H}_2(h)$ suffers a single error, resulting in the received word* $\mathbf{x}$, *then the syndrome* $S(\mathbf{x}) = \mathbf{x}H_2(h)^t$ *is the binary representation of the position in* $\mathbf{x}$ *of the error.* □

Proof If the error string is $\mathbf{e}_i$, then $S(\mathbf{x}) = S(\mathbf{e}_i) = \mathbf{e}_i H_2(h)^t$ is the ith column of $H_2(h)$, which is just the binary representation of the number i. ∎

Example 6.1.1 Using the binary Hamming code $\mathcal{H}_2(3)$, suppose $\mathbf{x} = 1011000$ is received. The syndrome is

$$
S(\mathbf{x}) = \begin{bmatrix} 1 & 0 & 1 & 1 & 0 & 0 & 0 \end{bmatrix}
\begin{bmatrix}
0 & 0 & 1 \\
0 & 1 & 0 \\
0 & 1 & 1 \\
1 & 0 & 0 \\
1 & 0 & 1 \\
1 & 1 & 0 \\
1 & 1 & 1
\end{bmatrix}
= \begin{bmatrix} 1 & 1 & 0 \end{bmatrix}
$$

which is the binary number $110_{binary} = 6_{decimal}$. Hence, if a single error occurred, it must have been in the 6th position of the codeword and so $\mathbf{c} = \mathbf{x} - \mathbf{e}_6 = 1011010$ is the nearest neighbor codeword.   □

In the nonbinary case, an error in the ith position produces an error string of the form $\alpha \mathbf{e}_i$, for some nonzero $\alpha \in \mathbb{Z}_p$. The syndrome of the received word $\mathbf{x} = \mathbf{c} + \alpha \mathbf{e}_i$ is thus

$$
S(\mathbf{x}) = \alpha \mathbf{e}_i H_p(h)^t
$$

which is α times the ith column of $H_p(h)$. But, since the most significant value of any column of $H_p(h)$ is equal to 1, the first nonzero entry in the syndrome is $\alpha 1 = \alpha$. Multiplying the syndrome by α^{-1} gives the ith column of $H_p(h)$. Comparing this with each column of $H_p(h)$ gives the value of i. Let us state this more formally in a theorem.

Theorem 6.1.3 *Suppose a codeword from the Hamming code $\mathcal{H}_p(h)$ suffers a single error, resulting in the received word $\mathbf{x}$, and let α be the most significant value of $S(\mathbf{x})$. If the column of $H_p(h)$ containing $\alpha^{-1} S(\mathbf{x})$ is the ith column, then the error string is $\alpha \mathbf{e}_i$ and the nearest neighbor codeword to $\mathbf{x}$ is $\mathbf{x} - \alpha \mathbf{e}_i$.*   □

Example 6.1.2 Using the ternary Hamming code $\mathcal{H}_3(3)$, the syndrome of the received word $\mathbf{x} = 1101112211201$ is

$$
\begin{bmatrix} 1 & 1 & 0 & 1 & 1 & 1 & 2 & 2 & 1 & 1 & 2 & 0 & 1 \end{bmatrix} H_3(3)^t
$$
$$
= \begin{bmatrix} 2 & 0 & 1 \end{bmatrix} = 2 \begin{bmatrix} 1 & 0 & 2 \end{bmatrix} = 2 \times (7th \text{ column of } H_3(3))
$$

It follows that the error string is $2\mathbf{e}_7$ and the nearest neighbor codeword is

$$
\mathbf{x} - 2\mathbf{e}_7 = 1101110211201
$$ □

The Hamming codes are indeed remarkable codes, especially when it comes to decoding. Also, the transmission rate of the Hamming code $\mathcal{H}_p(h)$ is

$$\mathcal{R}(\mathcal{H}_p(h)) = \frac{k}{n} = 1 - \frac{h}{n}$$

which tends to the maximum possible value of 1 as n gets large. Unfortunately, the error correction rate of $H_p(h)$ is

$$\delta(C) = \frac{\left\lfloor \frac{d-1}{2} \right\rfloor}{n} = \frac{1}{n}$$

which tends to the worst possible value of 0 as n gets large.

The Simplex Codes

Since the Hamming codes have some special properties, it is not surprising that their dual codes also have special properties (although these codes are much less well known). We will restrict attention to binary codes. The dual of the binary Hamming code $\mathcal{H}_2(h)$ is called the **simplex code** $\mathcal{S}(h)$. Since the rows of the matrix $H_2(h)$ are linearly independent, this matrix is a generator matrix for $\mathcal{S}(h)$.

The simplex code $\mathcal{S}(h)$ has length $n = 2^h - 1$ and dimension h. To determine the distance properties of the simplex codes, we observe that the generator matrix $H_2(h + 1)$ can be obtained from two copies of the matrix $H_2(h)$ as follows

$$H_2(h + 1) = \left[\begin{array}{ccc|c|ccc} 0 & \cdots & 0 & 1 & 1 & \cdots & 1 \\ \hline & & & 0 & & & \\ & H_2(h) & & \vdots & & H_2(h) & \\ & & & 0 & & & \end{array} \right] \tag{6.1.1}$$

Now, any codeword $\mathbf{c}$ in $\mathcal{S}(h + 1)$ is a sum of some of the rows of $H_2(h + 1)$. Hence, $\mathbf{c}$ has the form $\mathbf{c} = \mathbf{a}\alpha\mathbf{b}$, where $\mathbf{a}$ is a sum of rows of $H_2(h)$ (and possibly the zero row) and is therefore a codeword in $\mathcal{S}(h)$, α is either a 0 or a 1 and $\mathbf{b}$ is equal to $\mathbf{a}$ or $\mathbf{a} + \mathbf{1} = \mathbf{a}^c$, depending upon whether or not the first row of $H_2(h + 1)$ is included in the sum. Thus, we have two cases.

Case 1   If the sum constituting **c** does not involve the first row of $H_2(h + 1)$ then **c** = **a0a**, where **a** $\in$ $\mathcal{S}(h)$.

Case 2   If the sum constituting **c** does involve the first row of $H_2(h + 1)$ then **c** = **a1a**^{c}, where **a** $\in$ $\mathcal{S}(h)$.

These cases are summarized in the following theorem, which completely describes the simplex codes.

Theorem 6.1.4   *The simplex code* $\mathcal{S}(h)$ *can be described as follows.*

1. $\mathcal{S}(2) = \langle 011, 101 \rangle = \{000, 011, 101, 110\}$

2. *For any integer* $h \geq 2$,

$$\mathcal{S}(h + 1) = \{\mathbf{a0a} \mid \mathbf{a} \in \mathcal{S}(h)\} \cup \{\mathbf{a1a}^c \mid \mathbf{a} \in \mathcal{S}(h)\} \qquad \square$$

Theorem 6.1.4 can be used to establish a very interesting fact about the distances between codewords in a simplex code. We leave the proof as an exercise.

Theorem 6.1.5   *The simplex code* $\mathcal{S}(h)$ *is a* $[2^h - 1, h, 2^{h-1}]$-*code with the property that the distance between every pair of distinct codewords is* 2^{h-1}. $\qquad \square$

Thus, not only is the minimum distance of a simplex code $\mathcal{S}(h)$ equal to 2^{h-1}, which is rather large, but the distance between *every* pair of distinct codewords is 2^{h-1}. (This accounts for the small size of the simplex codes.) Figure 6.1.1 shows the simplex code $\mathcal{S}(2)$.

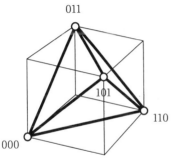

FIGURE 6.1.1   The simplex code $\mathcal{S}(2)$

Golay Codes

There are a total of four Golay codes—two binary codes and two ternary codes. We will define these codes by giving generating matrices, as did Marcel Golay in a one-page paper published in 1949. (Golay gave no hint as to how he obtained these generator matrices. However, there are other, more motivating ways to define these codes, although these methods require more mathematical structure than we have developed.)

The Binary Golay Code $\mathcal{G}_{24}$

The binary Golay code $\mathcal{G}_{24}$ is a $[24, 12, 8]$-code whose generator matrix has the form $G = [I_{12} \mid A]$, where

$$
A = \begin{bmatrix}
\cdot & 1 & 1 & 1 & 1 & 1 & 1 & 1 & 1 & 1 & 1 & 1 \\
1 & 1 & 1 & \cdot & 1 & 1 & 1 & \cdot & \cdot & \cdot & 1 & \cdot \\
1 & 1 & \cdot & 1 & 1 & 1 & \cdot & \cdot & \cdot & 1 & \cdot & 1 \\
1 & \cdot & 1 & 1 & 1 & \cdot & \cdot & \cdot & 1 & \cdot & 1 & 1 \\
1 & 1 & 1 & 1 & \cdot & \cdot & \cdot & 1 & \cdot & 1 & 1 & \cdot \\
1 & 1 & 1 & \cdot & \cdot & \cdot & 1 & \cdot & 1 & 1 & \cdot & 1 \\
1 & 1 & \cdot & \cdot & \cdot & 1 & \cdot & 1 & 1 & \cdot & 1 & 1 \\
1 & \cdot & \cdot & \cdot & 1 & \cdot & 1 & 1 & \cdot & 1 & 1 & 1 \\
1 & \cdot & \cdot & 1 & \cdot & 1 & 1 & \cdot & 1 & 1 & 1 & \cdot \\
1 & \cdot & 1 & \cdot & 1 & 1 & \cdot & 1 & 1 & 1 & \cdot & \cdot \\
1 & 1 & \cdot & 1 & 1 & \cdot & 1 & 1 & 1 & \cdot & \cdot & \cdot \\
1 & \cdot & 1 & 1 & \cdot & 1 & 1 & 1 & \cdot & \cdot & \cdot & 1
\end{bmatrix}
$$

and where dots have been used in place of 0s in the hope of increasing readability. We will show that $\mathcal{G}_{24}$ has minimum weight 8 through a series of simple facts.

Lemma 6.1.6　*The Golay code $\mathcal{G}_{24}$ is self-dual, that is, $\mathcal{G}_{24}^{\perp} = \mathcal{G}_{24}$.*　□

Proof　It is straightforward (but boring) to check that, if $\mathbf{r}$ and $\mathbf{s}$ are rows of G, then $\mathbf{r} \cdot \mathbf{s} = 0$. Hence, $\mathcal{G}_{24} \subseteq \mathcal{G}_{24}^{\perp}$. Since $\mathcal{G}_{24}$ and $\mathcal{G}_{24}^{\perp}$ both have dimension 12, they must be equal.　∎

Proof of the next result is left as an exercise.

Lemma 6.1.7　*The matrix $[A \mid I_{12}]$ is also a generator matrix for $\mathcal{G}_{24}$.*　□

Lemma 6.1.8   *The weight of every codeword in $\mathcal{G}_{24}$ is divisible by 4.*   □

Proof   It is easy to see by counting that the weight of every row of G is divisible by 4. If r and s are rows of G, then

$$w(\mathbf{r} + \mathbf{s}) = w(\mathbf{r}) + w(\mathbf{s}) - 2w(\mathbf{r} \cap \mathbf{s})$$

But $w(\mathbf{r} \cap \mathbf{s}) \equiv \mathbf{r} \cdot \mathbf{s} = 0 \bmod 2$, and so $w(\mathbf{r} + \mathbf{s})$ is also divisible by 4. It is now easy to construct a proof by induction that the weight of any sum of rows of G is divisible by 4. We leave details of this to the reader.   ■

Lemma 6.1.9   *The Golay code $\mathcal{G}_{24}$ has no codewords of weight 4.*   □

Proof   We take advantage of the two generating matrices $G_1 = [I_{12} \mid A]$ and $G_2 = [A \mid I_{12}]$ for $\mathcal{G}_{24}$. Suppose that $\mathbf{c} \in \mathcal{G}_{24}$ and consider the left half L and the right half R of $\mathbf{c}$. (Thus $c = LR$, where L and R are strings of length 12.) Since any nontrivial linear combination (i.e., sum) of the rows of G_1 has a left half of weight at least 1, we deduce that $w(L) \geq 1$. Similarly, using G_2, we get $w(R) \geq 1$. However, if $w(L) = 1$, then $\mathbf{c}$ must be a row of G_1, none of which has weight 4. Similarly, if $w(R) = 1$ then $\mathbf{c}$ cannot have weight 4. Assume that $w(L) \geq 2$ and $w(R) \geq 2$. Then $\mathbf{c}$ can have weight 4 if and only if $w(L) = 2$ and $w(R) = 2$. But this happens if and only if $\mathbf{c}$ is the sum of precisely two rows of G_1, which can be ruled out simply by checking that no sum of any two rows of G_1 has weight 4.   ■

Since the weight of any codeword in $\mathcal{G}_{24}$ is divisible by 4, but not equal to 4, we must have $w(\mathcal{G}_{24}) \geq 8$. On the other hand, the second row of G has weight 8 and so we have proved the following.

Theorem 6.1.10   *The binary Golay code $\mathcal{G}_{24}$ is a $[24, 12, 8]$-code.*   □

Decoding with the Binary Golay Code $\mathcal{G}_{24}$

Since $\mathcal{G}_{24}$ is a $[24, 12, 8]$-code, syndrome decoding would require that we construct (and check) a table with

$$\frac{2^{24}}{2^{12}} = 2^{12} = 4096$$

syndromes. On the other hand, using the structure of $\mathcal{G}_{24}$, we can considerably reduce the work involved in decoding.

Since $\mathcal{G}_{24}$ is self-dual, the matrices $G_1 = [I_{12} \mid A]$ and $G_2 = [A \mid I_{12}]$ are both parity check matrices. Suppose that 3 or fewer errors occur in the transmission of a codeword, and let $\mathbf{x}$ be the received word and $\mathbf{e}$ be the error vector. Thus, $w(\mathbf{e}) \le 3$. Let us write $\mathbf{e} = \mathbf{fg}$, where $\mathbf{f}$ and $\mathbf{g}$ have length 12. We can compute the syndromes of the received word using both parity check matrices as follows

$$S_1 = \mathbf{e}G_1^t = [\mathbf{f} \mid \mathbf{g}]\begin{bmatrix} I_{12} \\ \hline A \end{bmatrix} = \mathbf{f} + \mathbf{g}A$$

(we are mixing matrix and vector notation here) and similarly

$$S_2 = \mathbf{e}G_2^t = [\mathbf{f} \mid \mathbf{g}]\begin{bmatrix} A \\ \hline I_{12} \end{bmatrix} = \mathbf{f}A + \mathbf{g}$$

Now let us examine the possibilities.

1. If $w(\mathbf{f}) = 0$ and $1 \le w(\mathbf{g}) \le 3$, then $\mathbf{e} = \mathbf{0g} = \mathbf{0}S_2$ and

$$w(S_1) = w(\mathbf{g}A) \ge 5, w(S_2) = w(\mathbf{g}) \le 3$$

The first inequality following from the fact that the sum of any 3 or fewer rows of $[I_{12} \mid A]$ has weight at least 8, of which the contribution of the first half of the rows is at most 3.

2. If $1 \le w(\mathbf{f}) \le 3$ and $w(\mathbf{g}) = 0$, then $\mathbf{e} = \mathbf{f0} = S_1\mathbf{0}$ and

$$w(S_1) = w(\mathbf{f}) \le 3, w(S_2) = w(\mathbf{f}A) \ge 5$$

3. If $w(f) \ge 1$ and $w(g) \ge 1$, then $w(S_1) \ge 5$ and $w(S_2) \ge 5$.

Thus, if either syndrome has weight at most 3, we can easily recover the error vector $\mathbf{e}$. If $w(S_1)$ and $w(S_2)$ are both greater than 3, we know that one of the following holds

a) $w(\mathbf{f}) = 1$ and $w(\mathbf{g}) = 1$ or 2.

b) $w(\mathbf{f}) = 2$ and $w(\mathbf{g}) = 1$.

Let $\mathbf{e}_i$ be the vector of length 12 with a 1 in the ith position and zeros elsewhere. Under case a), we have $\mathbf{f} = \mathbf{e}_i$ for some i and so

$$\mathbf{y}_u = (\mathbf{x} + \mathbf{e}_u\mathbf{0})G_2^t = (\mathbf{e} + \mathbf{e}_u\mathbf{0})G_2^t = (\mathbf{e}_i\mathbf{g} + \mathbf{e}_u\mathbf{0})G_2^t = \mathbf{e}_iA + \mathbf{g} + \mathbf{e}_uA$$

Under case b), we have $\mathbf{f} = \mathbf{e}_i + \mathbf{e}_j$ for some $i \ne j$ and so

$$\mathbf{y}_u = (\mathbf{x} + \mathbf{e}_u\mathbf{0})G_2^t = (\mathbf{e} + \mathbf{e}_u\mathbf{0})G_2^t$$
$$= ((\mathbf{e}_i + \mathbf{e}_j)\mathbf{g} + \mathbf{e}_u\mathbf{0})G_2^t = \mathbf{e}_iA + \mathbf{e}_jA + \mathbf{g} + \mathbf{e}_uA$$

Suppose we make the above computations for all $u = 1, \ldots, 12$.

If case a) holds, then $w(\mathbf{y}_u) = w(\mathbf{g}) = 1$ or 2 precisely when $u = i$; otherwise, $w(\mathbf{y}_u) \geq 4$. On the other hand, if case b) holds, then $w(\mathbf{y}_u) \geq 4$ for all u. Thus, we can distinguish between the two cases and, if case a) holds, we can determine both the error position i (and thus $\mathbf{f}$) and the second half $\mathbf{g} = \mathbf{y}_i$ of the error string, by looking at the 12 strings $\mathbf{y}_1, \ldots, \mathbf{y}_{12}$.

If case b) holds, we make a similar computation using G_1^t. In this case, $\mathbf{f} = \mathbf{e}_i$ and $\mathbf{g} = \mathbf{e}_j$ and so

$$\mathbf{z}_u = (\mathbf{x} + \mathbf{0}\mathbf{e}_u)G_1^t = (\mathbf{e} + \mathbf{0}\mathbf{e}_u)G_1^t = (\mathbf{e}_i\mathbf{e}_j + \mathbf{0}\mathbf{e}_u)G_1^t = \mathbf{e}_i + \mathbf{e}_jA + \mathbf{e}_uA$$

which has weight $w(\mathbf{z}_u) = w(\mathbf{e}_i) = 1$ if $u = j$ and weight $w(\mathbf{z}_u) \geq 5$ for $u \neq j$. Thus, we may easily pick out both i and j in this case, by looking at the 12 strings $\mathbf{z}_1, \ldots, \mathbf{z}_{12}$.

In summary, if at most three errors occur, then we can decode correctly by computing at most the 26 syndromes

$$\mathbf{x}G_1^t, \ \mathbf{x}G_2^t, (\mathbf{x} + \mathbf{e}_1\mathbf{0})G_2^t, \ldots, (\mathbf{x} + \mathbf{e}_{12}\mathbf{0})G_2^t, (\mathbf{x} + \mathbf{0}\mathbf{e}_1)G_1^t, \ldots, (\mathbf{x} + \mathbf{0}\mathbf{e}_{12})G_1^t$$

(a task still best done by computer).

The Binary Golay Code $\mathcal{G}_{23}$

Let $\mathcal{G}_{23}$ be the code obtained simply by throwing away the last coordinate of every codeword in the Golay code $\mathcal{G}_{24}$. This process is called **puncturing** the code $\mathcal{G}_{24}$. The resulting punctured code has length 23 and, since the distance between codewords in $\mathcal{G}_{24}$ is greater than 1, all of the punctured codewords are distinct, so $\mathcal{G}_{23}$ has the same size as $\mathcal{G}_{24}$. It is clear that puncturing a code cannot increase the minimum distance nor decrease it by more than 1 and so $d(\mathcal{G}_{23}) = 7$ or 8. But the parameters $[23, 12, 7]$ satisfy the sphere-packing condition and so $d(\mathcal{G}_{23}) = 7$. Thus, the Golay code $\mathcal{G}_{23}$ is a perfect binary $[23, 12, 7]$-code.

The Ternary Golay Codes

The ternary Golay code $\mathcal{G}_{12}$ is the code with generating matrix $G = [I_6 \mid B]$, where

$$
B = \begin{bmatrix}
0 & 1 & 1 & 1 & 1 & 1 \\
1 & 0 & 1 & 2 & 2 & 1 \\
1 & 1 & 0 & 1 & 2 & 2 \\
1 & 2 & 1 & 0 & 1 & 2 \\
1 & 2 & 2 & 1 & 0 & 1 \\
1 & 1 & 2 & 2 & 1 & 0
\end{bmatrix}
$$

We will leave proof of the following as an exercise.

Theorem 6.1.11

1. *The ternary Golay code $\mathcal{G}_{12}$ is self-dual, that is, $\mathcal{G}_{12}^{\perp} = \mathcal{G}_{12}$.*
2. *The matrix B satisfies $B = B^t$.*
3. *$\mathcal{G}_{12}$ is a [12, 6, 6]-code.*
4. *The ternary code $\mathcal{G}_{11}$, obtained by puncturing $\mathcal{G}_{12}$ in its last coordinate position, is a perfect [11, 6, 5]-code.*   □

Uniqueness of the Golay Codes

In 1968, the coding theorist Vera Pless showed that any binary linear code with the same parameters as $\mathcal{G}_{24}$ must be equivalent to $\mathcal{G}_{24}$. In 1975, P. Delsarte and J.-M. Goethals extended this result, by showing that any binary code (linear or nonlinear) with the same parameters as $\mathcal{G}_{24}$ is equivalent to $\mathcal{G}_{24}$. (Actually, they showed that if such a code contains the zero codeword, then it must be linear, and so by the result of Pless, it must be equivalent to $\mathcal{G}_{24}$.) These authors also established the uniqueness of the other Golay codes. (In 1973, S. L. Snover also established the uniqueness of $\mathcal{G}_{23}$.) Let us state their results.

Theorem 6.1.12

1. *Any binary $(24, 2^{12}, 8)$-code is equivalent to the Golay code $\mathcal{G}_{24}$.*
2. *Any binary $(23, 2^{12}, 7)$-code is equivalent to the Golay code $\mathcal{G}_{23}$.*
3. *Any ternary $(12, 3^6, 6)$-code is equivalent to the Golay code $\mathcal{G}_{12}$*
4. *Any ternary $(11, 3^6, 5)$-code is equivalent to the Golay code $\mathcal{G}_{11}$.*   □

Theorem 6.1.12 can be summarized by saying that any code that has the parameters of a Golay code is equivalent to a Golay code.

Perfect Codes Revisited

We are now in a position to appreciate the following remarkable result (various parts of which are due to van Lint, Tietäväinen, Best, and Hong) concerning the existence of perfect codes. As we have seen, the code consisting of a single codeword, the entire space, and the repetition codes are all perfect. These are referred to as the **trivial codes**. All other codes are **nontrivial**.

Theorem 6.1.13

1. *For alphabets of prime power size, all nontrivial perfect codes C have the parameters of either a Hamming code or a Golay code. Furthermore,*

    (a) *if C has the paramteres of a Golay code, then it is equivalent to that Golay code.*

    (b) *if C is linear and has the parameters of a Hamming code, then it is equivalent to that Hamming code. However, there are nonlinear perfect codes with the Hamming parameters.*

2. *Over any alphabet, the only nontrivial t-error-correcting perfect code with $t \geq 3$ is the binary Golay code $\mathcal{G}_{23}$.* □

Notice that there are some gaps in Theorem 6.1.13. With regard to alphabets of prime power size, it is not known how many nonequivalent, nonlinear perfect codes there are with the Hamming parameters. (It is believed that there may be many thousands.) In 1962, Vasil'ev discovered a family of such codes, which we discuss in the exercises.

More generally, it is still not known whether there are perfect double-error-correcting codes over any alphabet whose size is not a power of a prime. (It is conjectured that there are none.) The issue of how many nonequivalent single-error-correcting perfect codes may exist seems to be extremely difficult.

The previous theorem can be rephrased from the point of view of error correcting capabilities as follows.

Theorem 6.1.14

1. *(For at least three-error-correcting codes) For $t \geq 3$, any nontrivial perfect t-error-correcting code is equivalent to the binary Golay code $\mathcal{G}_{23}$.*

2. *(For double-error-correcting codes) Over alphabets of prime power size, all nontrivial perfect double-error-correcting codes are equivalent to the ternary Golay code $\mathcal{G}_{11}$.*

3. *(For single-error-correcting codes) Any nontrivial perfect single-error-correcting code C over an alphabet of prime power size has the parameters of a Hamming code. Moreover, if C is linear, it is scalar multiple equivalent to a Hamming code. However, there are nonlinear codes with Hamming parameters that are not equivalent to a Hamming code.*   □

Exercises

1. Use the binary Hamming matrix $H_2(3)$ to decode the words a) 1111000, b) 1.

2. Use the ternary Hamming matrix $H_3(3)$ to decode the words a) 1111000222222, b) 2222100000000.

3. Write out the Hamming matrix $H_2(2)$ and use it to decode the word 101.

4. Write out the Hamming matrix $H_3(2)$ and use it to decode the words 0120 and 0010.

5. Write out the Hamming matrix $H_2(4)$ and use it to decode the word 111110000011111.

6. How many columns of the Hamming matrix $H_3(4)$ have a 0 in the first row? Write out these columns.

7. Show that the Hamming codes are perfect.

8. Show that the rows of the Hamming matrix $H_p(h)$ are linearly independent.

9. Apply elementary row operations to the parity check matrix $H_2(3)$ to obtain another parity check matrix in the right standard form $[A \mid I_3]$. Use this matrix to find a generator matrix for the Hamming code $\mathcal{H}_2(3)$.

10. Let M be the matrix obtained by adding an additional column of 0s to the left end of the Hamming matrix $H_2(h)$. Describe the pattern of 0s and 1s in the rows of M. Show that the Hamming code $\mathcal{H}_2(h)$ contains the codeword $\mathbf{1}$ and that the complement of every codeword in $\mathcal{H}_2(h)$ is also in $\mathcal{H}_2(h)$.

11. Show that if C is a linear code with the same parameters as that of $\mathcal{H}_p(h)$, then C is equivalent to $\mathcal{H}_p(h)$. Hint: let P be a parity check matrix for C with linearly independent rows. Multiply each column of P by the inverse of the value in the most significant position of that column. What can you say about the columns of the resulting matrix?

12. We remarked in the text that if C is a *linear* code with the same parameters as the Hamming code $\mathcal{H}_2(h)$, then C is equivalent to $\mathcal{H}_2(h)$. We now construct a binary *nonlinear* code $V(h)$ (first constructed by Vasil'ev in 1962) with the same parameters as the Hamming code $\mathcal{H}_2(h)$. Let $n = 2^h - 1$. Let $f : \mathcal{H}_2(h) \to \mathbb{Z}_2$ be the function defined by $f(\mathbf{1}) = 1$ and $f(\mathbf{c}) = 0$ for all $\mathbf{c} \neq \mathbf{1}$. Let $wt : \mathbb{Z}_2^n \to \mathbb{Z}_2$ be the function defined by $wt(\mathbf{x}) = 0$ if $\mathbf{x}$ has even weight and $wt(\mathbf{x}) = 1$ if $\mathbf{x}$ has odd weight. Now let

$$V(h) = \{\mathbf{x}(\mathbf{x} + \mathbf{c})(wt(\mathbf{x}) + f(\mathbf{c})) \mid \mathbf{x} \in \mathbb{Z}_2^n, \mathbf{c} \in \mathcal{H}_2(h)\}$$

Show that $V(h)$ is a binary $(\bar{n}, M, d)$-code, where $\bar{n}$, M, and d are the Hamming parameters

$$\bar{n} = 2^{h+1}, M = 2^{2n-h}, d = 3$$

Show also that $V(h)$ is nonlinear. Hint: consider the codewords that come from a Hamming codeword $\mathbf{c}$ and its complement $\mathbf{c}^c$. Show that $V(h)$ is not equivalent to $\mathcal{H}_2(h)$.

13. Use the fact that the Hamming code $\mathcal{H}_2(h)$ is perfect to determine the number of codewords of weight 3. Hint: consider the strings of weight 2 in $\mathbb{Z}_2^n$ and the packing spheres in which they lie.

14. Verify that

$$H_2(h+1) = \begin{bmatrix} 0 & \cdots & 0 & 1 & 1 & \cdots & 1 \\ & & & 0 & & & \\ & H_2(h) & & \vdots & & H_2(h) & \\ & & & 0 & & & \end{bmatrix}$$

15. Prove that the simplex code $\mathcal{S}(h)$ is an $[2^h - 1, h, 2^{h-1}]$-code with the property that the distance between every pair of distinct codewords is 2^{h-1}.

16. Prove Lemma 6.1.7. Hint: use the fact that $\mathcal{G}_{24}$ is self-dual.

17. How large is the syndrome table for the Golay code $\mathcal{G}_{24}$?

18. Prove Theorem 6.1.11. Hint for part 3): Let $\mathbf{c}$ be a codeword in $\mathcal{G}_{12}$ and let $\mathbf{c} = \mathbf{LR}$, where $\mathbf{L}$ and $\mathbf{R}$ are strings of length 6. Show that $G_1 = [I_6 \mid B]$ and $G_2 = [-B \mid I_6]$ are both generator matrices for $\mathcal{G}_{12}$. Now, if $w(\mathbf{c}) \leq 5$, then we must have $w(\mathbf{L}) \leq 2$ or $w(\mathbf{R}) \leq 2$. If, for instance, $w(\mathbf{L}) \leq 2$, then $\mathbf{c}$ is a linear combination of at most two rows of G_1. Is this possible?

The Nordstrom-Robinson Code

In the next exercises, we define and discuss the Nordstrom-Robinson code. This code has the interesting property that it has strictly larger minimum distance than any linear code with the same length and size. Hence, it shows that linear codes may not be as good as nonlinear codes. Incidentally, the Nordstrom-Robinson code was first defined by Alan Nordstrom using trial and error, when he was but a high-school student!

19. Recall that $\mathcal{G}_{24}$ has generator matrix $G = [I_{12} \mid A]$, where A is given earlier in this section. Show that, by permuting columns and using elementary row operations, the matrix G can be brought to the form

$$
G' = \left[
\begin{array}{ccccccc|cc}
1 & 0 & 0 & 0 & 0 & 0 & 0 & 1 & \\
0 & 1 & 0 & 0 & 0 & 0 & 0 & 1 & \\
0 & 0 & 1 & 0 & 0 & 0 & 0 & 1 & * \\
0 & 0 & 0 & 1 & 0 & 0 & 0 & 1 & \\
0 & 0 & 0 & 0 & 1 & 0 & 0 & 1 & \\
0 & 0 & 0 & 0 & 0 & 1 & 0 & 1 & \\
0 & 0 & 0 & 0 & 0 & 0 & 1 & 1 & \\
\hline
 & & & & & & & 0 & \\
 & & & & & & & 0 & \\
 & & & \text{0s} & & & & 0 & * \\
 & & & & & & & 0 & \\
 & & & & & & & 0 & \\
\end{array}
\right]
$$

where the asterisks represent some values. Note that the eighth column is the sum of the previous seven columns. Hint: Any seven columns of G are linearly independent and some eight columns of G are linearly dependent. Thus, the code C with generating matrix G' is equivalent to $\mathcal{G}_{24}$.

20. Show that there are $8 \times 32 = 256$ codewords in C whose first eight coordinates are one of

$$
\begin{array}{l}
0000\ 0000 \\
1000\ 0001 \\
0100\ 0001 \\
0010\ 0001 \\
0001\ 0001 \\
0000\ 1001 \\
0000\ 0101 \\
0000\ 0011
\end{array}
$$

21. The **Nordstrom-Robinson code** $\mathcal{N}$ is the code whose codewords are obtained from the 256 words from the previous exercise by deleting the first eight coordinate positions. Hence, $\mathcal{N}$ has length 16 and size 256. Show that $\mathcal{N}$ has minimum distance 6, and so is a $(16,256,6)$-code. (It is possible to show that $A_2(16, 6) = 256$ and so the code $\mathcal{N}$ is optimal. Hint: show that $d(\mathcal{N}) \geq 6$ and use Plotkin's bound to show that $d(\mathcal{N}) \leq 6$.

22. Show that there is no linear $(16,256,6)$-code. Hint: Suppose that C is such a code. Delete the last coordinate position to get a linear $(15,256,5)$-code C_1. The cross-section $x_1 = 0$ of this code is a linear $(14,128,5)$-code C_2. The cross-section $x_1 = 0$ of C_2 is a linear $(13,64,5)$-code C_3. Rearrange the coordinate positions if necessary to get a generating matrix of the form

$$
M = \begin{bmatrix}
1 & 1 & 1 & 1 & 1 & 0 & 0 & 0 & 0 & 0 & 0 & 0 & 0 \\
 & & G_1 & & & & & & & G_2 & & & \\
\end{bmatrix}
$$

Show that G_2 is the generator matrix of a linear $[8, 5, 3]$-code. Can such a code exist?

6.2   Reed-Muller Codes

Reed-Muller codes are one of the oldest families of codes, and have been widely used in applications. For each positive integer m, and each integer r satisfying $0 \leq r \leq m$, the rth order Reed-Muller code $\mathcal{R}(r, m)$ is a binary linear code with parameters

$$n = 2^m, M = 2^{1 + \binom{m}{1} + \cdots + \binom{m}{r}}, d = 2^{m-r}$$

We will restrict attention to the first order Reed-Muller codes $\mathcal{R}(m)$, which are binary linear $(2^m, 2^{m+1}, 2^{m-1})$-codes. As mentioned earlier, the code $\mathcal{R}(5)$ was used by Mariner 9 to transmit black and white photographs of Mars in 1972. There are many ways to define the Reed-Muller codes. We choose an inductive definition, since this definition makes it easy to establish the basic properties of the codes.

Definition   The **Reed-Muller codes** $\mathcal{R}(m)$ are binary codes defined, for all integers $m \geq 1$, as follows.

1. $\mathcal{R}(1) = \mathbb{Z}_2^2 = \{00, 01, 10, 11\}$

2. For $m \geq 1$,

$$\mathcal{R}(m + 1) = \{\mathbf{uu} \mid \mathbf{u} \in \mathcal{R}(m)\} \cup \{\mathbf{uu}^c \mid \mathbf{u} \in \mathcal{R}(m)\}$$

In words, the codewords in $\mathcal{R}(m + 1)$ are formed by juxtaposing each codeword in $\mathcal{R}(m)$ with itself and with its complement.   □

To demonstrate the virtues of an inductive definition, note that $\mathcal{R}(1)$ is a linear $(2^1, 2^2, 2^0)$-code in which every word except $\mathbf{0}$ and $\mathbf{1}$ has weight 2^0. We can easily extend this statement to the other Reed-Muller codes by induction.

Theorem 6.2.1   *For $m \geq 1$, the Reed-Muller code $\mathcal{R}(m)$ is a linear $(2^m, 2^{m+1}, 2^{m-1})$-code for which every codeword except $\mathbf{0}$ and $\mathbf{1}$ has weight 2^{m-1}.*   □

Proof   The result is true for $\mathcal{R}(1)$. Assume it is true for $\mathcal{R}(m)$ and consider the code $\mathcal{R}(m + 1)$. We leave it as an exercise to show that if $\mathcal{R}(m)$ is linear then so is $\mathcal{R}(m + 1)$. It is evident from the definition that the length of $\mathcal{R}(m + 1)$ is twice that of $\mathcal{R}(m)$ and so

$$\text{len}(\mathcal{R}(m + 1)) = 2 \cdot 2^m = 2^{m+1}$$

Moverover, since each of the sets $\{\mathbf{uu} \mid \mathbf{u} \in \mathcal{R}(m)\}$ and $\{\mathbf{uu}^c \mid \mathbf{u} \in \mathcal{R}(m)\}$ has size $|\mathcal{R}(m)|$ and since these sets are disjoint, it follows that

$$|\mathcal{R}(m+1)| = 2 \cdot |\mathcal{R}(m)| = 2 \cdot 2^{m+1} = 2^{m+2} .$$

Finally, assuming that all codewords in $\mathcal{R}(m)$ except $\mathbf{0}$ and $\mathbf{1}$ have weight 2^{m-1}, consider a codeword $\mathbf{c} \in \mathcal{R}(m+1)$. If $\mathbf{c} = \mathbf{uu}$ is different from $\mathbf{0}$ or $\mathbf{1}$, then $\mathbf{u} \neq \mathbf{0}$ or $\mathbf{1}$ and so

$$w(\mathbf{c}) = 2w(\mathbf{u}) = 2 \cdot 2^{m-1} = 2^m$$

If $\mathbf{c} = \mathbf{uu}^c$ then we must consider some cases. If $\mathbf{u} = \mathbf{0}$ then $\mathbf{c} = \mathbf{01}$, which has weight 2^m. If $\mathbf{u} = \mathbf{1}$ then $\mathbf{c} = \mathbf{10}$, which also has weight 2^m. Finally, if $\mathbf{u} \neq \mathbf{0}$ or $\mathbf{1}$ then, since $w(\mathbf{u}^c) = 2^m - 2^{m-1} = 2^{m-1} = w(\mathbf{u})$, we again have

$$w(\mathbf{c}) = 2w(\mathbf{u}) = 2 \cdot 2^{m-1} = 2^m$$

In all cases, $w(\mathbf{c}) = 2^m$. Thus, $\mathcal{R}(m+1)$ is a linear $(2^{m+1}, 2^{m+2}, 2^m)$-code, and the proof is complete.   ∎

Example 6.2.1 According to the definition, $\mathcal{R}(2) = \{0000, 0011, 0101,$ $0110, 1010, 1001, 1111, 1100\}$. Note that

$$R_2 = \begin{bmatrix} 0 & 0 & 1 & 1 \\ 0 & 1 & 0 & 1 \\ 1 & 1 & 1 & 1 \end{bmatrix}$$

is a generator matrix for $\mathcal{R}(2)$.   □

The inductive definition of $\mathcal{R}(m)$ also allows us to define generator matrices for these codes.

Theorem 6.2.2

1. *A generator matrix for $\mathcal{R}(1)$ is*

$$R_1 = \begin{bmatrix} 0 & 1 \\ 1 & 1 \end{bmatrix}$$

2. *If R_m is a generator matrix for $\mathcal{R}(m)$, then a generator matrix for $\mathcal{R}(m+1)$ is*

$$R_{m+1} = \left[\begin{array}{ccc|ccc} 0 & \cdots & 0 & 1 & \cdots & 1 \\ \hline & R_m & & & R_m & \end{array} \right] \tag{6.2.1}$$

☐

Proof   The first part is clear. Note that, if the rows of R_m are linearly independent, then so are the rows of R_{m+1}. For if any sum

$$\mathbf{s} = \mathbf{r}_{i_1} + \mathbf{r}_{i_2} + \cdots + \mathbf{r}_{i_u}$$

of the rows of R_{m+1} is $\mathbf{0}$, then this sum clearly cannot consist of just the first row. Hence, the sum of the left halves of the rows is a nontrivial sum (i.e., not all $\mathbf{0}$s) of rows of R_m that is equal to $\mathbf{0}$, which is not possible since the rows of R_m are assumed to be linearly independent. Thus, by induction, we conclude that each of the matrices R_{m+1} is a generator matrix (for some code).

Let C be the code whose generator matrix is the matrix on the right side of (6.2.1). We want to show that $C = \mathcal{R}(m + 1)$. If $\mathbf{r}$ is a row in R_m, then $\mathbf{rr}$ and $\mathbf{rr} + \mathbf{01} = \mathbf{rr}^c$ are both in C. It follows that if $\mathbf{u}$ is any sum of rows in R_m, that is, any codeword in $\mathcal{R}(m)$, then $\mathbf{uu}$ and $\mathbf{uu}^c$ are both in C. Hence, $\mathcal{R}(m + 1) \subseteq C$. But C has the same dimension as $\mathcal{R}(m + 1)$ and so $\mathcal{R}(m + 1) = C$, as desired.   ■

Example 6.2.2   Using the matrix R_2 from Example 6.2.1 (which also comes from R_1 using Theorem 6.2.2), we have

$$R_3 = \begin{bmatrix} 0 & 0 & 0 & 0 & 1 & 1 & 1 & 1 \\ 0 & 0 & 1 & 1 & 0 & 0 & 1 & 1 \\ 0 & 1 & 0 & 1 & 0 & 1 & 0 & 1 \\ 1 & 1 & 1 & 1 & 1 & 1 & 1 & 1 \end{bmatrix}$$

☐

Theorem 6.2.2 can be used to describe the generator matrices R_m directly, both in terms of their rows and their columns. We leave proof of the next theorem for the exercises.

Theorem 6.2.3

1. *The rows of R_m can be described as follows. The first row of R_m consists of a block of 2^{m-1} 0s followed by a block of 2^{m-1} 1s*

$$\overbrace{0 \cdots 0}^{2^{m-1}0s}\ \overbrace{1 \cdots 1}^{2^{m-1}1s}$$

The next row of R_m consists of alternating blocks of 0s and 1s of length 2^{m-2}

$$\overbrace{0 \cdots 0}^{2^{m-2}0s} \quad \overbrace{1 \cdots 1}^{2^{m-2}1s} \quad \overbrace{0 \cdots 0}^{2^{m-2}0s} \quad \overbrace{1 \cdots 1}^{2^{m-2}1s}$$

In general, the ith row of R_m consists of alternating blocks of 0s and 1s of length 2^{m-i}. The mth row of R_m thus consists of alternating 0s and 1s (blocks of length $2^{m-m} = 1$). The last row of R_m is a row of all 1s.

2. *The columns of R_m can be described as follows. Excluding the last row of R_m, the columns of R_m consist of all possible binary strings of length m which, when read from the top down as binary numbers, are $0, 1, \ldots, 2^m - 1$, in this order.*   □

The following table gives the parameters for $\mathcal{R}(m)$, for some small values of m. Note the large minimum distance, but small size, of these codes.

m	codeword length (2^m)	code size (2^{m+1})	minimum distance (2^{m-1})
2	4	8	2
3	8	16	4
4	16	32	8
5	32	64	16
6	64	128	32
7	128	256	64
8	256	512	128

It is interesting to compare the characteristics of the Reed-Muller codes with those of the Hamming codes. For approximately the same codeword length (2^m for Reed-Muller, $2^m - 1$ for Hamming), the code size of Reed-Muller (2^{m+1}) is significantly smaller than that of Hamming (2^{2^m-1-m}). With the Hamming codes, we pay for the large code size with a minimum distance of only 3 (and thus only single-error correction). For the Reed-Muller codes, the relatively large minimum distance grows along with the code size.

Decoding with Reed-Muller Codes

Since $\mathcal{R}(m)$ is a $(2^m, 2^{m+1}, 2^{m-1})$-code, it is capable of correcting

$$\left\lfloor \frac{2^{m-1} - 1}{2} \right\rfloor = 2^{m-2} - 1$$

errors. However, a standard array for $\mathcal{R}(m)$ has

$$\frac{2^{2^m}}{2^{m+1}} = 2^{2^m - m - 1}$$

rows. For instance, to correct seven errors, we need $m = 5$, in which case there are 67,108,864 cosets in a standard array! Thus, decoding using a syndrome table is time consuming, even for small values of m (such as the one used by Mariner 9). Fortunately, there are better ways to decode when using a Reed-Muller code.

We will describe a special type of *majority logic decoding*, called **Reed decoding**, that applies to Reed-Muller codes. The idea behind majority logic decoding is quite simple. Consider a linear code C, with basis $\mathcal{B} = \{\mathbf{b}_1, \mathbf{b}_2, \ldots, \mathbf{b}_k\}$. Suppose a codeword $\mathbf{c} = c_1 \cdots c_n$ is sent. Since $\mathcal{B}$ is a basis, there exist scalars $\alpha_i \in \mathbb{Z}_2$ for which

$$\mathbf{c} = \alpha_1 \mathbf{b}_1 + \alpha_2 \mathbf{b}_2 + \cdots + \alpha_k \mathbf{b}_k$$

Suppose we find a way to compute the coefficient α_1 directly from the coordinates c_i in, say, four different ways, and that each way uses *different* coordinates. For instance, suppose that (for $n = 8$)

$$\alpha_1 = c_1 + c_5$$
$$\alpha_1 = c_2 + c_6$$
$$\alpha_1 = c_3 + c_7$$
$$\alpha_1 = c_4 + c_8$$

Now imagine that the received word $\mathbf{x} = x_1 \cdots x_n$ has a single error. We attempt to compute α_1 four times, using the coordinates of $\mathbf{x}$, as follows

$$\alpha_1 = x_1 + x_5$$
$$\alpha_1 = x_2 + x_6$$
$$\alpha_1 = x_3 + x_7$$
$$\alpha_1 = x_4 + x_8$$

The point is that, if only one error has occurred, only one of the x_is will not equal the corresponding c_i and so only one of these four equations will not give the correct value of α_1. Put another way, the majority (in this case 3) of the equations will give the correct value of α_1. Doing this for each of the coefficients α_i enables us to recover the codeword $\mathbf{c}$. This is majority logic decoding.

Although we will discuss the general case R_m here, it might help to follow the procedure to keep an eye on the generator matrix

$$R_3 = \begin{bmatrix} 0 & 0 & 0 & 0 & 1 & 1 & 1 & 1 \\ 0 & 0 & 1 & 1 & 0 & 0 & 1 & 1 \\ 0 & 1 & 0 & 1 & 0 & 1 & 0 & 1 \\ 1 & 1 & 1 & 1 & 1 & 1 & 1 & 1 \end{bmatrix}$$

as we proceed.

Suppose a codeword $\mathbf{c} = c_1 \cdots c_n$ is sent. If the rows of R_m are denoted by $\mathbf{r}_1, \ldots, \mathbf{r}_{m+1}$, then

$$\mathbf{c} = \alpha_1 \mathbf{r}_1 + \cdots + \alpha_m \mathbf{r}_m + \alpha_{m+1} \mathbf{r}_{m+1}$$

for some scalars $\alpha_i \in \mathbb{Z}_2$. For reasons described earlier, we would like to find as many expressions for the coefficients α_i as possible. Suppose that $\mathbf{x}_i$ is a string that is orthogonal to every row of R_m except the ith row, and $\mathbf{x}_i \cdot \mathbf{r}_i = 1$. Then

$$\begin{aligned} \mathbf{x}_i \cdot \mathbf{c} &= \mathbf{x}_i \cdot (\alpha_1 \mathbf{r}_1 + \cdots + \alpha_{m+1} \mathbf{r}_{m+1}) \\ &= \alpha_1 (\mathbf{x}_i \cdot \mathbf{r}_1) + \cdots + \alpha_{m+1} (\mathbf{x}_i \cdot \mathbf{r}_{m+1}) \\ &= \alpha_i \end{aligned}$$

To find such a string $\mathbf{x}_i$, suppose that $\mathbf{r}_i = r_{i1} \cdots r_{in}$. Then

$$\mathbf{e}_u \cdot \mathbf{r}_i = (0 \cdots 010 \cdots 0) \cdot (r_{i1} \cdots r_{in}) = r_{iu}$$

and so,

$$(\mathbf{e}_u + \mathbf{e}_v) \cdot \mathbf{r}_i = (r_{iu} + r_{iv})(\bmod 2)$$

If $\mathbf{e}_u + \mathbf{e}_v$ is to be our candidate for $\mathbf{x}_i$, then we must have

$$(\mathbf{e}_u + \mathbf{e}_v) \cdot \mathbf{r}_k = (r_{ku} + r_{kv})(\bmod 2) = \begin{array}{ll} 1 & \text{if } k = i \\ 0 & \text{if } k \neq i \end{array}$$

Of course, the sum of two bits is equal to 1 if and only if the bits differ. Hence, this will happen if and only if the uth column and the vth column of R_m differ in the ith row and agree in all other rows.

For instance, in R_3, if $i = 1$ then consider the first and fifth columns. The entries in row 1 are different and so

$$(\mathbf{e}_1 + \mathbf{e}_5) \cdot \mathbf{r}_1 = \mathbf{e}_1 \cdot \mathbf{r}_1 + \mathbf{e}_5 \cdot \mathbf{r}_1 = 0 + 1 = 1$$

but the entries in any other row $\mathbf{r}_k (k \neq 1)$ are the same and so

$$(\mathbf{e}_1 + \mathbf{e}_5) \cdot \mathbf{r}_k = \mathbf{e}_1 \cdot \mathbf{r}_k + \mathbf{e}_5 \cdot \mathbf{r}_k = 0$$

Thus, for each row i, we want a pair of columns that are identical except in their ith row. We refer to such a pair of columns as a **good pair** for the ith row. Let us summarize.

Lemma 6.2.4　*If (u, v) are the column numbers of a good pair for the ith row of R_m, then for any codeword*

$$\mathbf{c} = \alpha_1 \mathbf{r}_1 + \cdots + \alpha_m \mathbf{r}_m + \alpha_{m+1} \mathbf{r}_{m+1}$$

we have (for $i = 1, \ldots, m$),

$$\alpha_i = (\mathbf{e}_u + \mathbf{e}_v) \cdot \mathbf{c} \qquad \qquad \square$$

Since the last row of R_m consists of all 1s, all pairs of entries in this row are equal and so the last row has no good pairs! (Do not dispair, we will deal with the last row later.) On the other hand, the last row will never give any trouble in finding good pairs for the other rows and so, for now, we can simply ignore the last row.

In fact, let R'_m be the matrix obtained from R_m by removing the last row. A good pair of columns for the ith row has the form

$$
\begin{bmatrix} \beta_1 \\ \vdots \\ \beta_{i-1} \\ 0 \\ \beta_{i+1} \\ \vdots \\ \beta_m \end{bmatrix}
\text{ and }
\begin{bmatrix} \beta_1 \\ \vdots \\ \beta_{i-1} \\ 1 \\ \beta_{i+1} \\ \vdots \\ \beta_m \end{bmatrix}
\qquad (6.2.2)
$$

We refer to the column on the left as the **0-half** of the good pair and the column on the right as the **1-half**. According to Theorem 6.2.3, the columns of R'_m consist of the binary representations of the numbers $0, 1, \ldots, 2^m - 1$, in this order. Hence, for any value of the β_is, there are two columns in R'_m of the form (6.2.2). Moreover, since the 1-half is a bigger number than the 0-half, the 1-half lies to the right of the 0-half in the matrix.

To determine how far to the right, observe that, as binary numbers,

$$\beta_1 \cdots \beta_{i-1} 1 \beta_{i+1} \cdots \beta_m = \beta_1 \cdots \beta_{i-1} 0 \beta_{i+1} \cdots \beta_m + 0 \cdots 010 \cdots 0$$
$$= \beta_1 \cdots \beta_{i-1} 0 \beta_{i+1} \cdots \beta_m + 2^{m-i}$$

and so the two halves have distance 2^{m-i} apart.

Thus, we can get all good pairs for the ith row as follows. Take all columns of R_m that have a 0 in the ith row as 0-halves. If a 0-half is in column j, then the corresponding 1-half is in column $j + 2^{m-i}$.

Testing this on the matrix R_3, we get the following good pairs:

For row 1, we have $2^{m-i} = 2^{3-1} = 4$ and so the good pairs are

(Col 1, Col 5), (Col 2, Col 6), (Col 3, Col 7), (Col 4, Col 8)

For row 2, we have $2^{m-i} = 2^{3-2} = 2$ and so the good pairs are

(Col 1, Col 3), (Col 2, Col 4), (Col 5, Col 7), (Col 6, Col 8)

For row 3, we have $2^{m-i} = 2^{3-3} = 1$ and so the good pairs are

(Col 1, Col 2), (Col 3, Col 4), (Col 5, Col 6), (Col 7, Col 8)

Thus, exactly half of the columns of R_m are 0-halves for the ith row and the other half of the columns are the corresponding 1-halves. In particular, there are exactly 2^{m-1} good pairs for each row.

Now imagine that a codeword

$$\mathbf{c} = \alpha_1 \mathbf{r}_1 + \cdots + \alpha_m \mathbf{r}_m + \alpha_{m+1} \mathbf{r}_{m+1}$$

is sent. Using the 2^{m-1} good pairs for row i, we get 2^{m-1} expressions for α_i (for $i \le m$). Specifically, if $\mathbf{c} = c_1 \cdots c_n$ and if (Col u, Col v) is a good pair for row i, then

$$\alpha_i = (\mathbf{e}_u + \mathbf{e}_v) \cdot \mathbf{c} = c_u + c_v$$

If (Col s, Col t) is another good pair for row i, then

$$\alpha_i = (\mathbf{e}_s + \mathbf{e}_t) \cdot \mathbf{c} = c_s + c_t$$

where s, t, u, and v are *distinct*, and so each of these 2^{m-1} expressions for α_i involves *different* positions in the codeword $\mathbf{c}$.

Thus, if no more than $2^{m-2} - 1$ errors occur, then at most $2^{m-2} - 1$ of the coordinates c_j are incorrect, and so at most $2^{m-2} - 1$ of the expressions for α_i are incorrect. This means that at least

$$2^{m-1} - (2^{m-2} - 1) = 2^{m-2} + 1$$

of these expressions give the correct value of α_i. In particular, a majority of the expressions are correct. It follows that we can get the correct value of α_i by computing the 2^{m-1} expressions for α_i and taking the majority value.

The final step is to obtain the coefficient α_{m+1}. If at most $2^{m-2} - 1$ errors have occurred in receiving $\mathbf{x}$, then the error string $\mathbf{e} = \mathbf{x} - \mathbf{c}$ has weight at most $2^{m-2} - 1$. Letting $\mathbf{d} = \alpha_1 \mathbf{r}_1 + \cdots + \alpha_m \mathbf{r}_m$, we have

$$\mathbf{x} - \mathbf{d} = \alpha_{m+1} \mathbf{r}_{m+1} + \mathbf{e} = \alpha_{m+1} \mathbf{1} + \mathbf{e}$$

There are two possibilities. If $\alpha_{m+1} = 0$ then $\mathbf{e} = \mathbf{x} - \mathbf{d}$ and if $\alpha_{m+1} = 1$ then $\mathbf{e} = (\mathbf{x} - \mathbf{d})^c$. Thus, if $w(\mathbf{x} - \mathbf{d}) \le 2^{m-2} - 1$, we decode α_{m+1} as 0 and if $w((\mathbf{x} - \mathbf{d})^c) \le 2^{m-2} - 1$, then we decode α_{m+1} as 1. (If neither of these cases occur, then more than $2^{m-2} - 1$ errors have occurred and we can just admit a decoding error.)

Thus, Reed decoding can correct up to $2^{m-2} - 1$ errors, as promised by the minimum distance of the code $\mathcal{R}(m)$. The decoded codeword is referred to as the *majority logic codeword* for the received word $\mathbf{x}$. Let us consider an example.

Example 6.2.2 Suppose that (unbeknownst to us) the codeword 11001100 from the $(8,16,4)$-Reed-Muller code $\mathcal{R}(3)$ is sent, but the received word is $\mathbf{x} = x_1 \cdots x_8 = 11011100$. (One error has occurred and so Reed decoding should correct it.) The good pairs for each row of R_3 were computed earlier:

Row 1: (Col 1, Col 5), (Col 2, Col 6), (Col 3, Col 7), (Col 4, Col 8)

Row 2: (Col 1, Col 3), (Col 2, Col 4), (Col 5, Col 7), (Col 6, Col 8)

Row 3: (Col 1, Col 2), (Col 3, Col 4), (Col 5, Col 6), (Col 7, Col 8)

Thus, if

$$\mathbf{c} = c_1 \cdots c_8 = \alpha_1 \mathbf{r}_1 + \alpha_2 \mathbf{r}_2 + \alpha_3 \mathbf{r}_3 + \alpha_4 \mathbf{r}_4$$

is the majority logic codeword, the expressions for α_1 are

$$\alpha_1 = c_1 + c_5$$

$$\alpha_1 = c_2 + c_6$$
$$\alpha_1 = c_3 + c_7$$
$$\alpha_1 = c_4 + c_8$$

The **majority logic equations** for α_i are found by replacing c_i (whose values we do not know) with x_i (whose values we do know) to get

$$\alpha_1 = x_1 + x_5 = 0$$
$$\alpha_1 = x_2 + x_6 = 0$$
$$\alpha_1 = x_3 + x_7 = 0$$
$$\alpha_1 = x_4 + x_8 = 1$$

Thus, the majority logic decision is $\alpha_1 = 0$. The majority logic equations for α_2 are

$$\alpha_2 = x_1 + x_3 = 1$$
$$\alpha_2 = x_2 + x_4 = 0$$
$$\alpha_2 = x_5 + x_7 = 1$$
$$\alpha_2 = x_6 + x_8 = 1$$

and so the majority logic decision is $\alpha_2 = 1$. The majority logic equations for α_3 are

$$\alpha_3 = x_1 + x_2 = 0$$
$$\alpha_3 = x_3 + x_4 = 1$$
$$\alpha_3 = x_5 + x_6 = 0$$
$$\alpha_3 = x_7 + x_8 = 0$$

and so the majority logic decision is $\alpha_3 = 0$. Thus,

$$\mathbf{x} - (\alpha_1 \mathbf{r}_1 + \alpha_2 \mathbf{r}_2 + \alpha_3 \mathbf{r}_3) = 11011100 - \mathbf{r}_2 = 11101111$$

Since the complement of this string has weight $1 \leq 2^{3-2} - 1$, we decode α_4 as 1. It follows that the majority logic codeword is

$$\mathbf{c} = \mathbf{r}_2 + \mathbf{1} = 11001100$$

which is indeed the codeword sent. $\square$

Exercises

1. Show using the definition that if $\mathcal{R}(m)$ is linear, then so is $\mathcal{R}(m + 1)$.

2. Show that $\mathcal{R}(m + 1) = \mathcal{R}(m) \oplus \mathrm{Rep}(2^m)$, where $\oplus$ is the $\mathbf{u}(\mathbf{u} + \mathbf{v})$-construction.

3. Show that the ith row of R_m (for $i \leq m$) consists of alternating blocks of 0s and 1s of length 2^{m-i} and that the last row of R_m is $\mathbf{1}$.

4. Show that, if we ignore the last row of R_m, the columns of R_m consist of all possible binary strings of length m which, when thought of as binary numbers, represent the numbers $0, 1, \ldots, 2^m - 1$ in increasing order.

5. Find a parity check matrix for $\mathcal{R}(2)$.

6. Show that the binary string $0 \cdots 010 \cdots 0$ of length m, with a 1 in the ith position *from the left*, is equal to 2^{m-i} when thought of as a binary number.

7. Find a parity check matrix for $\mathcal{R}(3)$.

8. Assuming the Reed-Muller code $\mathcal{R}(3)$ is used, decode the received word 01111100.

9. Assuming the Reed-Muller code $\mathcal{R}(3)$ is used, decode the received word 11000001.

10. Assuming the Reed-Muller code $\mathcal{R}(3)$ is used, decode the received word 01101001.

11. What do you expect to get from the majority logic equations if you apply Reed decoding to a codeword?

12. Find all good pairs for the rows of R_4. Decode the received word 0111 0110 1110 0010 (the spaces are for readability).

13. The higher order Reed-Muller codes can be defined as follows.

    (a) The **zero order Reed-Muller codes** are defined, for $m \geq 0$, by $\mathcal{R}_0(m) = \mathrm{Rep}_2(2^m)$.

    (b) The **first order Reed-Muller codes** are defined, as in the text, for $m \geq 1$, by $\mathcal{R}_1(m) = \mathcal{R}(m)$. (This is just to set the notation.)

    (c) For any $r \geq 2$, then **rth order Reed-Muller codes** are defined, for $m \geq r$, by

    $$\mathcal{R}_r(r) = \mathbb{Z}_2^{2^m}$$

$$\mathcal{R}_r(m + 1) = \mathcal{R}_r(m) \oplus \mathcal{R}_{r-1}(m)$$

where $\oplus$ is the $\mathbf{u}(\mathbf{u} + \mathbf{v})$-construction.

Show that $\mathcal{R}_r(m)$ has parameters

$$[2^m, 1 + \binom{m}{1} + \cdots + \binom{m}{r}, 2^{m-r}]$$

where the sum for the dimension is taken to be 1 if $r = 0$.

6.3   Some Decimal Codes

In this section, we discuss some interesting decimal codes, that is, codes over the alphabet $\mathbb{Z}_{10}$. Since 10 is not a prime number, the set $\mathbb{Z}_{10}$ is not a field. To work around this problem, we first define codes over $\mathbb{Z}_{11}$, and then "reduce" them to decimal codes.

The Single-Error-Detecting ISBN Code

The well-known ISBN code is used on nearly all recently published books. (As we will see, the ISBN code is not quite a decimal code.) Every book has a number associated with it, known as its *International Standard Book Number*, or ISBN. (This number generally appears on the back cover of the book.) An ISBN is a 10-digit number, such as

$$0-387-97812-7$$

The first digit of an ISBN indicates the language of the book, which in this case is 0, for English. The next group of digits (387) stands for the publisher, which in this case is Springer-Verlag. The next group of numbers (97812) is the book number, assigned by the publisher. The final digit is a redundant check digit, designed to detect errors. Since the hyphens do not have any relevance to error detection, we will ignore them and treat an ISBN as a string of length 10 over $\mathbb{Z}_{11}$.

Given the first 9 digits of an ISBN, say $\mathbf{x} = x_1 \cdots x_9$, the tenth digit x_{10}, called the **check digit**, is the solution, in $\mathbb{Z}_{11}$, of the equation

$$x_1 + 2x_2 + 3x_3 + \cdots + 9x_9 + 10x_{10} = 0$$

In fact, since $-10 = 1$, solving for the check digit gives (in $\mathbb{Z}_{11}$)

$$x_{10} = x_1 + 2x_2 + 3x_3 + \cdots + 9x_9$$

For instance, in the ISBN given above, the check digit is

$$1 \cdot 0 + 2 \cdot 3 + 3 \cdot 8 + 4 \cdot 7 + 5 \cdot 9 + 6 \cdot 7 + 7 \cdot 8 + 8 \cdot 1 + 9 \cdot 2 = 7$$

If the check digit is equal to 10, the letter X is used in its place. Thus, for example, 0-201-02988-X is a valid ISBN. Note that all other digits in an ISBN are decimal digits (by definition), but because the last digit may be X, an ISBN is not really a decimal codeword.

In the language of coding theory, we construct the ISBN code as follows. Let C be the linear code in $\mathbb{Z}_{11}^{10}$ with parity check matrix

$$P = \begin{bmatrix} 1 & 2 & 3 & 4 & 5 & 6 & 7 & 8 & 9 & 10 \end{bmatrix}$$

This code is a [10, 9, 2]-code over $\mathbb{Z}_{11}$, of size 11^9. Let $\mathcal{I}$ be the code obtained from C by removing all codewords that have a "10" in any position, except possibly the last. The codewords in $\mathcal{I}$ are precisely the ISBNs and so we refer to $\mathcal{I}$ as the **ISBN code**. We leave it as an exercise to show that $\mathcal{I}$ is a nonlinear $(10, 10^9, 2)$-code.

We now show that the ISBN code $\mathcal{I}$ can detect any single error, as well as any transposition of two digits in a codeword. As to the former, suppose that an ISBN codeword $\mathbf{c}$ is sent, but that a string $\mathbf{x}$, containing a single error, is received. If the error is in the ith position, then the error string is $u\mathbf{e}_i$ where $u \neq 0$ and $\mathbf{x} = \mathbf{c} + u\mathbf{e}_i$. The syndrome of $\mathbf{x}$ is

$$S(\mathbf{x}) = S(u\mathbf{e}_i) = uS(\mathbf{e}_i) = u \cdot i$$

which is the product of two nonzero elements of the field $\mathbb{Z}_{11}$ and is therefore nonzero. Thus, a single error is detected by the presence of a nonzero syndrome. (It is interesting to note that, if the size of the error is known, then its position can be determined from the syndrome. Similarly, if the position is known, then the size can be determined.)

The fact that the ISBN code can detect transpositions can be seen as follows. Suppose that an ISBN $\mathbf{c} = c_1 c_2 \cdots c_{10}$ is sent, but that c_i and c_j (with $c_i \neq c_j$) get transposed during transmission. Then the received string is

$$\mathbf{x} = c_1 \cdots c_{i-1} c_j c_{i+1} \cdots c_{j-1} c_i c_{j+1} \cdots c_{10}$$

(We have assumed that $i < j$.) Hence, the error is

$$\mathbf{e} = 0 \cdots 0(c_j - c_i)0 \cdots (c_i - c_j)0 \cdots 0$$

and the syndrome of x is

$$S(\mathbf{x}) = \mathbf{e}P^t = i \cdot (c_j - c_i) + j \cdot (c_i - c_j) = (c_j - c_i)(i - j)$$

But since $i \neq j$ and $c_i \neq c_j$, again we get a nonzero syndrome in $\mathbb{Z}_{11}$. Thus, a transposition of digits is also detected by means of a nonzero syndrome.

A Single-Error-Correcting Decimal Code

Next, we discuss a single-error-correcting decimal code that has a particularly nice method for correcting errors—one that does not require a syndrome table.

When we think of nice decoding in a single-error-correcting code, we think of Hamming codes—so let us start with the Hamming code $\mathcal{H}_{11}(2)$, which is an 11-ary linear [12, 10, 3]-code with parity check matrix

$$H_{11} = \begin{bmatrix} 0 & 1 & 1 & 1 & 1 & 1 & 1 & 1 & 1 & 1 & 1 & 1 \\ 1 & 0 & 1 & 2 & 3 & 4 & 5 & 6 & 7 & 8 & 9 & 10 \end{bmatrix}$$

Now we shorten this code in the first and last coordinates, taking the double cross-section $x_1 = 0$, $x_{11} = 0$, which has parity check matrix

$$P = \begin{bmatrix} 1 & 1 & 1 & 1 & 1 & 1 & 1 & 1 & 1 & 1 \\ 0 & 1 & 2 & 3 & 4 & 5 & 6 & 7 & 8 & 9 \end{bmatrix}$$

and is a linear [10, 8, 3]-code. Let D be the decimal code obtained from this code by removing all codewords that have a 10 in any position. We leave it as an exercise to show that D is a nonlinear code with minimum distance 3 over $\mathbb{Z}_{10}$. It is possible to show, using a formula known as the Principle of Inclusion-Exclusion that D has size 82,644,629. (See the exercises.)

Let us see how we might correct single errors in D. Suppose that a codeword $\mathbf{c}$ is sent, but that an error occurs of the form $u\mathbf{e}_i$, where $u \neq 0$. Denoting the syndrome (taken modulo 11) of the received word $\mathbf{x} = \mathbf{c} + u\mathbf{e}_i$ by $S(\mathbf{x}) = [s\ t]$, we have

$$\begin{bmatrix} s & t \end{bmatrix} = S(\mathbf{x}) = u\mathbf{e}_i P^t = \begin{bmatrix} u & u(i-1) \end{bmatrix}$$

Thus, assuming a single error, we see immediately that the magnitude u of the error is the first coordinate s. To determine the position of the error, we solve the equation $t = u(i - 1)$ for i (in $\mathbb{Z}_{11}$), which gives $i = u^{-1}t + 1 = s^{-1}t + 1$. Thus, the error string in this case is $s\mathbf{e}_{s^{-1}t+1}$.

Example 6.3.1 Suppose that $\mathbf{x} = 1274235110$ is received. The syndrome of $\mathbf{x}$ is $S(\mathbf{x}) = 48$ and so $s = 4$, $t = 8$ and $s^{-1}t + 1 = 3$. Hence, the error string is $4\mathbf{e}_3$ and the codeword is $\mathbf{x} - 4\mathbf{e}_3 = 1234235110$.   □

While the decimal code D is single-error-correcting and double-error-detecting, it is not *simultaneously* single-error-correcting and double-error-detecting. However, it is capable of simultaneously correcting single errors and detecting transposition errors.

To see this, suppose that a codeword $\mathbf{c} = c_1c_2\cdots c_{10}$ is sent, but that c_i and c_j (with $c_i \neq c_j$) get transposed during transmission. Then the received string is

$$\mathbf{x} = c_1 \cdots c_{i-1}c_jc_{i+1} \cdots c_{j-1}c_ic_{j+1} \cdots c_{10}$$

(We have assumed that $i < j$.) Hence, the error is

$$\mathbf{e} = 0\cdots 0(c_j - c_i)0 \cdots (c_i - c_j)0 \cdots 0$$

and the syndrome of $\mathbf{x}$ is

$$\begin{bmatrix} s & t \end{bmatrix} = \mathbf{e}P^t = \begin{bmatrix} 0 & (c_j - c_i)(i - j) \end{bmatrix}$$

Note that the first component is 0 and, since $i \neq j$ and $c_i \neq c_j$, the second component is nonzero.

Thus, to decode with D, we first compute the syndrome $S(\mathbf{x}) = [s\,t]$, taken modulo 11, and then proceed as follows.

1. If $s = t = 0$, assume that no error has occurred.

2. If $s \neq 0$ and $t \neq 0$, assume that the error string is $s\mathbf{e}_{s^{-1}t+1}$.

3. If $s = 0$ and $t \neq 0$, assume that a transposition error has occurred.

This procedure guarantees that all single errors are corrected *and* all transposition errors are detected.

A Double-Error-Correcting Decimal Code

We now turn our attention to constructing a double-error-correcting decimal code. Naturally, as the error-correcting capabilities of the codes increase, so does their complexity. We begin with a definition.

Definition For $m \geq 2$, let $a_1, a_2, \ldots, a_m$ be elements of a field. The matrix

$$V(a_1, a_2, \ldots, a_m) = \begin{bmatrix} 1 & 1 & \cdots & 1 \\ a_1 & a_2 & \cdots & a_m \\ a_1^2 & a_2^2 & \cdots & a_m^2 \\ \vdots & \vdots & \vdots & \vdots \\ a_1^{m-1} & a_2^{m-1} & \cdots & a_m^{m-1} \end{bmatrix}$$

whose ith column consists of the successive powers of the element a_i, is called a **Vandermonde matrix**. □

Vandermonde matrices play an important role in many areas of mathematics. We are primarily interested in these matrices because, when the elements a_i are distinct, the columns (and rows) of the matrix are linearly independent.

The easiest way to prove this is to use a standard result from linear algebra involving determinants, which we state without proof. (If you are not familiar with determinants, just read the statements of the next two theorems.

Theorem 6.3.1 *If A is an $m \times m$ matrix with nonzero determinant, then the columns (and rows) of A are linearly independent.* □

Theorem 6.3.2 *The determinant of the Vandermonde matrix $V(a_1, a_2, \ldots, a_m)$ is*

$$D(a_1, a_2, \ldots, a_m) = \prod_{i<j} (a_j - a_i)$$

Hence, if the a_is are distinct, the determinant is nonzero and the columns (and rows) of $V(a_1, a_2, \ldots, a_m)$ are linearly independent. □

Proof First note that, if two of the a_is agree, then two columns of the Vandermonde matrix are identical and the determinant is 0. So let us assume that the a_is are distinct. Consider the matrix formed by replacing the element a_m by a variable x,

$$V(a_1, a_2, \ldots, a_{m-1}, x) = \begin{bmatrix} 1 & 1 & \cdots & 1 \\ a_1 & a_2 & \cdots & x \\ a_1^2 & a_2^2 & \cdots & x^2 \\ \vdots & \vdots & \vdots & \vdots \\ a_1^{m-1} & a_2^{m-1} & \cdots & x^{m-1} \end{bmatrix}$$

Expanding the determinant of this matrix along the last column gives

$$D(a_1, a_2, \ldots, a_{m-1}, x) = A_0 + A_1 x + A_2 x^2 + \cdots + A_{m-1} x^{m-1}$$

where the coefficients A_i do not involve x. Hence, $D(a_1, a_2, \ldots, a_{m-1}, x)$ is a polynomial in x of degree at most $m - 1$. Notice also that the coefficient A_{m-1} is obtained by striking out the last row and column of $V(a_1, a_2, \ldots, a_{m-1}, x)$ and taking the determinant of the remaining matrix. But this matrix is $V(a_1, a_2, \ldots, a_{m-1})$ and so

$$A_{m-1} = D(a_1, a_2, \ldots, a_{m-1})$$

Now, if we replace x by a_i for any $i \leq m - 1$, then the ith column and the mth column of $V(a_1, a_2, \ldots, a_{m-1}, x)$ will agree and so the determinant will be 0. In symbols,

$$D(a_1, a_2, \ldots, a_{m-1}, a_i) = 0$$

for all $i = 1, \ldots, m-1$. This shows that the polynomial $D(a_1, a_2, \ldots, a_{m-1}, x)$ has $m - 1$ distinct roots $a_1, \ldots, a_{m-1}$ and therefore must have degree exactly $m - 1$. It follows that $a_1, \ldots, a_{m-1}$ is a *complete* set of the roots of $D(a_1, a_2, \ldots, a_{m-1}, x)$ and so

$$D(a_1, a_2, \ldots, a_{m-1}, x) = A_{m-1}(x - a_1)(x - a_2) \cdots (x - a_{m-1})$$
$$= D(a_1, a_2, \ldots, a_{m-1})(x - a_1)(x - a_2) \cdots (x - a_{m-1})$$

Setting $x = a_m$ gives

$$D(a_1, a_2, \ldots, a_{m-1}, a_m) = D(a_1, a_2, \ldots, a_{m-1})(a_m - a_1)(a_m - a_2) \cdots (a_m - a_{m-1})$$

An inductive argument can now be used to complete the proof. (Details of this are left as an exercise.) ∎

Now consider the matrix

$$P = \begin{bmatrix} 1 & 1 & 1 & \cdots & 1 \\ 1 & 2 & 3 & \cdots & 10 \\ 1 & 2^2 & 3^2 & \cdots & 10^2 \\ 1 & 2^3 & 3^3 & \cdots & 10^3 \end{bmatrix}$$

Since any four columns of P form a Vandermonde matrix with distinct elements, any four columns of P are linearly independent. However, the first five columns are linearly dependent. Since the rows of P are linearly independent (exercise), it follows that P is the parity check matrix of a linear $[10, 6, 5]$-code over $\mathbb{Z}_{11}$. Let E be the code obtained by removing all codewords that have a 10 in any position. Then E is a nonlinear code of length 10 and minimum distance 5 over $\mathbb{Z}_{10}$ (exercise) and so it is double-error-correcting. We leave it as an exercise to show (again using the Principle of Inclusion-Exclusion) that E has size 683,024.

As to decoding with E, suppose a codeword $\mathbf{c}$ is sent and at most 2 errors occur. Thus, the error string has the form $u\mathbf{e}_i + v\mathbf{e}_j$, where $u, v \in \mathbb{Z}_{10}$, with $i < j$. The syndrome of the received word $\mathbf{x} = \mathbf{c} + u\mathbf{e}_i + v\mathbf{e}_j$ is

$$S(\mathbf{x}) = (u\mathbf{e}_i + v\mathbf{e}_j)P^t = \begin{bmatrix} u + v & ui + vj & ui^2 + vj^2 & ui^3 + vj^3 \end{bmatrix}$$

(All computations are made over $\mathbb{Z}_{11}$.) Letting $S(\mathbf{x}) = [s_1\ s_2\ s_3\ s_4]$, we get the system

$$u + v = s_1 \qquad\qquad (6.3.1)$$
$$ui + vj = s_2$$
$$ui^2 + vj^2 = s_3$$
$$ui^3 + vj^3 = s_4$$

In order to solve these equations, we multiply each of the first three equations by i and subtract the equation following it, to get

$$v(i - j) = is_1 - s_2$$
$$vj(i - j) = is_2 - s_3$$
$$vj^2(i - j) = is_3 - s_4$$

Notice that squaring the middle equation gives

$$v^2 j^2 (i - j)^2 = (is_2 - s_3)^2$$

and multiplying the first and third equations gives

$$v^2 j^2 (i-j)^2 = (is_1 - s_2)(is_3 - s_4)$$

Equating these two expressions for $v^2 j^2 (i-j)^2$, we get

$$(is_2 - s_3)^2 = (is_1 - s_2)(is_3 - s_4)$$

which is equivalent to

$$(s_2^2 - s_1 s_3)i^2 + (s_1 s_4 - s_2 s_3)i + s_3^2 - s_2 s_4 = 0 \qquad (6.3.2)$$

Going back to (6.3.1) and multiplying by j instead of i leads to the same equation. Hence, the positions i and j of the errors can be found by solving (6.3.2). Once we know the values of i and j, the values of u and v can be found by solving the first two equations in (6.3.1).

For convenience, let us write

$$A = s_2^2 - s_1 s_3, \quad B = s_1 s_4 - s_2 s_3, \quad C = s_3^2 - s_2 s_4$$

so that the error locations i and j are the solutions to the equation

$$Ax^2 + Bx + C = 0 \qquad (6.3.3)$$

The polynomial $Ax^2 + Bx + C$ is called the **error locator polynomial**.

Now let us consider the possibilities.

Case 1   If no errors occur, then the syndrome $S(\mathbf{x})$ will be zero.

Case 2   If exactly one error occurs, then the syndrome is

$$S(\mathbf{x}) = \begin{bmatrix} u & ui & ui^2 & ui^3 \end{bmatrix}$$

which is nonzero. Also, $A = 0$, $B = 0$ and $C = 0$ and so the error locator polynomial is the zero polynomial.

Case 3   If exactly two errors occur, then $u, v \neq 0$ and $i \neq j$. It follows that the error locator polynomial cannot be the zero polynomial. To see this, suppose that $A = 0$ and $C = 0$ (which implies that $B = 0$). A little manipulation shows that this is equivalent to

$$\frac{s_4}{s_3} = \frac{s_3}{s_2} = \frac{s_2}{s_1} = \lambda$$

for some constant λ. Hence, $s_4 = \lambda^3 s_1$, $s_3 = \lambda^2 s_1$ and $s_2 = \lambda s_1$ and equations (6.3.1) become

$$u + v = s_1$$
$$ui + vj = \lambda s_1$$
$$ui^2 + vj^2 = \lambda^2 s_1$$
$$ui^3 + vj^3 = \lambda^3 s_1$$

Multiplying the first of these equations by λ and equating with the second gives $u\lambda + v\lambda = ui + vj$, or

$$u(\lambda - i) = v(j - \lambda)$$

Similarly, multiplying the second equation by λ and equating with the third gives $ui\lambda + vj\lambda = ui^2 + vj^2$, or

$$ui(\lambda - i) = vj(j - \lambda)$$

Comparing these equations shows that $i = j$, a contradiction. Hence, one of A or C is not 0 and the error locator polynomial is nonzero.

We can now describe a double-error-correcting decoding procedure. Let the syndrome of the received word be $S(\mathbf{x}) = [s_1\ s_2\ s_3\ s_4]$.

Step 1   If $S(\mathbf{x}) = \mathbf{0}$, assume no errors have occurred.

Step 2   If $S(\mathbf{x}) \neq \mathbf{0}$ but the error locator polynomial is the zero polynomial, assume that one error has occurred, whose magnitude is s_1 and whose location is $s_1^{-1}s_2$. That is, the error string is $s_1\mathbf{e}_{s_1^{-1}s_2}$.

Step 3   If $S(\mathbf{x}) \neq \mathbf{0}$ and the error locator polynomial is nonzero, assume that two errors have occurred, whose locations are given by the roots of the error locator polynomial. After finding these roots, use the first two equations in (6.3.1) to find the magnitudes of the errors. If the error locator polynomial has no roots in $\mathbb{Z}_{11}$, then more than two errors must have occurred.

Before looking at an example, we emphasize that the error locator equation must be solved in the field $\mathbb{Z}_{11}$, not in the field of real numbers! Fortunately, since $\mathbb{Z}_{11}$ has size 11, the roots can be found simply by trying all 11 possibilities. Alternatively, since the quadratic formula holds in any field *except* $\mathbb{Z}_2$, we have

$$i, j = \frac{-B \pm \sqrt{B^2 - 4AC}}{2A}$$

To use this formula, a table of square roots in $\mathbb{Z}_{11}$, created by squaring each element of $\mathbb{Z}_{11}$, is very handy. (For example, since $5^2 = 3$, we have $\sqrt{3} = 5$. We leave computation of such a table to the reader.

Example 6.3.2 Suppose $\mathbf{x} = 3235556411$ is received. The syndrome is $S(\mathbf{x}) = [2\ 8\ 10\ 7]$. Since $A = 0$, $B = 0$ and $C = 0$, we assume that there is one error, with magnitude 2 and location $2^{-1} \cdot 8 = 6 \cdot 8 = 4$. Hence, the error string is $2\mathbf{e}_4$ and the decoded word in $\mathbf{c} = \mathbf{x} - 2\mathbf{e}_4 = 3233556411$.

Now suppose $\mathbf{x} = 4739688119$ is received. The syndrome in this case is $S(\mathbf{x}) = [1\ 7\ 10\ 10]$ and so $A = 6$, $B = 6$, $C = 8$. Hence, the error locator equation is

$$6x^2 + 6x + 8 = 0$$

whose solutions are the error locations

$$i, j = \frac{-B \pm \sqrt{B^2 - 4AC}}{2A} = \frac{5 \pm \sqrt{9}}{1} = 5 \pm 3 = 2, 8$$

The magnitudes of the errors are given by the solutions to the equations

$$u + v = 1$$
$$2u + 8v = 7$$

which are $u = 2$, $v = 10$. Thus, the error string is $2\mathbf{e}_2 + 10\mathbf{e}_8$ and the decoded codeword is $\mathbf{c} = \mathbf{x} - (2\mathbf{e}_2 + 10\mathbf{e}_8) = 4539688219$.

Finally, suppose $\mathbf{x} = 1111037407$ is received. The syndrome is $S(\mathbf{x}) = [5\ 10\ 4\ 4]$ and so $A = 3$, $B = 2$, $C = 9$. Hence, the error locator equation is

$$3x^2 + 2x + 9 = 0$$

whose solutions are the error locations

$$i, j = \frac{-B \pm \sqrt{B^2 - 4AC}}{2A} = \frac{9 \pm \sqrt{6}}{6}$$

However, the number 6 does not have a square root in $\mathbb{Z}_{11}$, as can be seen by squaring all elements of $\mathbb{Z}_{11}$. Hence, more than two errors have occurred. $\square$

It is probably worth mentioning that, even though this decoding procedure may *sometimes* tell us that more than two errors have occurred, it is not simultaneously double-error-correcting and triple-error-detecting. For it is possible that three errors errors may occur but the resulting received word has distance 2 from the wrong codeword.

Generalizations

The last two codes presented here can be generalized. Actually, these codes are very special cases of codes known as **BCH codes**, named after their discoverers R.C. Bose, D.K. Ray-Chaudhuri, and A. Hocquenghem. The BCH codes form an extremely important class of codes for several reasons. For instance, they have good error-correcting properties when the length is not too great, they can be encoded and decoded relatively easily and they provide a good basis upon which to build other families of codes. In any case, to define the BCH codes requires fairly sophisticated algebraic tools and so we will not do so here. However, we do want to take at least one step in that direction.

Let $d \leq n \leq p - 1$, where p is a prime. We want to construct a linear code B of length n and minimum distance d over $\mathbb{Z}_p$. To do so, consider the matrix

$$P = \begin{bmatrix} 1 & 1 & 1 & \cdots & 1 \\ 1 & 2 & 3 & \cdots & n \\ 1 & 2^2 & 3^2 & \cdots & n^2 \\ \vdots & \vdots & \vdots & \cdots & \vdots \\ 1 & 2^{d-2} & 3^{d-2} & \cdots & n^{d-2} \end{bmatrix}$$

where all elements are in $\mathbb{Z}_p$. Notice that, for $p = 11$, $n = 10$ and $d = 5$, we get the parity check matrix used to define the double-error-correcting code E. Let $\mathcal{B}_p(n, d)$ be the code over $\mathbb{Z}_p$ with parity check matrix P. Thus, the code E is the set of all codewords in $\mathcal{B}_{11}(10, 5)$ that have coordinates in $\mathbb{Z}_{10}$.

Since any $d - 1$ columns of P form a Vandermonde matrix with *distinct* elements (this is why we need $n \leq p - 1$), we conclude that any $d - 1$ columns of P are linearly independent. On the other hand, since P has only $d - 1$ rows, we may appeal to a result of linear algebra that says that the maximum number of linearly independent columns is the same as the maximum number of linearly independent rows to conclude that no d columns of P are linearly independent. Hence, $B_p(n, d)$ is a linear $[n, n-d+1, d]$-code over $\mathbb{Z}_p$. Since the parameters of the code $B_p(n, d)$ give equality in Singleton's bound for $A_p(n, d)$ we get the following theorem.

Theorem 6.3.3 *If p is a prime and $d \leq n \leq p - 1$, then the code $B_p(n, d)$ defined above is an optimal $[n, n - d + 1, d]$-code over $\mathbb{Z}_p$ and so*

$$A_p(n, d) = p^{n-d+1} \qquad \square$$

Actually, the above construction can be extended to any prime power p and so Theorem 6.3.3 holds for any prime power p as well.

Exercises

1. Prove that the ISBN code $\mathcal{I}$ is a nonlinear $(10, 10^9, 2)$-code.

2. Compute the ISBN check digits for the numbers
   a) 0-387-94180   b) 0-19-859678   c) 0-13-283796

3. Show that the single-error-correcting decimal code D is nonlinear and has minimum distance 3.

4. Finish the proof of Theorem 6.3.2.

5. Assuming that the single-error-correcting code D is used, decode the received words
   a) 0118246792   b) 4511156214   c) 4535797952

6. Show that the first five columns of the matrix

$$P = \begin{bmatrix} 1 & 1 & 1 & \cdots & 1 \\ 1 & 2 & 3 & \cdots & 10 \\ 1 & 2^2 & 3^2 & \cdots & 10^2 \\ 1 & 2^3 & 3^3 & \cdots & 10^3 \end{bmatrix}$$

   are linearly dependent.

7. Why do you suppose it is important to consider trasnposition errors, especially in a decimal code?

8. Let P be the parity check matrix that gives rise to the code E. Verify that the first five columns of P and the rows of P are linearly independent.

9. Show that the code E has minimum distance 5.

10. Show that the code D has size 82,644,629. Hint: If $S_1, \ldots, S_n$ are finite sets, the Principle of Inclusion-Exclusion gives a formula for the size of their union $U = S_1 \cup \cdots \cup S_n$ in terms of the sizes of various intersections of these sets. In particular,

$$|U| = \sum_i |S_i| - \sum_{i<j} |S_i \cap S_j| + \sum_{i<j<k} |S_i \cap S_j \cap S_k| - \cdots$$

$$\cdots + (-1)^{n+1}|S_1 \cap \cdots \cap S_n|$$

(Notice that the sum alternates in sign.) Now, let C denote the linear [10, 8]-code over $\mathbb{Z}_{11}$ with parity check

$$P = \begin{bmatrix} 1 & 1 & 1 & 1 & 1 & 1 & 1 & 1 & 1 & 1 \\ 0 & 1 & 2 & 3 & 4 & 5 & 6 & 7 & 8 & 9 \end{bmatrix}$$

that was used to obtain the code D. Let S_i be the set of all codewords in C that have a 10 in the ith position. Show that, in this case, the sum is

$$|U| = \binom{10}{1} 11^7 - \binom{10}{2} 11^6 + \cdots - \binom{10}{8}$$

and that this sum is equal to 131,714,252. Since U is the set of all codewords in C that have a 10 in *at least* one position, the size of D is $|C| - |U|$.

11. Use the hint for the previous exercise to compute the size of the double-error-correcting code E.

12. Construct a table of square roots in $\mathbb{Z}_{11}$.

13. Assuming the double-error-correcting code E is used, decode the following words   a) 1208680845 b) 7189385648 c) 1251913347
d) 1211837417 e) 5149608219

14. Show that the linear code that gave rise to the code D is optimal by considering the Singleton bound. Hence, $A_{11}(10, 3) = 11^8$.

15. Show that the linear code that gave rise to the code E is optimal by considering the Singleton bound. Hence, $A_{11}(10, 5) = 11^6$.

16. Show that the linear code B is optimal by considering the Singleton bound.

6.4   Codes from Latin Squares

There is a strong connection between coding theory and a branch of mathematics known as combinatorics, which can be (very) loosely defined as the study of finite sets and their subsets. Let us explore one aspect of this connection. Throughout this section, we will let S_q denote a set of size q.

Consider codes of length 4 and minimum distance 3 over S_q. Suppose C is such a code, of size M. Since $d(C) = 3$, no two codewords can agree in their first two coordinate positions. Since there are q^2 possibilities for the first two coordinates, we have $M \leq q^2$. (This argument is precisely the one used to prove the Singleton bound.) Thus,

$$A_q(4, 3) \leq q^2$$

If there exists a $(4, q^2, 3)$-code C, then the first two coordinates must yield all of the q^2 pairs of elements of S_q, so we can write C in the form

$$C = \{ija_{ij}b_{ij} \mid \text{all } i, j \in S_q\}$$

In order for C to have minimum distance 3, every pair of codewords must have distance at least 3. Put another way, every pair of coordinate positions must yield q^2 distinct pairs. In particular,

1. the ordered pairs (i, a_{ij}) for all $i, j \in S_q$, must all be distinct.
2. the ordered pairs (j, a_{ij}) for all $i, j \in S_q$, must all be distinct.
3. the ordered pairs (i, b_{ij}) for all $i, j \in S_q$, must all be distinct.
4. the ordered pairs (j, b_{ij}) for all $i, j \in S_q$, must all be distinct.
5. the ordered pairs (a_{ij}, b_{ij}) for all $i, j \in S_q$, must all be distinct.

Let us consider the first two of these conditions. Let $A = (a_{ij})$ be the $q \times q$ matrix whose (i, j)th entry is a_{ij}. The index i is the row number of the entry a_{ij} and the index j is the column number. To say that the ordered pairs $(i_1, a_{i_1j_1})$ and $(i_2, a_{i_2j_2})$ are distinct is to say that if $i_1 = i_2$ then $a_{i_1j_1} \neq a_{i_2j_2}$. Put another way, if two entries $a_{i_1j_1}$ and $a_{i_2j_2}$ of A lie in the same row ($i_1 = i_2$), then they must be distinct. Similarly, condition 2) is equivalent to saying that if two entries of A lie in the same column, they must be distinct. We are now ready for a definition.

Definition   A **Latin square** of order q over S_q is a $q \times q$ matrix over S_q with the property that the elements in each row are distinct and the elements in each column are distinct. This is equivalent to saying that each row contains all q of the elements of S_q (in some order) and similarly for each column. □

For instance, the matrix

$$\begin{bmatrix} 0 & 1 & 2 \\ 1 & 2 & 0 \\ 2 & 0 & 1 \end{bmatrix}$$

is a Latin square of order 3 over $\mathbb{Z}_3 = \{0, 1, 2\}$, since every row and every column of the array contains all three numbers 0, 1, and 2. Similarly, the matrix

$$\begin{bmatrix} 0 & 1 & 2 & 3 \\ 1 & 2 & 3 & 0 \\ 2 & 3 & 0 & 1 \\ 3 & 0 & 1 & 2 \end{bmatrix}$$

is a Latin square of order 4 over $\mathbb{Z}_4 = \{0, 1, 2, 3\}$.

Thus, we have shown that the ordered pairs (i, a_{ij}) are distinct if and only if the matrix $A = (a_{ij})$ is a Latin square. In the language of Latin squares, we can now say that

$$C = \{ija_{ij}b_{ij} \mid \text{all } i, j \in S_q\}$$

is an optimal $(4, q^2, 3)$-code over S_q if and only if

1. the matrix $A = (a_{ij})$ is a Latin square over S_q.
2. the matrix $B = (b_{ij})$ is a Latin square over S_q.
3. the ordered pairs (a_{ij}, b_{ij}) are distinct.

Latin squares have applications in various areas of applied mathematics, such as in the theory of statistical experiments. For example, imagine a drug company wants to test the effects of three drugs on human subjects. In order to get the most reliable information possible, the company wants to test each drug on each subject over a period of several days. For simplicity, let us assume that there are three subjects in the experiment. Then one possible schedule is shown in the following table, where the three drugs are simply denoted by 0, 1, and 2.

	Day 1	Day 2	Day 3
Subject 1	0	1	2
Subject 2	1	2	0
Subject 3	2	0	1

Notice that, precisely because this table is a Latin square, on each day (column) all three drugs are tested, and each subject (row) tests all three drugs.

The pattern used to create the previous two examples of Latin squares can be generalized to create larger Latin squares as follows. Take the first row to be the numbers $0, 1, 2, \ldots, q - 1$, in order

$$012 \cdots q - 1$$

The second row is obtained from the first by rotating it one position to the right, wrapping the last number around to the first column

$$\begin{array}{ccccc} 0 & 1 & 2 & \cdots & q - 1 \\ q - 1 & 0 & 1 & \cdots & q - 2 \end{array}$$

Each subsequent row is created in a similar manner, to get

$$\begin{bmatrix} 0 & 1 & 2 & \cdots & q - 1 \\ q - 1 & 0 & 1 & \cdots & q - 2 \\ q - 2 & q - 1 & 0 & \cdots & q - 3 \\ \vdots & \vdots & \vdots & \cdots & \vdots \\ 1 & 2 & 3 & \cdots & 0 \end{bmatrix}$$

This construction proves the following theorem.

Theorem 6.4.1 *For all positive integers q, there exists a Latin square of order q.* □

Now we need to deal with the last condition for the existence of an optimal $(4, q^2, 3)$-code, namely, that the ordered pairs (a_{ij}, b_{ij}) for $i, j \in S_q$, are distinct.

Definition Two latin squares $L_1 = (a_{ij})$ and $L_2 = (b_{ij})$ over S_q are said to be **mutually orthogonal Latin squares** (abbreviated **MOLS**) if the q^2 ordered pairs (a_{ij}, b_{ij}) are distinct. □

Put another way, two Latin squares are orthogonal if, when we superimpose the squares to form an array whose (i, j)th entry is the ordered pair (a_{ij}, b_{ij}), the resulting array has distinct entries.

Example 6.4.1 The Latin squares

$$\begin{bmatrix} 0 & 1 & 2 \\ 1 & 2 & 0 \\ 2 & 0 & 1 \end{bmatrix} \quad \text{and} \quad \begin{bmatrix} 0 & 1 & 2 \\ 2 & 0 & 1 \\ 1 & 2 & 0 \end{bmatrix}$$

are orthogonal, since the nine corresponding ordered pairs are

$$\begin{bmatrix} (0,0) & (1,1) & (2,2) \\ (1,2) & (2,0) & (0,1) \\ (2,1) & (0,2) & (1,0) \end{bmatrix} \quad (6.4.1)$$

and these ordered pairs are distinct. □

Mutually orthogonal latin squares also have important applications. For instance, referring to our earlier example involving drugs, suppose that a drug company wants to test three types of decongestants and three kinds of antihistamines. In order to observe any combined effects of the two types of drugs, the company would like to test all possible combinations on its subjects. One possible schedule would be to employ table (6.4.1) as follows

	Day 1	Day 2	Day 3
Subject 1	(0,0)	(1,1)	(2,2)
Subject 2	(1,2)	(2,0)	(0,1)
Subject 3	(2,1)	(0,2)	(1,0)

For instance, on Day 2, subject 2 receives decongestant number 2 and antihistamine number 0. In this way, each subject tests each drug, each drug is tested each day, and all pairs of drugs are tested in combination.

We can now state the following theorem.

Theorem 6.4.2 *The code*

$$C = \{ija_{ij}b_{ij} \mid \text{all } i,j \in S_q\}$$

is an optimal $(4, q^2, 3)$-*code over* S_q *if and only if* $A = (a_{ij})$ *and* $B = (b_{ij})$ *is a pair of mutually orthogonal Latin squares.* □

Corollary 6.4.3 *A q-ary* $(4, q^2, 3)$-*code exists if and only if a pair of mutually orthogonal Latin squares of order q exists.* □

We are now left with the question of whether or not a pair of MOLS exists for every positive integer q. The answer is "almost." We first settle the question for prime powers $q = p^m$. We will leave it as an exercise to show that there is no pair of MOLS of order 2. However, that is the only exception as far as prime powers go.

Theorem 6.4.4 *If q is a prime power, but $q \neq 2$, then a pair of MOLS of order q exists.*   □

Proof   Let $F_q = \{\lambda_1, \ldots, \lambda_q\}$ be the field of size q. (Recall that for any prime power q, there exists an essentially unique field of size q.) Let α and β be distinct nonzero elements of F_q, and consider the $q \times q$ matrices $A = (a_{ij})$ and $B = (b_{ij})$ defined by

$$a_{ij} = \lambda_i + \alpha\lambda_j \text{ and } b_{ij} = \lambda_i + \beta\lambda_j$$

where i, j range over all elements of F_q. Let us show that $L_1 = (a_{ij})$ and $L_2 = (b_{ij})$ are MOLS.

To show that L_1 is a Latin square, observe that if two entries $\lambda_i + \alpha\lambda_j$ and $\lambda_i + \alpha\lambda_{j'}$ in the same row of L_1 but different columns ($j \neq j'$) are equal, then

$$\lambda_i + \alpha\lambda_j = \lambda_i + \alpha\lambda_{j'}$$

Subtracting λ_i and then multiplying both sides by α^{-1} (α is assumed nonzero), gives $\lambda_j = \lambda_{j'}$, which implies that $j = j'$, a contradiction. A similar argument shows that entries in the same column are distinct. In a similar way, L_2 is a Latin square.

To see that L_1 and L_2 are mutually orthogonal, consider the q^2 ordered pairs $(a_{ij}, b_{ij}) = (\lambda_i + \alpha\lambda_j, \lambda_i + \beta\lambda_j)$. If $(a_{ij}, b_{ij}) = (a_{kl}, b_{kl})$, then

$$(\lambda_i + \alpha\lambda_j, \lambda_i + \beta\lambda_j) = (\lambda_k + \alpha\lambda_\ell, \lambda_k + \beta\lambda_\ell)$$

This is equivalent to the system of equations

$$\lambda_i + \alpha\lambda_j = \lambda_k + \alpha\lambda_\ell$$
$$\lambda_i + \beta\lambda_j = \lambda_k + \beta\lambda_\ell$$

Subtracting the second equation from the first gives

$$(\alpha - \beta)\lambda_j = (\alpha - \beta)\lambda_\ell$$

or

$$(\alpha - \beta)(\lambda_j - \lambda_\ell) = 0$$

But since $\alpha \neq \beta$, we get

$$\lambda_j - \lambda_\ell = 0$$

and so $j = \ell$. Using this in the first equation of the system gives $i = k$. Hence, the ordered pairs (a_{ij}, b_{ij}) are distinct, and L_1 and L_2 are MOLS. ∎

Example 6.4.2 Let $F_q = \mathbb{Z}_5$ and let $\alpha = 1$, $\beta = 2$. Then, according to the previous theorem,

$$L_1 = (i + j) = \begin{bmatrix} 0 & 1 & 2 & 3 & 4 \\ 1 & 2 & 3 & 4 & 0 \\ 2 & 3 & 4 & 0 & 1 \\ 3 & 4 & 0 & 1 & 2 \\ 4 & 0 & 1 & 2 & 3 \end{bmatrix}$$

and

$$L_2 = (i + 2j) = \begin{bmatrix} 0 & 2 & 4 & 1 & 3 \\ 1 & 3 & 0 & 2 & 4 \\ 2 & 4 & 1 & 3 & 0 \\ 3 & 0 & 2 & 4 & 1 \\ 4 & 1 & 3 & 0 & 2 \end{bmatrix}$$

are MOLS. The 5-ary (4,25,3)-code associated with these MOLS is

$$C = \{0000, 0112, 0224, 0331, 0443, 1011, 1123, \ldots, 4320, 4432\} \qquad \square$$

The next theorem shows how to go from prime powers to other integers.

Theorem 6.4.5 *If a pair of MOLS of order m and a pair of MOLS of order n exist, then a pair of MOLS of order mn also exists.* $\qquad \square$

Proof The idea of the proof is a lot simpler than its precise description. Let A_1, A_2 be MOLS of order m and B_1, B_2 be MOLS of order n. We construct MOLS C_1, C_2 of order mn. Since C_2 is constructed from A_2 and B_2 in exactly the same way as C_1 is constructed from A_1 and B_1, we will just deal with the latter. Let $A_1 = (a_{ij})$, $B_1 = (b_{ij})$ and $C_1 = (c_{ij})$.

It will help to make a definition. By $a_{ij} \odot B_1$, we mean the matrix formed by replacing each entry b_{uv} in B_1 by the ordered pair (a_{ij}, b_{uv}).

Thus,

$$
a_{ij} \odot B_1 = \begin{bmatrix}
(a_{ij}, b_{11}) & (a_{ij}, b_{12}) & (a_{ij}, b_{13}) & \cdots \\
(a_{ij}, b_{21}) & (a_{ij}, b_{22}) & (a_{ij}, b_{23}) & \cdots \\
(a_{ij}, b_{31}) & (a_{ij}, b_{32}) & (a_{ij}, b_{33}) & \cdots \\
\vdots & \vdots & \vdots & \ddots
\end{bmatrix}
$$

Now we form C_1 in blocks by "expanding" each entry a_{ij} of A_1 into the block $a_{ij} \odot B_1$. An example may help to make this clear.

For the sake of illustration only, let

$$
A_1 = \begin{bmatrix} \alpha & \beta \\ \gamma & \delta \end{bmatrix} \text{ and } B_1 = \begin{bmatrix} a & b & c \\ d & e & f \\ g & h & i \end{bmatrix}
$$

Then C_1 is the 6×6 matrix

$$
\begin{bmatrix}
(\alpha, a) & (\alpha, b) & (\alpha, c) & (\beta, a) & (\beta, b) & (\beta, c) \\
(\alpha, d) & (\alpha, e) & (\alpha, f) & (\beta, d) & (\beta, e) & (\beta, f) \\
(\alpha, g) & (\alpha, h) & (\alpha, i) & (\beta, g) & (\beta, h) & (\beta, i) \\
(\gamma, a) & (\gamma, b) & (\gamma, c) & (\delta, a) & (\delta, b) & (\delta, c) \\
(\gamma, d) & (\gamma, e) & (\gamma, f) & (\delta, d) & (\delta, e) & (\delta, f) \\
(\gamma, g) & (\gamma, h) & (\gamma, i) & (\delta, g) & (\delta, h) & (\delta, i)
\end{bmatrix}
$$

We will leave it as an exercise to explain why C_1 is a Latin square and why C_1, C_2 is a pair of MOLS. ∎

Since any positive integer q can be written in the form

$$
q = p_1^{e_1} p_2^{e_2} \cdots p_k^{e_k}
$$

where $p_1 < p_2 < \cdots < p_k$ are prime numbers, Theorems 6.4.5 and 6.4.4 imply that there is a pair of MOLS of order any positive integer q with the exception that, if $p_1 = 2$, then e_1 must be greater than 1. In other words, q must have the property that if $2 \mid q$ then $4 \mid q$. (The symbol $\mid$ stands for divides.) We leave it as an exercise to show a positive integer q has this property if and only if it has the form $q = 4u + r$, where $r = 0$, 1 or 3, that is, if and only if $q \equiv 0$, 1 or 3 modulo 4. Thus, we have established the following result.

Theorem 6.4.6 *If $q \equiv 0$, 1 or 3 (mod 4), then there exists a pair of MOLS of order q.* □

This still leaves open the question of existence of a pair of MOLS for $q \equiv 2$ (mod 4), that is, for $q = 2, 6, 10, 14, \ldots$. It was not until 1960 that three mathematicians named Bose, Shrikhande, and Parker settled the remaining case as follows.

Theorem 6.4.7   *There exists a pair of MOLS of order q if and only if $q \neq 2$ and $q \neq 6$.*   □

The proof of Theorem 6.4.7 is rather complicated, and we will not go into it here. However, this result, together with Corollary 6.4.3 gives the following.

Theorem 6.4.8

1. *There exists a q-ary $(4, q^2, 3)$-code for all q except $q = 2$ and $q = 6$.*

2. *$A_q(4, 3) = q^2$ for all q except $q = 2$ and $q = 6$.*   □

We leave it as an exercise to show that $A_2(4, 3) = 2$. It has also been shown that $A_6(4, 3) = 34$. We thus have all values of $A_q(4, 3)$.

Theorem 6.4.9

1. $A_q(4, 3) = q^2$ for all $q \neq 2, 6$

2. $A_2(4, 3) = 2$

3. $A_6(4, 3) = 34$ □

Exercises

1. Prove that there is no pair of MOLS of size 2×2.

2. With reference to Theorem 6.4.5, explain why C_1 is a Latin square and why C_1, C_2 is a pair of MOLS.

3. Prove that, for a positive integer q, the properties (1) if 2 divides q then so does 4 and (2) $q \equiv 0, 1$ or 3 modulo 4 are equivalent. Hint: any number q can be written in the form $q = 4u + r$, where the remainder $r = 0, 1, 2,$ or 3.

4. Show that $A_2(4, 3) = 2$.

5. Construct a Latin square of order 5.

6. Show that the matrices

$$A = \begin{bmatrix} 1 & 2 & 3 & 4 \\ 4 & 3 & 2 & 1 \\ 2 & 1 & 4 & 3 \\ 3 & 4 & 1 & 2 \end{bmatrix} \text{ and } B = \begin{bmatrix} 1 & 2 & 3 & 4 \\ 3 & 4 & 1 & 2 \\ 4 & 3 & 2 & 1 \\ 2 & 1 & 4 & 3 \end{bmatrix}$$

form a pair of MOLS. Write down the codewords in the (4,16,3)-code associated with this pair of MOLS.

7. Construct a pair of MOLS of order 5. Describe the (4,25,3)-code associated with this pair of MOLS.

8. Use a pair of MOLS of order 3 to construct a pair of MOLS of order 9.

9. A **transversal** of a Latin square L of order q is a set of q locations, no two of which are in the same row or column, that have distinct entries. For example, in the Latin square

$$\begin{bmatrix} 2 & 1 & 3 \\ 3 & 2 & 1 \\ 1 & 3 & 2 \end{bmatrix}$$

the underlined locations (3,1), (2,2), and (3,1) form a transversal. Find two more transversals in this matrix. (It is a theorem that a Latin square has an orthogonal mate if and only if it has q disjoint transversals, as in this case.)

10. Show that $A_q(3, 2) = q^2$ using Latin squares.

11. We can generalize the results of this section as follows. Consider an $(n, q^2, n - 1)$-code C over S_q of the form

$$C = \{ ija_{ij}^{(1)} a_{ij}^{(2)} \cdots a_{ij}^{(n-2)} \mid \text{all } i, j \in S_q \}$$

The Singleton bound gives $A_q(n, n - 1) \le q^2$ and so such a code (if it exists) would be optimal. A **set of MOLS** of order q is a set of Latin squares of order q, every pair of which is a pair of MOLS.

(a) Show that C exists if and only if there exists a set of $n - 2$ MOLS, that is, a set of $n - 2$ Latin squares with the property that every pair of these Latin squares is mutually orthogonal.

(b) Show that, if q is a prime power, then there exists a set of $q - 1$ MOLS. Hint: Let $F_q = \{\lambda_1, \ldots, \lambda_q\}$ be the field of size q. Generalizing the proof of Theorem 6.4.4, consider the $q \times q$ matrices

$A_k = (a_{ij}^{(k)})$ and $B = (b_{ij})$ defined by

$$a_{ij}^{(k)} = \lambda_i + \lambda_k \lambda_j$$

(c) Show that each pair of the A_ks is a pair of MOLS.

(d) Show that, if q is a prime power and $n \leq q + 1$, then $A_q(n, n-1) = q^2$.

(e) Prove that there are at most $q - 1$ Latin squares in any set of MOLS of order q. Hint: Suppose that $\{L_1, \ldots, L_r\}$ is a set of MOLS. Explain why we may relabel any of the Latin squares L_i without affecting the orthogonality of the set. Thus, we may assume that the first row of each L_i is $[1 \; 2 \; 3 \; \cdots \; q]$. Consider the (2,1)-th entry in each L_i. Can any of these entries be a 1? Can any of these entries be the same? (Incidentally, part b) shows that there are the maximum number $q - 1$ of MOLS of order q when q is a prime power. It has not yet been determined what the maximum possible number of MOLS of order q is when q is not a prime power.)

12. Another type of combinatorial object that gives rise to codes is the so-called projective plane. Here is an example. The sets of points and lines shown in Figure 6.4.1 is referred to as the projective plane of order 2 or the **Fano plane**. It is an example of a combinatorial design. Let us denote the lines in this figure by $\ell_1 = \overline{14}$, $\ell_2 = \overline{16}$, $\ell_3 = \overline{46}$, $\ell_4 = \overline{15}$, $\ell_5 = \overline{26}$, $\ell_6 = \overline{34}$, $\ell_7 = \overline{25}$. Notice that each line is incident with exactly 3 points, and each pair of lines meets at exactly 1 point.

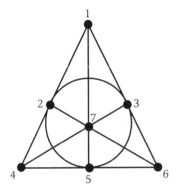

FIGURE 6.4.1

The **incidence matrix** of this plane is the matrix $A = (a_{ij})$ for which

$$a_{ij} = \begin{cases} 1 & \text{if line } \ell_i \text{ contains point } j \\ 0 & \text{otherwise} \end{cases}$$

This matrix is the 7×7 matrix

$$A = \begin{bmatrix} 1 & 1 & 0 & 1 & 0 & 0 & 0 \\ 1 & 0 & 1 & 0 & 0 & 1 & 0 \\ 0 & 0 & 0 & 1 & 1 & 1 & 0 \\ 1 & 0 & 0 & 0 & 1 & 0 & 1 \\ 0 & 1 & 0 & 0 & 0 & 1 & 1 \\ 0 & 0 & 1 & 1 & 0 & 0 & 1 \\ 0 & 1 & 1 & 0 & 1 & 0 & 0 \end{bmatrix}$$

Show that the binary code consisting of the rows $\mathbf{r}_1, \ldots, \mathbf{r}_7$ of A and the complements of the rows of A is a nonlinear $(7,16,3)$-code. Show also that it is perfect.

7

CHAPTER

An Introduction to Cyclic Codes

7.1 Cyclic Codes

An introduction to coding theory would not be complete without some discussion of cyclic codes, which form the most important class of linear codes. Indeed, the binary Hamming and all Golay codes are equivalent to cyclic codes, as are many of the important known codes. However, a thorough study of cyclic codes requires more abstract algebra than we wish to assume of the reader and so we will tread lightly here.

The definition of cyclic code is very simple.

Definition The **right cyclic shift** of a string $\mathbf{x} = x_1 \cdots x_n$ is the string $x_n x_1 \cdots x_{n-1}$ obtained by shifting each element to the right one position, wrapping the last element around to the first position.

A linear code C is **cyclic** if whenever $\mathbf{c} \in C$ then the right cyclic shift of $\mathbf{c}$ is also in C. □

To conserve ink, we will often refer to the right cyclic shift of a word $\mathbf{x}$ simply as a right shift of $\mathbf{x}$.

As an immediate consequence of this definition, if C is a cyclic code and $\mathbf{c} \in C$, then the string obtained by shifting the elements of $\mathbf{c}$ any number of positions (with wrapping) is also a codeword in C. Also, a code is cyclic if and only if whenever $\mathbf{c} = c_0 c_1 \cdots c_{n-1}$ is a codeword, so is the word $c_1 \cdots c_{n-1} c_0$, obtained by a left cyclic shift.

Example 7.1.1 The binary code $C = \{000, 101, 011, 110\}$ is cyclic, since the right cyclic shift of each codeword is another codeword.

Note that the binary code $D = \{0000, 1001, 0110, 1111\}$ is *not* cyclic, since shifting 1001 gives 1100, which is not in C. However, D is *equivalent* to a cyclic code, for if we interchange the third and fourth positions, the result is an equivalent code $E = \{0000, 1010, 0101, 1111\}$, which is cyclic. Thus, a noncyclic code may be equivalent to a cyclic code.   □

Example 7.1.2 We mentioned that the binary Hamming codes are equivalent to cyclic codes. Consider the Hamming code $\mathcal{H}_2(3)$, with parity check matrix

$$H_2(3) = \begin{bmatrix} 0 & 0 & 0 & 1 & 1 & 1 & 1 \\ 0 & 1 & 1 & 0 & 0 & 1 & 1 \\ 1 & 0 & 1 & 0 & 1 & 0 & 1 \end{bmatrix}$$

It is clear that $\mathbf{c} = 1000011$ is in $\mathcal{H}_2(3)$ since $\mathbf{c}H_2(3)^t = 0$. However, the shift 1100001 is not in $\mathcal{H}_2(3)$, since it is not orthogonal to the first row of $H_2(3)$. Thus, $\mathcal{H}_2(3)$ is not cyclic. However, consider the linear code C whose parity check matrix is obtained by reordering the columns of $H_2(3)$ to get

$$P = \begin{bmatrix} 1 & 0 & 0 & 1 & 0 & 1 & 1 \\ 0 & 1 & 0 & 1 & 1 & 1 & 0 \\ 0 & 0 & 1 & 0 & 1 & 1 & 1 \end{bmatrix}$$

We leave it as an exercise to find a generator matrix for this code and show that C is cyclic. Hence, $\mathcal{H}_2(3)$ is equivalent to a cyclic code.

We will see a bit later that the ternary Hamming code $\mathcal{H}_3(2)$ is not cyclic, nor is it equivalent to a cyclic code.   □

Codewords as Polynomials

To get a better understanding of cyclic codes, it pays to think of strings as polynomials. In particular, to each string $\mathbf{c} = c_0 c_1 c_2 \cdots c_{n-1}$ over $\mathbb{Z}_p$, we associate a polynomial, with coefficients in $\mathbb{Z}_p$, as follows

$$c_0 c_1 c_2 \cdots c_{n-1} \leftrightarrow c_0 + c_1 x + c_2 x^2 + \cdots + c_{n-1} x^{n-1}$$

Note that, as is customary, we start the indexing of the elements of the string at 0, rather than 1, so that the element c_k is the coefficient of the kth power of the variable x.

Note that addition and scalar multiplication of strings corresponds to the analogous operations for polynomials. That is, if the string $\mathbf{c}$ corresponds to the polynomial $p(x)$ and if $\mathbf{d}$ corresponds to the polynomial $q(x)$, then the sum $\mathbf{c} + \mathbf{d}$ corresponds to the polynomial sum $p(x) + q(x)$. In symbols,

$$\mathbf{c} + \mathbf{d} \leftrightarrow (c_0 + d_0) + (c_1 + d_1)x + (c_2 + d_2)x^2 + \cdots + (c_{n-1} + d_{n-1})x^{n-1}$$
$$= p(x) + q(x)$$

Similarly, if α is a scalar, then

$$\alpha\mathbf{c} \leftrightarrow (\alpha c_0) + (\alpha c_1)x + (\alpha c_2)x^2 + \cdots + (\alpha c_{n-1})x^{n-1} = \alpha p(x) \qquad (7.1.1)$$

Because of these facts, we may simply think of strings as polynomials, and vice versa.

To be more formal, the set $\mathcal{P}_n(\mathbb{Z}_p)$ of all polynomials of degree _less than_ n, with coefficients in $\mathbb{Z}_p$, is a vector space, under the operations of addition of polynomials and scalar multiplication of polynomials by elements of the alphabet $\mathbb{Z}_p$. Moreover, if C is a linear code of length n, then the association (7.1.1) associates with each codeword $\mathbf{c} \in C$ a polynomial $p(x) \in \mathcal{P}_n(\mathbb{Z}_p)$. Moreover, this association _preserves_ the operations of addition and scalar multiplcation and so we may think of C as a subspace of $\mathcal{P}_n(\mathbb{Z}_p)$. (For those who have studied abstract algebra, the association in (7.1.1) defines an isomorphism of C onto a subspace of $\mathcal{P}_n(\mathbb{Z}_p)$ and so C is isomorphic to a subspace of $\mathcal{P}_n(\mathbb{Z}_p)$.)

Thus, we may think of a codeword of length n as a polynomial of degree less than n and a linear code C of length n over $\mathbb{Z}_p$ as a subspace of $\mathcal{P}_n(\mathbb{Z}_p)$.

Example 7.1.3 Referring to the codes in Example 7.1.1, the binary code C takes the form $C = \{0, 1 + x^2, x + x^2, 1 + x\}$, where 0 is the zero polynomial. The code E takes the form $E = \{0, 1 + x^2, x + x^3, 1 + x + x^2 + x^3\}$. $\square$

Since addition and scalar multiplication of codewords is addition and scalar multiplication of polynomials, it is natural to ask how we might express the process of performing a right cyclic shift in terms of operations on polynomials. Notice that multiplying a codeword

$$p(x) = c_0 + c_1 x + \cdots + c_{n-1} x^{n-1}$$

by x gives

$$xp(x) = c_0 x + c_1 x^2 + \cdots + c_{n-1} x^n$$

which has some resemblence to a right cyclic shift, and indeed would be a right cyclic shift if we replaced x^n by $x^0 (= 1)$.

But we can do exactly this by dividing $xp(x)$ by $x^n - 1$ and keeping only the remainder; that is, by taking the product $xp(x)$ modulo $x^n - 1$.

Definition   In the set $P_n(\mathbb{Z}_p)$ of all polynomials of degree less than n, the operation of **multiplication modulo** $x^n - 1$ is defined by taking the ordinary product, then dividing by $x^n - 1$ and keeping only the remainder.   □

Let us denote by $R_n(\mathbb{Z}_p)$ the set of all polynomials of degree less than n, with coefficients from the set $\mathbb{Z}_p$, and with the operations of addition of polynomials, scalar multiplication of a polynomial by an element of $\mathbb{Z}_p$ and multiplication modulo $x^n - 1$. If we do not need to emphasize the alphabet $\mathbb{Z}_p$, we will simply denote this by R_n. Note that taking the product modulo $x^n - 1$ is very easy, since we simply take the ordinary product and then replace x^n by 1. As an example, in $R_4(\mathbb{Z}_2)$,

$$\begin{aligned}
(x^3 + x^2 + 1)(x^2 + 1) &= x^5 + x^4 + x^3 + 1 \\
&= x^4 \cdot x + x^4 + x^3 + 1 \\
&= 1 \cdot x + 1 + x^3 + 1 \\
&= x^3 + x
\end{aligned}$$

It is important to note that, since the polynomials in $R_n(\mathbb{Z}_p)$ have coefficients in a finite field, their properties are different from those of the polynomials with coefficients in the field of real numbers. For instance, in $R_4(\mathbb{Z}_2)$, since any multiple of 2 is equal to 0 and since $-1 = +1$, we have

$$(x - 1)^4 = x^4 - 4x^3 + 6x^2 - 4x + 1 = x^4 + 1 = 0$$

Thus, the product of nonzero polynomials may equal the zero polynomial. This cannot happen when the coefficients come from the field of real numbers.

We can now think of a linear code C over an alphabet $\mathbb{Z}_p$ as a subspace of the vector space $\mathcal{R}_n(\mathbb{Z}_p)$. In addition, if $p(x) \in C$, then the right cyclic shift of $p(x)$ is the polynomial $xp(x)$. Shifting the elements of $p(x)$ twice is equivalent to multiplying by x^2, to get $x^2p(x)$. In general, applying k right cyclic shifts is equivalent to multiplying the polynomial (codeword) by x^k.

We can now say that a linear code $C \subseteq \mathcal{R}_n$ is cyclic if and only if $p(x) \in C$ implies $xp(x) \in C$. Proof of the following result is left as an exercise.

Theorem 7.1.1 *A linear code $C \subseteq \mathcal{R}_n$ is cyclic if and only if $p(x) \in C$ implies that $f(x)p(x) \in C$ for any polynomial $f(x) \in \mathcal{R}_n$.* □

In the language of abstract algebra, the set $\mathcal{R}_n$, together with the operations of addition, scalar multiplication, and multiplication modulo $x^n - 1$, is called an algebra and any subset C of $\mathcal{R}_n$ that is a vector subspace and also has the property described in the previous theorem is called an ideal of $\mathcal{R}_n$. That is, the cyclic codes in $\mathcal{R}_n$ are precisely the ideals of $\mathcal{R}_n$.

The Generator Polynomial of a Cyclic Code

It happens that the cyclic codes in $\mathcal{R}_n$ can be described very simply. Before doing so, we need a bit of notation. The **leading coefficient** of a polynomial $p(x)$ is the coefficient of the largest power of x that appears in the polynomial (with nonzero coefficient). If the leading coefficient of $p(x)$ is equal to 1, we say that $p(x)$ is **monic**.

Now suppose that C is a cyclic code. Let $g(x)$ be a polynomial of smallest degree in C, among all nonzero polynomials in C. Multiplying $g(x)$ by a scalar if necessary, we can get a monic polynomial in C, so assume that $g(x)$ is also monic. If $p(x) \in C$ then we may divide $p(x)$ by $g(x)$ to get

$$p(x) = q(x)g(x) + r(x)$$

where the remainder $r(x)$ is either the zero polynomial or else has degree less than that of $g(x)$. However, since $r(x) = p(x) - q(x)g(x) \in C$, and since $g(x)$ has the smallest degree of any nonzero polynomial in C, we deduce that $r(x)$ must be the zero polynomial. Thus, $p(x) = q(x)g(x)$; that is, $p(x)$ is a multiple of $g(x)$. Conversely, any multiple of $g(x)$ is a codeword in C and so C consists precisely of the multples of $g(x)$. We will use the notation $\langle\langle g(x)\rangle\rangle$ to denote the set of all multiples of $g(x)$. Thus,

$$C = \langle\langle g(x)\rangle\rangle = \{f(x)g(x) \mid f(x) \in \mathcal{R}_n\}$$

Moreover, the polynomial $g(x)$ is unique, in the sense that there is only one polynomial in C that is both monic and has the smallest degree among all nonzero polynomials in C. For if $h(x)$ also has these properties, then by what we have just learned, there is a polynomial $q(x)$ for which $h(x) = q(x)g(x)$. But since $h(x)$ and $g(x)$ have the same degree, the polynomial $q(x)$ must be a constant and since $h(x)$ and $g(x)$ are both monic, we get $q(x) = 1$, whence $h(x) = g(x)$. Let us summarize.

Theorem 7.1.2   *Let C be a cyclic code in $\mathcal{R}_n$. Then there is a unique polynomial $g(x)$ in C that is both monic and has the smallest degree among all nonzero polynomials in C. Moreover, $C = \langle\langle g(x)\rangle\rangle$.* $\square$

In the language of abstract algebra, the set $\langle\langle g(x)\rangle\rangle$ is called the ideal generated by $g(x)$ and so the previous theorem says that any cyclic code C is generated by a single polynomial $g(x)$. This polynomial is called the **generator polynomial** of C. Also, a cyclic code has one and only one generator polynomial.

Example 7.1.4   Referring to the cyclic code $C = \{0, 1 + x^2, x + x^2, 1 + x\}$ in Example 7.1.3, we have (since $x^3 = 1$)

$$0 = 0 \cdot (1 + x)$$
$$1 + x^2 = x^2 \cdot (1 + x)$$
$$x + x^2 = x \cdot (1 + x)$$
$$1 + x = 1 \cdot (1 + x)$$

and so $C = \langle\langle 1 + x\rangle\rangle$. Since $1 + x$ has minimum degree in C, it is the generator polynomial for C.

Notice also that

$$0 = 0 \cdot (1 + x^2)$$
$$1 + x^2 = 1 \cdot (1 + x^2)$$
$$x + x^2 = x^2 \cdot (1 + x^2)$$
$$1 + x = x \cdot (1 + x^2)$$

and so C is also generated by the polynomial $1 + x^2$. However, since this polynomial does not have minimum degree in C, it is *not* the generator polynomial for C. □

The moral of the previous example is that a cyclic code may be generated by different polynomials, but has only one generator polynomial. *From now on, whenever we write $C = \langle\langle g(x)\rangle\rangle$, we will mean that $g(x)$ is the generator polynomial for C.*

It is very easy to characterize those polynomials that are generator polynomials.

Theorem 7.1.3 *A monic polynomial $p(x) \in \mathcal{R}_n(\mathbb{Z}_p)$ is the generator polynomial of a cyclic code in $\mathcal{R}_n(\mathbb{Z}_p)$ if and only if it divides $x^n - 1$.* □

Proof Suppose first that $p(x)$ is the generator polynomial for a cyclic code C. Dividing $p(x)$ into $x^n - 1$ gives

$$x^n - 1 = q(x)p(x) + r(x)$$

where the remainder $r(x)$ is either the zero polynomial or has degree less than that of $p(x)$. However, in $\mathcal{R}_n$, the polynomial $x^n - 1$ is the zero polynomial and so, interpreting this equation as being in $\mathcal{R}_n$, we have

$$r(x) = -q(x)p(x) \in C$$

It follows that $r(x)$ must be the zero polynomial, since no nonzero polynomial in C has degree less than that of $p(x)$. Thus, $p(x) \mid x^n - 1$.

For the converse, suppose that $p(x) \mid x^n - 1$. Thus, $x^n - 1 = a(x)p(x)$ for some polynomial $a(x)$ in $\mathcal{R}_n$. Note that $\deg a(x) = n - \deg p(x)$. Let C be the cyclic code generated by $p(x)$. Suppose further that $C = \langle\langle g(x)\rangle\rangle$, that is, $g(x)$ is the generator polynomial for C. Since $g(x) \in C$, we have

$$g(x) = f(x)p(x)$$

for some polynomial $f(x)$ in $\mathcal{R}_n$. Multiplying by $a(x)$ gives, in $\mathcal{R}_n$,

$$a(x)g(x) = a(x)f(x)p(x) = f(x)(x^n - 1) = 0$$

However, if $\deg p(x) > \deg g(x)$, then $\deg a(x) = n - \deg p(x) < n - \deg g(x)$, that is, $\deg a(x)g(x) < n$ and so it cannot be equal to 0 in $\mathcal{R}_n$ (since there is no reduction modulo $x^n - 1$). It follows that $\deg p(x) = \deg g(x)$ and since $p(x)$ is monic, we actually have $p(x) = g(x)$, whence $p(x)$ is the generator polynomial for C. ■

Theorem 7.1.3 is very important, for it tells us that there is precisely one cyclic code in $\mathcal{R}_n$ for each factor of $x^n - 1$, and that this accounts for all cyclic codes. Thus, we can find all cyclic codes in $\mathcal{R}_n$ by factoring $x^n - 1$. Before considering an example, we need a definition.

Definition　A *nonconstant* polynomial $p(x)$ in $\mathcal{R}_n$ is **irreducible** if whenever $p(x) = a(x)b(x)$ for polynomials $a(x)$, $b(x)$ in $\mathcal{R}_n$, then at least one of $a(x)$ or $b(x)$ must be a constant polynomial.　　□

In other words, a nonconstant polynomial in $\mathcal{R}_n$ is irreducible if it cannot be factored into nonconstant polynomials. Thus, by factoring $x^n - 1$ into monic irreducible factors, we can get all monic divisors of $x^n - 1$.

Example 7.1.5　The polynomial $x^3 - 1$ factors into irreducible factors over $\mathbb{Z}_2$ as follows

$$x^3 - 1 = (1 + x)(1 + x + x^2)$$

Hence, the factors of $x^3 - 1$ are 1, $1 + x$, $1 + x + x^2$ and $x^3 - 1$ itself. Thus, a complete list of binary cyclic codes of length 3 is

$$C_0 = \langle\langle 1 \rangle\rangle = \mathcal{R}_3 = \mathbb{Z}_2^3$$
$$C_1 = \langle\langle 1 + x \rangle\rangle = \{0, 1 + x, x + x^2, 1 + x^2\} = \{000, 110, 011, 101\}$$
$$C_2 = \langle\langle 1 + x + x^2 \rangle\rangle = \{0, 1 + x + x^2\} = \{000, 111\}$$
$$C_3 = \langle\langle x^3 - 1 \rangle\rangle = \{0\} = \{000\}$$

□

We have seen that if $g(x)$ is the generator polynomial for a cyclic code C, then C consists of all polynomial multiples of $g(x)$. We can easily obtain a basis for C from $g(x)$.

Theorem 7.1.4　*Let C be a nonzero cyclic code in $\mathcal{R}_n$ with generator polynomial $g(x) = g_0 + g_1 x + \cdots + g_k x^k$, of degree k. Then C has basis*

$$\mathcal{B} = \{g(x), xg(x), \ldots, x^{n-k-1}g(x)\}$$

and generator matrix

$$G = \begin{bmatrix} g_0 & g_1 & g_2 & \cdots & & g_k & 0 & 0 & \cdots & 0 \\ 0 & g_0 & g_1 & g_2 & \cdots & & g_k & 0 & \cdots & 0 \\ 0 & 0 & g_0 & g_1 & g_2 & \cdots & & g_k & & \vdots \\ \vdots & \vdots & & \ddots & \ddots & & & & \ddots & 0 \\ 0 & 0 & \cdots & 0 & g_0 & g_1 & g_2 & \cdots & & g_k \end{bmatrix}$$

whose $n - k$ rows each consist of a right cyclic shift of the row above. Note also that $\dim(C) = n - \deg g(x)$. □

Proof We know that any codeword in C has the form $c(x) = a(x)g(x)$ for some polynomial $a(x)$. Since $g(x) \mid x^n - 1$, we may write $x^n - 1 = h(x)g(x)$. Dividing $a(x)$ by $h(x)$ gives

$$a(x) = f(x)h(x) + r(x)$$

where, as usual, the remainder $r(x)$ is either 0 or has degree less than that of $h(x)$. Multiplying this by $g(x)$ and noting that $h(x)g(x) = 0$ in $\mathcal{R}_n$, we get

$$c(x) = a(x)g(x) = f(x)h(x)g(x) + r(x)g(x) = r(x)g(x)$$

and so, if $c(x)$ is not the zero polyomial, then $r(x) \neq 0$ and so $c(x) = r(x)g(x)$ is a multiple of $g(x)$, where the polynomial $r(x)$ has degree at most $\deg h(x) - 1 = n - k - 1$. It follows that $r(x) = r_0 + r_1 x + \cdots + r_{n-k-1}x^{n-k-1}$ and so

$$c(x) = r_0 g(x) + r_1 x g(x) + \cdots + r_{n-k-1}x^{n-k-1}g(x)$$

is a linear combination of the elements of $\mathcal{B}$. Hence, $\mathcal{B}$ spans C. We leave it as an exercise to show that $\mathcal{B}$ is linearly independent, whence it is a basis for C.

Since the string versions of the polynomials in $\mathcal{B}$ are the rows of G, it follows that G is a generator matrix for C and since $|\mathcal{B}| = n - k$, it follows that $\dim(C) = n - k$. ■

Example 7.1.6 Let us find all ternary cyclic codes of length 4. Over $\mathbb{Z}_3$, the polynomial $x^4 - 1$ factors into irreducible factors as follows

$$x^4 - 1 = (x^2 - 1)(x^2 + 1) = (-1 + x)(1 + x)(1 + x^2)$$

Incidentally, to see that $1 + x^2$ is irreducible, we note that if not, then it would factor into linear factors and thus would have at least one root in $\mathbb{Z}_3 = \{0, 1, 2\}$. But none of the elements of $\mathbb{Z}_3$ is a root of $1 + x^2$ and so $1 + x^2$ is irreducible over $\mathbb{Z}_3$.

It follows that there are a grand total of $2^3 = 8$ factors of $x^4 - 1$ over $\mathbb{Z}_3$, corresponding to eight distinct cyclic codes. Here is a list of the factors and their corresponding generator matrices.

1. Factor: 1
   Generator matrix: I_4

2. Factor: $-1 + x$

Generator matrix: $\begin{bmatrix} -1 & 1 & 0 & 0 \\ 0 & -1 & 1 & 0 \\ 0 & 0 & -1 & 1 \end{bmatrix}$

3. Factor: $1 + x$

Generator matrix: $\begin{bmatrix} 1 & 1 & 0 & 0 \\ 0 & 1 & 1 & 0 \\ 0 & 0 & 1 & 1 \end{bmatrix}$

4. Factor: $1 + x^2$

Generator matrix: $\begin{bmatrix} 1 & 0 & 1 & 0 \\ 0 & 1 & 0 & 1 \end{bmatrix}$

5. Factor: $(-1 + x)(1 + x) = -1 + x^2$

Generator matrix: $\begin{bmatrix} -1 & 0 & 1 & 0 \\ 0 & -1 & 0 & 1 \end{bmatrix}$

6. Factor: $(-1 + x)(1 + x^2) = -1 + x - x^2 + x^3$

Generator matrix: $\begin{bmatrix} -1 & 1 & -1 & 1 \end{bmatrix}$

7. Factor: $(1 + x)(1 + x^2) = 1 + x + x^2 + x^3$

Generator matrix: $\begin{bmatrix} 1 & 1 & 1 & 1 \end{bmatrix}$

8. Factor: $-1 + x^4$

$C = \{0\}$

It is clear from this exhaustive list that the two cyclic codes of dimension 3 have minimum distance 2 and so the ternary Hamming $[4, 3, 3]$-code $\mathcal{H}_3(1)$ is not equivalent to a cyclic code, as mentioned earlier. □

We have seen that the cyclic codes in $\mathcal{R}_n$ correspond precisely to the factors of $x^n - 1$ in $\mathcal{R}_n$. Thus, from a theoretical standpoint, we have solved the problem of finding all cyclic codes. However, from a practical viewpoint, the problem of efficiently factoring $x^n - 1$ is a complicated one, which we cannot go into here. For reference, we include the following short list of factorizations of $x^n - 1$ over $\mathbb{Z}_2$.

Factorization of $x^n - 1$ over $\mathbb{Z}_2$

$x^3 - 1 = (x + 1)(x^2 + x + 1)$

$x^5 - 1 = (x + 1)(x^4 + x^3 + x^2 + x + 1)$

$$x^7 - 1 = (x + 1)(x^3 + x + 1)(x^3 + x^2 + 1)$$
$$x^9 - 1 = (x + 1)(x^2 + x + 1)(x^6 + x^3 + 1)$$
$$x^{11} - 1 = (x + 1)(x^{10} + x^9 + \cdots + 1)$$
$$x^{13} - 1 = (x + 1)(x^{12} + x^{11} + \cdots + 1)$$
$$x^{15} - 1 = (x + 1)(x^2 + x + 1)(x^4 + x + 1)(x^4 + x^3 + 1)(x^4$$
$$+ x^3 + x^2 + x + 1)$$
$$x^{17} - 1 = (x + 1)(x^8 + x^5 + x^4 + x^3 + 1)(x^8 + x^7 + x^6 + x^4 + x^2 + x + 1)$$
$$x^{19} - 1 = (x + 1)(x^{18} + x^{17} + \cdots + 1)$$
$$x^{21} - 1 = (x + 1)(x^2 + x + 1)(x^3 + x^2 + 1)(x^3 + x + 1)(x^6 + x^4$$
$$+ x^2 + x + 1)(x^6 + x^5 + x^4 + x^2 + 1)$$
$$x^{23} - 1 = (x + 1)(x^{11} + x^9 + x^7 + x^6 + x^5 + x + 1)(x^{11} + x^{10}$$
$$+ x^6 + x^5 + x^4 + x^2 + 1)$$
$$x^{25} - 1 = (x + 1)(x^4 + x^3 + x^2 + x + 1)(x^{20} + x^{15} + x^{10} + x^5 + 1)$$

Note that further factorizations can be obtained from the fact that, over $\mathbb{Z}_2$,

$$x^{2^k n} - 1 = (x^n - 1)^{2^k}$$

Thus, for instance,

$$x^6 - 1 = x^{2 \cdot 3} - 1 = (x^3 - 1)^2 = (x + 1)^2 (x^2 + x + 1)^2$$

The Check Polynomial of a Cyclic Code

We have seen that the generator polynomial $g(x)$ of a cyclic code $C \subseteq \mathcal{R}_n$ divides $x^n - 1$. Hence, we may write

$$x^n - 1 = h(x)g(x)$$

where $h(x) \in \mathcal{R}_n$. The polynomial $h(x)$, which has degree equal to the dimension of C, is referred to as the **check polynomial** of C. Since the generator polynomial $g(x)$ is unique, so is the check polynomial. The following theorem shows why the check polynomial is important.

Theorem 7.1.5 *Let C be a cyclic code in $\mathcal{R}_n$, with check polynomial $h(x)$. Then a polynomial $p(x) \in \mathcal{R}_n$ is in C if and only if $p(x)h(x) = 0$ in $\mathcal{R}_n$.* $\qquad \square$

Proof   Note first that $h(x)g(x) = 0$ in $\mathcal{R}_n$. If $p(x)$ is a codeword, then $p(x) = f(x)g(x)$ for some polynomial $f(x)$ and so $p(x)h(x) = f(x)g(x)h(x) = 0$ in $\mathcal{R}_n$. Conversely, if $p(x)h(x) = 0$ in $\mathcal{R}_n$ then let us write

$$p(x) = q(x)g(x) + r(x)$$

where $r(x) = 0$ or $\deg r(x) < \deg g(x)$. Multiplying by $h(x)$ gives (since $h(x)g(x) = 0$)

$$0 = p(x)h(x) = q(x)g(x)h(x) + r(x)h(x) = r(x)h(x)$$

and since $\deg[r(x)h(x)] < \deg[g(x)h(x)] = \deg(x^n - 1) = n$, we deduce that $r(x)$ must be the zero polynomial. Hence, $p(x) = q(x)g(x) \in C$.   ■

From the check polynomial $h(x)$ of a cyclic code, we may obtain a parity check matrix. Suppose that C is a cyclic $[n, k]$-code, with check polynomial $h(x)$ of degree k. The problem is to relate the polynomial product $p(x)h(x)$ in $\mathcal{R}_n$, which is 0 precisely for codewords $p(x)$, to the scalar product of strings. Since we are dealing with a product in $\mathcal{R}_n$, the easiest coefficients of $p(x)h(x)$ to determine are those of $x^k, \ldots, x^{n-1}$, since they do not involve any "wrap around." In particular, for a codeword $c(x) = c_0 + c_1 x + \cdots + c_{n-1}x^{n-1}$, these coefficients must be 0, and so we get

$$c_0 h_k + \quad c_1 h_{k-1} + \cdots + \quad c_k h_0 = 0$$
$$c_1 h_k + \quad c_2 h_{k-1} + \cdots + \quad c_{k+1} h_0 = 0$$
$$\vdots$$
$$c_{n-k-1}h_k + c_{n-k-2}h_{k-1} + \cdots + c_{n-1}h_0 = 0$$

In string language, any codeword $\mathbf{c} = c_0 c_1 \cdots c_{n-1}$ is orthogonal to the string $\mathbf{h} = h_k h_{k-1} \cdots h_0 0 \cdots 0$ and to the first $n - k - 1$ right cyclic shifts of $\mathbf{h}$. In symbols, c is orthogonal to the rows of the $(n - k) \times n$ matrix

$$H = \begin{bmatrix} h_k & h_{k-1} & h_{k-2} & \cdots & & h_0 & 0 & 0 & \cdots & 0 \\ 0 & h_k & h_{k-1} & h_{k-2} & \cdots & & h_0 & 0 & \cdots & 0 \\ 0 & 0 & h_k & h_{k-1} & h_{k-2} & \cdots & & h_0 & & \vdots \\ \vdots & \vdots & & \ddots & \ddots & & & & \ddots & 0 \\ 0 & 0 & \cdots & 0 & h_k & h_{k-1} & h_{k-2} & \cdots & & h_0 \end{bmatrix}$$

Hence, these rows lie in $C^\perp$. Since the rows of H are linearly independent (because $h_k = 1$), it follows that H is a generator matrix for a code D for which $D \subseteq C^\perp$. However, $\dim(D) = n - (n - k) = k$ and so $D = C^\perp$.

Thus, H is a generator matrix for $C^\perp$ and hence a parity check matrix for C. Let us summarize.

Theorem 7.1.6 _If C is a cyclic code with check polynomial $h(x)$, then the matrix H defined above is a parity check matrix for C._ ☐

Example 7.1.7 Referring to Example 7.1.6, let C be the ternary cyclic code with generator polynomial $g(x) = -1 + x$ and generator matrix

$$G = \begin{bmatrix} -1 & 1 & 0 & 0 \\ 0 & -1 & 1 & 0 \\ 0 & 0 & -1 & 1 \end{bmatrix}$$

The check polynomial is

$$h(x) = \frac{x^4 - 1}{-1 + x} = 1 + x + x^2 + x^3$$

and so a parity check matrix for C is

$$H = \begin{bmatrix} 1 & 1 & 1 & 1 \end{bmatrix}$$

For the code with generator polynomial $g(x) = 1 + x^2$, the check polynomial is $h(x) = -1 + x^2$ and so a parity check matrix is

$$H = \begin{bmatrix} -1 & 0 & 1 & 0 \\ 0 & -1 & 0 & 1 \end{bmatrix} \qquad ☐$$

Hamming Codes as Cyclic Codes

We have already seen that the binary Hamming code $\mathcal{H}_2(3)$ is equivalent to a cyclic code and that the ternary Hamming code $\mathcal{H}_3(1)$ is not. In order to show that all binary Hamming codes are equivalent to cyclic codes, we need to take a slightly different look at cyclic codes. Since this is not the place to go into a detailed discussion, we will be deliberately sketchy.

Let n be an odd positive integer and consider the polynomial $x^n - 1$ over $\mathbb{Z}_2$. It is possible to show that $x^n - 1$ can have at most n roots in any field containing $\mathbb{Z}_2$ as a subfield. Moreover, there is a field $\mathcal{F}$ containing $\mathbb{Z}_2$ as a subfield that contains all of the roots of $x^n - 1$. These roots are known as the nth roots of unity over $\mathbb{Z}_2$.

It is not hard to show that the nth roots of unity are distinct, that is, that $x^n - 1$ has no multiple roots. For if not, there would be a root $r \in \mathcal{F}$

for which

$$x^n - 1 = (x - r)^2 p(x)$$

Taking the derivative of both sides (this is legitimate even for polynomials over finite fields) gives (since $2 = 0$ in any field containing $\mathbb{Z}_2$ as a subfield)

$$nx^{n-1} = 2(x - r)p(x) + (x - r)^2 p'(x) = (x - r)^2 p'(x)$$

Making the substitution $x = r$ results in $nr^{n-1} = 0$, which is not possible unless n is divisible by 2. Since we are assuming that n is odd, it follows that $x^n - 1$ has no multiple roots.

Thus, there exists a field $\mathcal{F}$ containing $\mathbb{Z}_2$ as a subfield that contains all n distinct roots of the polynomial $x^n - 1$. Moreover, it can be shown that there is a root β with the property that all of its powers, together with the zero element, constitute the entire field $\mathcal{F}$, in symbols,

$$\mathcal{F} = \{0, 1, \beta, \beta^2, \ldots, \beta^k\}$$

where $|\mathcal{F}| = k + 2$. The determination of k is a bit beyond our scope. However, when n has the form $2^r - 1$, it turns out that $k = 2^r$. The element β is called a **primitive field element** of $\mathcal{F}$.

Now let us consider the set C of all polynomials in $\mathcal{R}_n$ for which β is a root

$$C = \{p(x) \in \mathcal{R}_n \mid p(\beta) = 0\}$$

If $x^n - 1$ can.be factored into a product of irreducible polynomials

$$x^n - 1 = m_1(x) \cdots m_s(x),$$

then β is a root of one of these polynomials, say $m_i(x)$. Hence, any $p(x) \in C$ shares a common root (namely β) with the irreducible polynomial $m_i(x)$. It follows that $p(x)$ is divisible by $m_i(x)$ and that, moreover, C is precisely the set of polynomials that are divisible by $m_i(x)$. In other words, C is the binary cyclic code with generator polynomial $m_i(x)$.

Of course, the length of C is $n = 2^r - 1$. It is possible to show that the degree of the generator polynomial $m_i(x)$ is r and so the dimension of C is $n - r$. Finally, it is easy to see that no polynomial $p(x)$ for which $p(\beta) = 0$ can have weight less than 3 (as a binary string). Hence, the minimum distance of C is at least 3, which implies that it must be equal to 3 (by the sphere-packing condition). Hence, C is a linear code with the same parameters as the Hamming code $\mathcal{H}_2(r)$ and is therefore equivalent to $\mathcal{H}_2(r)$. This shows that the binary Hamming codes are equivalent to cyclic codes.

Exercises

1. Let C be a linear code with basis $\mathcal{B} = \{\mathbf{b}_1, \ldots, \mathbf{b}_k\}$. Show that if the right cyclic shift of each basis codeword $\mathbf{b}_i$ is also in C, then C is cyclic.

2. Show that a code is cyclic if and only if whenever $\mathbf{c} = c_0 c_1 \cdots c_{n-1}$ is a codeword, so is the word $c_1 \cdots c_{n-1} c_0$, obtained by a left cyclic shift.

3. Find a generator matrix for the code C in Example 7.1.2. (Hint: note that P has the form $P = [I \mid A]$.) Show that the code C is cyclic.

4. Prove that a linear code $C \subseteq \mathcal{R}_n$ is cyclic if and only if $p(x) \in C$ implies that $f(x)p(x) \in C$ for any polynomial $f(x) \in \mathcal{R}_n$.

5. Find the generator polynomial for the binary cyclic code $E = \{0, 1 + x^2, x + x^3, 1 + x + x^2 + x^3\}$. Find all polynomials in E that generate E.

6. With reference to Theorem 7.1.4, show that $\mathcal{B}$ is linearly independent.

7. Find all distinct binary cyclic codes of length 4.

8. Find the check polynomial $h(x)$ for each of the codes in Example 7.1.6 and verify that $h(x)c(x) = 0$ for all codewords $c(x)$. (You can take advantage of the basis for each code given in the example.)

9. Prove that if C is a cyclic code in $\mathcal{R}_n$ with check polynomial $h(x)$ of degree k, then the polynomial $h^r(x) = x^k h(x^{-1})$, called the **reciprocal polynomial** of $h(x)$, is the generator polynomial of the dual code $C^\perp$ (that is, after multiplying by a suitable constant to make it monic).

10. Find the check polynomial and a parity check matrix for each code in Example 7.1.6.

11. Determine all of the binary cyclic codes of length 5.

12. Show that, for any prime p, the p-ary repetition code $\text{Rep}_p(n)$ is cyclic. Find the generator polynomial for this code.

13. Show that the binary code E_n consisting of all codewords of length n that have even weight is a cyclic code. Find the generator polynomial for E_n.

14. If $g(x)$ is the generator polynomial of a cyclic code, show that $g(x)$ has nonzero constant term. Does the polynomial $x^k g(x)$ also generate the code? Is it a generator polynomial?

15. Let $C = \langle\langle g(x)\rangle\rangle$ be a binary cyclic code. If C contains at least one word of odd weight, show that the set E of all codewords in C with even weight is also a cyclic code. What is the generator polynomial for E?

16. How many cyclic codes of length n are there over $\mathbb{Z}_p$? Express the answer in terms of the number of irreducible factors of $x^n - 1$.

17. Let $h(x)$ be the check polynomial of a cyclic code C. Is $C^\perp = \langle\langle h(x)\rangle\rangle$? If not, what can you say?

Answers to Selected Odd-Numbered Exercises

Section 1.1

1.

$\oplus_3$	0	1	2
0	0	1	2
1	1	2	0
2	2	0	1

$\odot_3$	0	1	2
0	0	0	0
1	0	1	2
2	0	2	1

3. a) 1 b) 1 c) 0 d) 0 e) 1 f) 2

5. $\binom{8}{4}4^4 = 17{,}920$; $\binom{8}{3}\binom{5}{2}3^3 = 15{,}20$

7. $\binom{6}{0}2^6 + \binom{6}{1}2^5 + \binom{6}{2}2^4 = 496$

9. $\binom{n}{k}(r-1)^k$

11. Since addition is performed elementwise, we need only consider four possibilities in each case.

For part a),

$(0 + 0)^c = 0^c = 1 = 0 + 0^c$
$(1 + 0)^c = 1^c = 0 = 1 + 0^c$
$(0 + 1)^c = 1^c = 0 = 0 + 1^c$
$(1 + 1)^c = 0^c = 1 = 1 + 1^c$

For part b),

$0^c + 0^c = 1 + 1 = 0 = 0 + 0$
$1^c + 0^c = 0 + 1 = 1 = 1 + 0$
$0^c + 1^c = 1 + 0 = 1 = 0 + 1$
$1^c + 1^c = 0 + 0 = 0 = 1 + 1$

13. a) $d(\mathbf{x}^c, \mathbf{y}) = n - d(\mathbf{x}, \mathbf{y})$ b) $d(\mathbf{x}^c, \mathbf{y}^c) = d(\mathbf{x}, \mathbf{y})$

Section 1.2

1. Since $2^6 = 64$ and $2^7 = 128$, the minimum length is 7.
3. There are only two encoding functions: a) $f(a) = 0, f(b) = 1$ and b) $g(a) = 1, g(b) = 0$.
5. $n!$
7. The codes of maximum codeword length n over a set A of size r are precisely the subsets of A_n. Since a set of size k has 2^k subsets and since $|A_n| = \frac{1-r^{n+1}}{1-r}$ the answer is

$$2^{\frac{1-r^{n+1}}{1-r}}$$

For $r = 2$ and $n = 5$, this number is $2^{63} \approx 9 \times 10^{18}$.

Section 1.3

1. Yes. Let $\mathbf{x}$ be a sequence of codewords. Read $\mathbf{x}$ from left to right.
     a) If we encounter a 0, this must represent the codeword 0.
     b) If we encounter a 1 followed by a 0, this must represent 10.
     c) If we encounter a 1 followed by another 1, the sequence 11, together with the next 2 elements, uniquely determines which codeword is represented.
3. Yes. Let $\mathbf{x}$ be a sequence of codewords. Read $\mathbf{x}$ from right to left.
a) If we encounter a 0, it must be the codeword 0. If we encounter a 1, count the number of 1s before encountering a 0. Any string of five 1s must be the codeword 11111. Otherwise, the number of 1s before the 0 uniquely determines the codeword.
5. No. The string 110110 can be interpreted as itself or as the sequence 1101 10
7. The codeword lengths do not satisfy Kraft's inequality, so there is no such code.
9. a) $N(1) = 1, N(2) = 3$.
b) The $N(k)$ strings of codewords of length k bits can be divided into two disjoint groups — those that begin with a codeword of length 1, namely with the codeword 0 and those that begin with a codeword of length 2, namely 10 or 11. Since there is only one possibility for the former and two

possibilities for the latter, we get

$$N(k) = N(k-1) + 2N(k-2)$$

c) Letting $N(k) = \alpha^k$ leads to the quadratic equation $\alpha^2 - \alpha - 2 = 0$, whose solutions are $\alpha = -1, 2$. Checking in the original equation, we find that $(-1)^k$ and 2^k are both solutions and that $a(-1)^k + b2^k$ is a solution for numbers a and b, to be determined by the facts that $N(1) = 1$ and $N(2) = 3$. This gives $-a + 2b = 1$ and $a + 4b = 3$, whose solution is $a = \frac{1}{3}$, $b = \frac{2}{3}$. Hence, $N(k) = \frac{1}{3}(-1)^k + \frac{2}{3}(2^k)$.
d) $N(1) = 1, N(2) = 2, N(3) = 5, N(4) = 9, N(5) = 18$.

Section 1.4

1. Suppose that C is not uniquely decipherable. Then there is a string **x** over the code alphabet A that can be written in two different ways as a sequence of codewords

$$\mathbf{x} = \mathbf{c}_1\mathbf{c}_2\cdots\mathbf{c}_n = \mathbf{d}_1\mathbf{d}_2\cdots d_m$$

We may assume that $\mathbf{c}_1 \neq \mathbf{d}_1$, for otherwise, we can just remove this codeword and consider the remaining portion of the message. (Sooner or later, we must run into a situation where $\mathbf{c}_i \neq \mathbf{d}_i$, since the two representations of **x** are assumed to be different.) But then either $\text{len}(\mathbf{c}_1) < \text{len}(\mathbf{d}_1)$, in which case $\mathbf{c}_1$ is a prefix of $\mathbf{d}_1$, or $\text{len}(\mathbf{c}_1) > \text{len}(\mathbf{d}_1)$, in which case $\mathbf{d}_1$ is a prefix of $\mathbf{c}_1$. In either case, we conclude that C fails to have the prefix property and so is not instantaneous.

3. No, since 10 is a prefix of 1011.

5. No. Each codeword of length 5 must begin with 11. But there are only $2^3 = 8$ binary strings of length 5 that start with 11.

7. A word of length n has n prefixes, including itself.

9. No, by Kraft's Theorem.

11. Yes. $C = \{00, 01, 100, 101, 1100, 1101, 11110, 11111\}$

13. Yes. $C = \{0, 1, 2, 3, 40, 41, 42, 43, 440, 441, 442, 4430, 4431, 4432\}$

15. Twelve additional codewords of length 6. $\{0, 10, 1100, 1101xy, 1110xy, 1111xy\}$, where $x, y \in \mathbb{Z}_2$.

Section 2.1

1. Scheme 1 has average codeword length 2.9, whereas scheme 2 has average codeword length 3. Hence, scheme 1 is more efficient.

3. $\frac{1}{10}(1 + 2 + \cdots + 9 + 9) = 5.4$

5. By minimizing the sum of the lengths of all codewords.

Section 2.2

1. $A \to 1100$, $B \to 111$, $C \to 0$, $D \to 10$, $E \to 1101$; len $= 2.2$, sav $= 27\%$

3. $A \to 1100$, $B \to 10$, $C \to 0$, $D \to 1101$, $E \to 1110$, $F \to 1111$; len $= 2.4$, sav $= 20\%$

5. $A \to 101$, $B \to 100$, $C \to 011$, $D \to 010$, $E \to 001$, $F \to 000$, $G \to 1111$, $H \to 1110$, $I \to 1101$, $J \to 1111$; len $= 3.4$, sav $= 15\%$

9. The Huffman tree must have exactly 3 levels. The probability distribution $\{p_1, p_2, p_3, p_4\}$ must satisfy $p_i + p_j \geq p_k$ and $p_i + p_j \geq p_\ell$ where $\{i, j, k, \ell\} = \{1, 2, 3, 4\}$.

11. Suppose there are n source symbols. If we combine s nodes into 1 on the first step and then we reduce the number of nodes on the top level by $s - 1$. Suppose we need u additional steps to get to exactly r nodes on the top level. Each of these u steps will reduce the number of nodes on the top level by $r - 1$. Hence, we must have

$$n - (s - 1) - u(r - 1) = r$$

or

$$s = n - (u + 1)(r - 1)$$

Since $2 \leq s \leq r$, the first reduction size s is uniquely determined by the condition

$$s \equiv n \bmod (r - 1), \quad 2 \leq s \leq r$$

Once s is determined, we may determine u from the first equation above to be

$$u = \frac{n - s + 1 - r}{r - 1} = \frac{n - s}{r - 1} - 1$$

The total number of steps to construct the Huffman tree is $u + 2$.

13. a) The average codeword length of (C, f) is

$$\text{AveCodeLenHuff}_2(C, f) = \frac{1}{n} \sum_{i=1}^{n} \text{len}(f(s_i)) = \frac{1}{n} T$$

and since $\text{AveCodeLenHuff}_2(C, f)$ is minimum among all instantaneous 2 encodings, so is T.

b) It is clear from the construction of a Huffman tree that the bottom level of the tree contains only sibling pairs associated with source symbols. Let $\mathbf{c}$ and $\mathbf{d}$ be the codewords of sibling pairs on the bottom level of the Huffman tree. Since a source symbol on level k has codeword length $k - 1$, we conclude that $\mathbf{c}$ and $\mathbf{d}$ have length L and the corresponding nodes lie on level $L + 1$. Moreover, since they are sibling pairs, they differ only in their last positions.

c) Let $\mathbf{c} = c_1 \cdots c_{L-1}0$ and $\mathbf{d} = c_1 \cdots c_{L-1}1$ be the codewords of maximum length guaranteed by the previous part. Suppose (for the purposes of contradiction) that there is a codeword $\mathbf{e} = e_1 \cdots e_m$ of length $m \leq L - 2$. Let D be the code derived from C by deleting the codewords $\mathbf{c}$, $\mathbf{d}$, and $\mathbf{e}$ and adding the codewords $\mathbf{f} = c_1 \cdots c_{L-1}$, $\mathbf{e}0$ and $\mathbf{e}1$. Since deleting codewords cannot destroy the prefix property, we need only be concerned about the additions. Adding $\mathbf{f}$ does not destroy the prefix property since a prefix of $\mathbf{f}$ is also a prefix of $\mathbf{c}$ and so is not a codeword. Adding $\mathbf{e}0$ and $\mathbf{e}1$ does not destroy the prefix property because we removed $\mathbf{e}$. Thus, D has the prefix property and is therefore instantaneous. However, the next change in the sum of the codeword lengths is

$$-L - L - m + L - 1 + m + 1 + m + 1 = m + 1 - L < 0$$

and so there has been a net decrease in total codeword length, which is impossible in view of part a). Hence, all codewords in C have length L or $L - 1$.

d) Let the top level of a complete binary tree be level 0. Then there are at most 2^k nodes on level k, for $k = 0, 1, 2, \ldots$. Thus, since

$$2^k < \alpha 2^k \leq 2^{k+1}$$

we see that the source nodes all lie on levels k and $k + 1$ and while there may be none on level k, there is at least one on level $k + 1$. It follows that $L = k + 1$.

Since there are n nodes, we have $u + v = n = \alpha 2^k$. Also, if we were to add an additional node pair to the bottom of each of the source nodes on level k, then we would get $2u$ nodes on level $k + 1$ and so $2u + v = 2^L = 2^{k+1}$. It follows that $u = 2^{k+1} - \alpha 2^k = 2^k(2 - \alpha)$ and $v = n - u = 2^{k+1}(\alpha - 1)$.

e) $k + 2 - \frac{2}{\alpha}$

Section 3.1

1. 0.918

3. $\lg a - \frac{4}{a}$

5. Since the sum that defines the entropy consists only of nonnegative terms, the entropy is 0 if and only if each term is 0, which happens if and only if $\mathcal{P}(s) = 1$ for some source symbol s.

7. The entropies are a) lg 12 for tossing a coin and rolling a die, b) lg 8 for tossing three coins, and c) lg 16 for tossing four coins. Thus, the most information comes from tossing four coins and the least from tossing three coins.

9. $H = \frac{7}{24} \lg \frac{24}{7} + \frac{11}{24} \lg \frac{24}{11} + \frac{5}{24} \lg \frac{24}{5} + \frac{1}{24} \lg 24$

Section 3.2

1. a) If $0 < x < 1$, then we apply the mean value theorem to the interval $[x, 1]$ to get

$$\frac{\ln x - \ln 1}{x - 1} = \frac{1}{z} > 1$$

for some $z \in (x, 1)$. Multiplying both sides by the negative number $x - 1$ gives the desired inequality. For $x > 1$, the mean value theorem gives

$$\frac{\ln x - \ln 1}{x - 1} = \frac{1}{z} < 1$$

for some $z \in (1, x)$. Multiplying by $x-1$ gives the desired inequality. (The result is clearly true for $x = 1$.) As for equality, the function $f(x) = \ln x - x + 1$ has $f(1) = 0$ and also has $f'(x) < 0$ for $x > 1$ and $f'(x) > 0$ for $x < 1$. Hence, it can be 0 nowhere else.

b) This follows from part a) and the change of base formula $\lg x = (\ln x)/(\ln 2)$.

3. $H(\frac{1}{2} + x) = (\frac{1}{2} + x) \lg \frac{1}{(\frac{1}{2}+x)} + (\frac{1}{2} - x) \lg \frac{1}{(\frac{1}{2}-x)} = H(\frac{1}{2} - x)$

5. The change of base formula for logarithms is $\log_r x = \frac{\log_s x}{\log_s r}$ and so

$$H_r(\mathcal{S}) = \sum_{i=1}^{q} p_i \log_r \frac{1}{p_i} = \sum_{i=1}^{q} p_i \frac{\log_s \frac{1}{p_i}}{\log_s r} = \frac{1}{\log_s r} \sum_{i=1}^{q} p_i \log_s \frac{1}{p_i} = \frac{H_s(\mathcal{S})}{\log_s r}$$

7. We may first write the lemma in the form

$$\sum_{i=1}^{q} p_i \lg p_i \geq \sum_{i=1}^{q} p_i \lg r_i$$

Using both sides as exponents of 2, we get

$$\prod_{i=1}^{q} p_i^{p_i} \geq \prod_{i=1}^{q} r_i^{p_i}$$

Now let $r_i = p_i x_i / \sum p_j x_j$. Since $\sum r_i = 1$, we may substitute into the previous inequality to get

$$\prod_{i=1}^{q} p_i^{p_i} \geq \prod_{i=1}^{q} \frac{(p_i x_i)^{p_i}}{(\sum_j p_j x_j)^{p_i}} = \frac{1}{\sum_j p_j x_j} \prod_{i=1}^{q} (p_i x_i)^{p_i} = \frac{1}{\sum_j p_j x_j} \prod_{i=1}^{q} p_i^{p_i} \prod_{i=1}^{q} x_i^{p_i}$$

Canceling the common factor and rearranging gives the desired inequality. Equality holds throughout if and only if it holds in the lemma, that is, if and only if $r_i = p_i$ for all i. This is equivalent to $x_i = \sum_j p_j x_j$ for all i, which is equivalent to saying that all of the x_is are equal.

Section 3.3

1. For S, we have $a \to 0, b \to 1$, with average codeword length 1. For S^2, we have

$$aa \to 010, ab \to 011, ba \to 00, bb \to 1$$

with an average codeword length per source symbol of $\frac{27}{32} = 0.84375$. For S^3, we have

$$aaa \to 11100, aab \to 11101, aba \to 11110, baa \to 11111$$

$$abb \to 100, bab \to 101, bba \to 110, bbb \to 0$$

with an average codeword length per source symbol of $\frac{158}{192} = 0.82292$.

Section 3.4

1. $H_2(S) = \frac{3}{2}$. A Huffman encoding is $a \to 00, b \to 01, c \to 1$ with average codeword length of $\frac{3}{2}$. Hence, $H_2(S) = \text{MinAveCodeLen}_2(S)$.

3. The 100th extension S^{100}. We would need $2^{100} \approx 1.3 \times 10^{30}$ codewords!

5. Yes.

Section 4.1

1.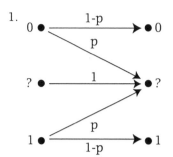

3. Because the input completely determines the output.

5. Because if a 0 is received, it is equally likely that a 0 or a 1 was sent, and similarly if a 1 is received. Thus, knowledge of the output tells us nothing about the input.

7. Reverse the outputs, that is, change a 0 to a 1 and a 1 to a 0.

Section 4.2

1. Let E_k be the event that $\mathbf{c}_k$ is sent over the channel and let A be the event that $\mathbf{x}$ is received. Apply Bayes' Theorem.

3. a) $P(010 \text{ received}|\ 000 \text{ sent}) = (0.99)^2(0.01) = 0.009801$
   $P(010 \text{ received}|\ 001 \text{ sent}) = (0.99)(0.01)^2 = 0.000099$
   $P(010 \text{ received}|\ 111 \text{ sent}) = (0.99)(0.01)^2 = 0.000099$
   Decode as 000.
   b) $P(110 \text{ received}|\ 000 \text{ sent}) = (0.99)(0.01)^2 = 0.000099$
   $P(110 \text{ received}|\ 001 \text{ sent}) = (0.01)^3 = 0.000001$
   $P(110 \text{ received}|\ 111 \text{ sent}) = (0.99)^2(0.01) = 0.009801$
   Decode as 111.

5. a) $P(010 \text{ received}|\ 000 \text{ sent}) = (0.99)^2(0.008) = 0.0078408$
   $P(010 \text{ received}|\ 001 \text{ sent}) = (0.99)(0.008)^2 = 0.00006336$
   $P(010 \text{ received}|\ 111 \text{ sent}) = (0.008)^2(0.99) = 0.00006336$
   Decode as 000.
   b) $P(10? \text{ received}|\ 000 \text{ sent}) = (0.008)(0.99)(0.002) = 0.00001584$
   $P(10? \text{ received}|\ 001 \text{ sent}) = (0.008)(0.99)(0.002) = 0.00001584$
   $P(10? \text{ received}|\ 111 \text{ sent}) = (0.99)(0.008)(0.002) = 0.00001584$
   A three-way tie!
   c) $P(??0 \text{ received}|\ 000 \text{ sent}) = (0.002)^2(0.99) = 0.00000396$

P(??0 received| 001 sent) $= (0.002)^2(0.008) = 0.000000032$
P(??0 received| 111 sent) $= (0.002)^2(0.008) = 0.000000032$
·Decode as 000

Section 4.3

1. a) 5 b) 8

3. a) 11200 b) tie c) 00111 d) 01221

5. If $i \neq j$ then $P(a_i \mid a_j) = \frac{1-p}{r-1}$. If a codeword $\mathbf{c}$ and a received word $\mathbf{x}$, both of length n, differ in exactly k places, then

$$P(\mathbf{x} \text{ received } \mid \mathbf{c} \text{ sent }) = p^{n-k} \left(\frac{1-p}{r-1} \right)^k$$

Since $p > \frac{1-p}{r-1}$, this probability is larger for larger values of $n - k$, that is, for smaller values of k. Hence, the probability is maximized when k, the Hamming distance, is as small as possible.

7. Channel alphabet $\{0,1\}$. Let $P(0 \text{ received } \mid 0 \text{ sent}) = 0.1$, $P(1 \text{ received}\mid 0 \text{ sent}) = 0.9$, $P(1 \text{ received}\mid 1 \text{ sent}) = P(0 \text{ received}\mid 1 \text{ sent}) = 0.5$. Code $C = \{001, 011\}$. Suppose that 000 is received. Then

$$P(000 \text{ received } \mid 001 \text{ sent }) = (0.1)^2(0.5) = 0.005$$

$$P(000 \text{ received } \mid 011 \text{ sent }) = (0.1)(0.5)^2 = 0.025$$

Thus, it is more probable that the codeword farther from 000 was sent. The key here is that it is more likely that a 0 is changed into a 1 than not changed at all!

Section 4.4

1. a) $d(C) = 3$ b) 10010, 11100, tie (decoding error)

3. Length $= n$, size $= 2^{n-1}$, $d = 2$.

5. First we prove part 1). Recall that Theorem 4.4.2 says that $d(C) \geq 2v + 1$ if and only if C is v-error-correcting. If C is exactly v-error-correcting then, by Theorem 4.4.2, we have $d(C) \geq 2v + 1$. However, if $d(C) \geq 2v + 3$ then Theorem 4.4.2 would imply that C is $(v + 1)$-error-correcting, which is not

the case. Hence, $2v + 1 \le d(C) \le 2v + 2$. For the converse, suppose that $d(C) = 2v + 1$ or $d(C) = 2v + 2$. Then Theorem 4.4.2 says that C is v-error-correcting. However, if C were $(v + 1)$-error-correcting, then Theorem 4.4.2 would imply that $d(C) \ge 2v + 3$, which is not true. Hence, C is exactly v-error-correcting. We may rephrase the first part of Corollary 4.4.3 by saying that a code C with odd minimum distance is exactly v-error-correcting if and only if $d(C) = 2v + 1$ and a code with even minimum distance is v-error-correcting if and only if $d(C) = 2v + 2$. Setting $s = 2v + 1$ and $t = 2v + 2$, this is equivalent to saying that a code C with odd minimum distance is exactly $\frac{s-1}{2}$-error-correcting if and only if $d(C) = s$ and a code with even minimum distance is $\frac{t-2}{2}$-error-correcting if and only if $d(C) = t$. But, since s is odd, we have $\lfloor \frac{s-1}{2} \rfloor = \frac{s-1}{2}$ and since t is even, $\lfloor \frac{t-1}{2} \rfloor = \frac{t-2}{2}$. We can now combine the two cases into one—which is precisely the statement of part 2) of the corollary.

7. $P(\text{decode error}) \le 1 - (0.999)^{15} - 15(0.001)(0.999)^{14} = 0.000104094$

9. No. We can assume that any such code contains the zero codeword $\mathbf{0} = 0000000$. Then any other codeword must have at least five 1s. We may also assume that one of the codewords is $\mathbf{c} = 11111xy$. But any other codeword $\mathbf{d}$ with at least five 1s will have at least three positions in which there is a 1 in common with $\mathbf{c}$. Hence, $d(c) \le d(\mathbf{c}, \mathbf{d}) \le 4$, a contradiction. Hence, the best we can do (size-wise) is a (7,2,5)-code.

11. Suppose that C is maximal, but that, for some string $\mathbf{x} \notin C$, we have $d(\mathbf{x}, \mathbf{c}) \ge d$ for all codewords $\mathbf{c}$ in C. Then we may add the string $\mathbf{x}$ to the code C and still have a code with minimum distance d. On the other hand, if for all words $\mathbf{x} \notin C$, there is a codeword $\mathbf{c}$ with the property that $d(\mathbf{x}, \mathbf{c}) < d$ and if we try to add a word $\mathbf{x}$ to C, the minimum distance of the code would be less than d.

Section 4.5

1. a) 6   b) 51   c) 175   d) 256   e) 5^{10} (no computation necessary)

3. Suppose that C is an $(n, M, 2e)$-code. Let $\mathbf{c}$ and $\mathbf{d}$ be codewords of minimum distance $2e$ apart and suppose (as we may) that they differ in the first $2e$ positions. Let $\mathbf{x}$ be the word that agrees with $\mathbf{c}$ in the first e positions, agrees with $\mathbf{d}$ in the next e positions, and agrees with both $\mathbf{c}$ and $\mathbf{d}$ in the remaining positions. Then $d(\mathbf{c}, \mathbf{x}) = d(\mathbf{d}, \mathbf{x}) = e$. The triangle inequality can be used to show that no other codeword is closer to $\mathbf{x}$. However, the packing radius of C is $pr(C) = \lfloor \frac{2e-1}{2} \rfloor = e - 1 < e$ and so no packing sphere contains $\mathbf{x}$, whence C is not perfect.

5. Plugging into the left side of the sphere-packing condition gives

$$r^n \cdot 1 = r^n$$

which is equal to the right side.

7. Plugging into the left side of the sphere-packing condition gives (with the help of the binomial theorem)

$$2 \cdot \sum_{k=0}^{m} \binom{2m + 1}{k}$$

Now, the binomial theorem says that the "complete" sum is

$$\sum_{k=0}^{2m+1} \binom{2m + 1}{k} = 2^{2m+1}$$

but since, for any k satisfying $0 \leq k \leq m$, we have

$$\binom{2m + 1}{k} = \binom{2m + 1}{(2m + 1) - k}$$

and as k runs through 0 to m, the expression $2m + 1 - k$ runs (backwards) through $m + 1$ to $2m + 1$, we see that

$$\sum_{k=0}^{m} \binom{2m + 1}{k} = \sum_{k=0}^{m} \binom{2m + 1}{2m + 1 - k} = \sum_{j=m+1}^{2m+1} \binom{2m + 1}{j}$$

Hence, the "top half" of the complete sum is equal to the "bottom half", whence

$$\sum_{k=0}^{m} \binom{2m + 1}{k} = \sum_{k=0}^{2m+1} \binom{2m + 1}{k} = 2^{2m+1} = 2^n$$

as desired.

9. Let $C = \{\mathbf{c}, \mathbf{d}\}$ be a binary $(2m + 1, 2, 2m + 1)$-code. Since C has minimum distance equal to its length, the two codewords must differ in every position. By interchanging corresponding bits in the two codewords in any given position, we can arrange it so that one of the resulting words is the zero word $\mathbf{0} = 0 \cdots 0$. Hence, the other word must be $\mathbf{1} = 1 \cdots 1$ and the new code is $\text{Rep}_2(2m + 1)$. Thus, C is just the repetition code after some "relabeling" of the symbols in fixed positions. (There are different codes with the same parameters where you cannot relabel the symbols and turn one code into the other.)

Section 4.6

1. $\overline{C}$ = {000000, 111001, 001111, 110110}. Parameters are (6,4,4).

3. Let $\mathbf{c}$ and $\mathbf{d}$ be codewords in C. Suppose that a) $\mathbf{c}$ and $\mathbf{d}$ both have 1s in α positions, b) $\mathbf{c}$ and $\mathbf{d}$ both have 0s in β positions, c) $\mathbf{c}$ has a 0 and $\mathbf{d}$ has a 1 in γ positions, and d) $\mathbf{c}$ has a 1 and d has a 0 in δ positions. Then $d(\mathbf{c},\mathbf{d}) = \gamma + \delta$. Also, $\alpha + \delta = w(\mathbf{c})$ is even and $\alpha + \gamma = w(d)$ is even. Since $\alpha + \delta$ and $\alpha + \gamma$ are both even, δ and γ must have the same parity and so $d(\mathbf{c},\mathbf{d}) = \gamma + \delta$ is even. This proves that the distance between every pair of codewords is even. For the rest, suppose that $d(\mathbf{c},\mathbf{d}) = d(C)$. Let $\overline{c}$ be the word obtained from $\mathbf{c}$ by adding an even parity check and similarly for $\overline{\mathbf{d}}$. Thus, $d(\overline{\mathbf{c}},\overline{\mathbf{d}}) = d(\mathbf{c},\mathbf{d})$ or $d(\overline{\mathbf{c}},\overline{\mathbf{d}}) = d(\mathbf{c},\mathbf{d}) + 1$. First suppose that $d(C)$ is even. Thus $d(\mathbf{c},\mathbf{d})$ is even and since all distances in $\overline{C}$ are even, we must have $d(\overline{\mathbf{c}},\overline{\mathbf{d}}) = d(\mathbf{c},\mathbf{d})$, whence $d(\overline{C}) = d(C)$. Now suppose that $d(C)$ is odd. Then $d(\mathbf{c},\mathbf{d})$ is odd and since $d(\overline{\mathbf{c}},\overline{\mathbf{d}})$ is even, we must have $d(\overline{\mathbf{c}},\overline{\mathbf{d}}) = d(\mathbf{c},\mathbf{d}) + 1$, whence $d(\overline{C}) = d(C) + 1$.

5. This cross-section is $D = \{000, 001, 010, 011, 110, 111\}$ with parameters (3,6,1).

7. Since M_i is the number of codewords in C that have an i in the first position, if we sum M_i over all possible values of i, we must get the total number of codewords in C. Thus, $\sum M_i = M$. For the rest, if D_i is the code formed from C by taking only those codewords with an i in the first position, then D_i is a subset (subcode) of C and so its minimum distance e_i cannot be smaller than that of C, that is, $e_i \geq d$. But, every codeword in D_i has the same element in the first position (namely, an i). Therefore, removing this position does not affect the distances between words and so $d_i = e_i \geq d$.

9. {0000000,0000111,0111000,0111111,1101010,1010101}. (Build the smaller code first.)

11. $C + 1 \oplus C_2$ = {000000000000000, 0000000011111000,
              0000000000011111, 1100000011000000,
              1100000000111000, 1100000011011111

13. For part a), since $d(\mathbf{c},\mathbf{d}) + d(\mathbf{c},\mathbf{d}^c)$ is the number of places where $\mathbf{c}$ and $\mathbf{d}$ differ plus the number of places where $\mathbf{c}$ and $\mathbf{d}$ agree, we get $d(\mathbf{c},\mathbf{d}) + d(\mathbf{c},\mathbf{d}^c) = n$. For part b), let $D = C \cup C^c$. We have $d(C) = d$ and since taking the complement of all words does not change any distances between words, we also have $d(C^c) = d$. Thus, the only way that $d(D)$ could be different from $d(C)$ is if $d(\mathbf{c},\mathbf{d}^c) < d(C)$ for some pair of codewords $\mathbf{c}$ and $\mathbf{d}$ in C. But by part a),

$$\min\{d(\mathbf{c},\mathbf{d}^c)\} = \min\{n - d(\mathbf{c},\mathbf{d})\} = n - \max\{d(\mathbf{c},\mathbf{d})\} = n - d_{\max}$$

Thus, if $n - d_{max} \geq d(C)$ we have $d(D) = d(C)$ and if $n - d_{max} < d(C)$ then $d(D) = n - d_{max}$. Hence, $d(D) = \min\{d, n - d_{max}\}$.

15. $C_1 \odot C_2$ has parameters $(16,6,2)$, which is not as good as $C_1 \oplus C_2$, since $C_1 \oplus C_2$ is single-error-correcting but $C_1 \odot C_2$ does not correct any errors!

17. First note that $C_2 = \{0000, 0011, 0101, 0110, 1001, 1010, 1100, 1111\}$ and $\mathrm{Rep}_2(4) = \{0000, 1111\}$. Hence, $C_3 = \{00000000, 00001111, 00110011, 00111100, 01010101, 01011010, 01100110, 01101001, 10011001, 10010110, 10101010, 10100101, 11001100, 11000011, 11111111, 11110000\}$ is a $(8,16,4)$-code. Furthermore, C_4 is a $(16,32,8)$-code and C_5 is a $(32,64,16)$-code. More generally, since $\mathrm{Rep}_2(2^k)$ is a $(2^k, 2, 2^k)$-code, if C_k is a $(2^k, 2^{k+1}, 2^{k-1})$-code then $C_{k+1} = C_k \oplus \mathrm{Rep}_2(2^k)$ is a $(2^k + 2^k, 2^{k+1} \cdot 2, \min\{2 \cdot 2^{k-1}, 2^k\})$-code, that is, a $(2^{k+1}, 2^{k+2}, 2^k)$-code.

19. Let C_1 and C_2 be equivalent codes, where

$$c_1 c_2 \cdots c_n \in C_1 \text{ if and only if } \pi_1(c_{\sigma(1)})\pi_2(c_{\sigma(2)}) \cdots \pi_n(c_{\sigma(n)}) \in C_2$$

It is clear from the definition that the equivalent codes have the same length n. Moreover, since the functions π_i and σ are one-to-one, the codes have the same size. In particular, $c_1 c_2 \cdots c_n = d_1 d_2 \cdots d_n$ if and only if $c_i = d_i$ for all i, which happens if and only if

$$c_{\sigma(1)} c_{\sigma(2)} \cdots c_{\sigma(n)} = d_{\sigma(1)} d_{\sigma(2)} \cdots d_{\sigma(n)}$$

and this holds if and only if

$$\pi_1(c_{\sigma(1)})\pi_2(c_{\sigma(2)}) \cdots \pi_n(c_{\sigma(n)}) = \pi_1(d_{\sigma(1)})\pi_2(d_{\sigma(2)}) \cdots \pi_n(d_{\sigma(n)})$$

Since permuting the positions or the code alphabet has no effect on distances, we have $d(C_1) = d(C_2)$.

Section 4.7

1. a) 4   b) 4   c) 4

3. Note that $\mathbf{x} + \mathbf{y}$ has a 0 in the kth position if and only if $\mathbf{x}$ and $\mathbf{y}$ have the same value in the kth position. Hence, the weight of $\mathbf{x} + \mathbf{y}$ is precisely the number of positions in which $\mathbf{x}$ and $\mathbf{y}$ differ, which is $d(\mathbf{x}, \mathbf{y})$.

5. $\mathcal{R}(\mathcal{R}(r, m)) = \left[1 + \binom{m}{1} + \cdots + \binom{m}{r}\right]/2^m$, $\delta(\mathcal{R}(r, m)) = (2^{m-r-1} - 1)/2^m$ for $r < m$, 0 for $r = m$.

7. Let C be a binary $(7, M, 5)$-code with $\mathbf{0} \in C$. Then all other codewords must have weight at least 5. If $\mathbf{1} \in C$ then $C = \{\mathbf{0}, \mathbf{1}\}$. If $\mathbf{1} \notin C$ then note that the farthest apart two strings of length 7 and weight at least 5 can be is 4. Hence, C can have at most one other (nonzero) codeword and so $M \leq 2$.

9. Let C be a binary $(6, M, 3)$-code and consider the cross-sections C_i defined by $x_1 = i$ $(i = 0, 1)$. These codes have length 5 and minimum distance at least 3, and since $A_2(5, 3) = 4$, and $A_2(5, 4) = A_2(5, 5) = 2$ they have size at most 4. Hence, $A_2(6, 3) \leq 4 + 4 = 8$. To find an optimal code, let us first try to construct the aforementioned cross-sections in such a way that words from different cross-sections are "reasonably" far apart. We start with an optimal $(5, 4, 3)$-code $D_1 = \{00000, 11100, 00111, 11011\}$. Next, create D_2 by switching the first elements of each codeword in D_1 (change 0 to 1 and 1 to 0) and also switching the last elements. Thus, $D_2 = \{10001, 01101, 10110, 01010\}$. The resulting code has the same parameters as D_1 (corresponding positions in two codewords of D_2 are different if and only if they are different in D_1) and so is also optimal. In addition, the distance between any codeword in D_1 and any codeword in D_2 is at least 2. Therefore, if we adjoin a 0 to the front of each codeword in D_1 and a 1 to the front of each codeword in D_2, the resulting eight words forms a code with parameters $(6, 8, 3)$. Thus, $A_2(6, 3) \geq 8$.

11. Certainly $A_r(n, 1) \leq r^n$. But the code $\mathbb{Z}_r^n$ has length n, size r^n and minimum distance 1.

13. No. The binary code $C = \{000, 111\}$ has minimum distance 3 but there is no way to add words to C to get a code with minimum distance 2.

15. According to Theorem 4.7.7 (with some slight modifications), if d is even then a binary $(n - 1, M, d - 1)$-code exists if and only if a binary (n, M, d)-code exists. Hence, if C is an optimal $(n, A_2(n, d), d)$-code, there is an $(n - 1, A_2(n, d), d-1)$-code D, whence $A_2(n, d) \leq A_2(n-1, d-1)$. Similarly, if C is an optimal $(n-1, A_2(n-1, d-1), d-1)$-code, there is an $(n, A_2(n-1, d-1), d)$-code D, whence $A_2(n - 1, d - 1) \leq A_2(n, d)$. Putting the two inequalities together gives the result.

17. If there exists a binary (n, M, d)-code C with d even, then by Theorem 4.7.7 there is a binary $(n - 1, M, d - 1)$-code D. Add an overall parity check to D to get a binary (n, M, d)-code with all codewords of even weight.

19. a) Yes. If you cannot add a new word to C without decreasing its minimum distance then it is, by definition, maximal. Moreover, since A^n is a finite set, the process of adding new codewords must stop, and the result will be a maximal code. b) No. Recall that $A_2(6, 3) = 8$. Now consider the code $C = \{000000, 011100, 000111, 011011, 100000, 111111\}$. (We formed this code by adjoining a 0 to the front of an optimal $(5, 4, 3)$-code and then including the codewords $(100000, 111111)$ designed to "spoil" the resulting code.) The code C has minimum distance 3 but cannot be enlarged to an optimal code. For we cannot add any new codewords that begin with a 0 (and preserve the minimum distance) the cross-section $x_1 = 0$ is an optimal $(5, 4, 3)$-code and we cannot add any additional words that begin with a 1 (and preserve

the minimum distance) since any such words would have to have weight at least 4 (because of 100000) and at most 3 (because of 111111). c) The code C above is maximal but not optimal.

Section 4.8

1. Setting the inequality sign in the sphere-packing bound to an equal sign gives the sphere-packing condition.

3. covering bound 5; packing bound 16

5. covering bound 3; packing bound 21

7. covering bound 4; packing bound 56

9. If $n = r + 1$ and $d = 3$, then

$$V_r^n\left(\left\lfloor \frac{d-1}{2} \right\rfloor\right) = V_r^{r+1}(1) = 1 + \binom{r+1}{1}(r-1) = r^2$$

and so the sphere-packing bound is

$$A_r(r+1, 3) \leq \frac{r^{r+1}}{r^2} = r^{r-1}$$

Section 4.9

7. The function $f(x) = x(M - x)$, defined for all real numbers x, has a global maximum at $x = \frac{M}{2}$. If M is even, this is also the maximum over the integers from 0 to M, whence $f(k) \leq f(\frac{M}{2}) = (\frac{M}{2})^2$. If M is odd, since the graph of f is symmetric about the line $x = \frac{M}{2}$, we deduce that $f(\frac{M-1}{2}) = f(\frac{M+1}{2}) = \frac{M^2-1}{4}$ is the desired maximum.

9. Let us write $a = \frac{d}{2d-n}$. Then (4.9.2) implies that

$$M \leq \begin{array}{ll} \lfloor 2a \rfloor & \text{for } M \text{ even} \\ \lfloor 2a \rfloor - 1 & \text{for } M \text{ odd} \end{array}$$

Suppose first that $k \leq a < k + \frac{1}{2}$ for some integer k. Then $\lfloor 2a \rfloor = 2k$ and $2\lfloor a \rfloor = 2k = \lfloor 2a \rfloor$. Thus, regardless of the parity of M, we have $M \leq 2\lfloor a \rfloor$, as desired. Now suppose that $k + \frac{1}{2} \leq a < k + 1$. Then $\lfloor 2a \rfloor = 2k + 1$ and $2\lfloor a \rfloor = 2k$. If M is odd, then we have $M \leq \lfloor 2a \rfloor - 1 = 2k = 2\lfloor a \rfloor$ and if M is even, then $M \leq \lfloor 2a \rfloor = 2k + 1$ implies that $M \leq 2k = 2\lfloor a \rfloor$, as desired.

11. If d is odd, then $A_2(2d + 1, d) = A_2(2d + 2, d + 1)$ and we may apply the results of part a) (with d replaced by $d + 1$) to get

$$A_2(2d + 1, d) = A_2(2d + 2, d + 1) \le 4(d + 1)$$

as desired.

Section 5.1

1. If $\alpha = 0$ then we have

$$\alpha\mathbf{0} = (\alpha + \alpha)\mathbf{0} = \alpha\mathbf{0} + \alpha\mathbf{0}$$

and so $\alpha\mathbf{0} = \alpha\mathbf{0} - \alpha\mathbf{0} = \mathbf{0}$. If $\alpha \ne 0$, then for any $\mathbf{x} \in \mathbb{Z}_p^n$ we have

$$\alpha\mathbf{0} + \mathbf{x} = \alpha\mathbf{0} + \alpha(\alpha^{-1}\mathbf{x}) = \alpha(\mathbf{0} + \alpha^{-1}\mathbf{x}) = \alpha(\alpha^{-1}\mathbf{x}) = \mathbf{x}$$

and similarly $\mathbf{x} + \alpha\mathbf{0} = \mathbf{x}$. It follows from the definition of $\mathbf{0}$ that $\alpha\mathbf{0} = \mathbf{0}$. For the second statement, if $\alpha\mathbf{x} = \mathbf{0}$ and $\alpha \ne 0$, then we may multiply both sides by α^{-1} and use the properties of scalar multiplication to get

$$\mathbf{x} = \alpha^{-1}(\alpha\mathbf{x}) = \alpha^{-1}\mathbf{0} = \mathbf{0}$$

Hence, if $\alpha \ne 0$ then $\mathbf{x} = \mathbf{0}$. Thus, one of α or $\mathbf{x}$ must be zero.

3. Since $\mathbf{0} + \mathbf{0} = \mathbf{0}$ and $\alpha\mathbf{0} = \mathbf{0}$ for all scalars α (by an earlier exercise), we deduce that $\{\mathbf{0}\}$ is closed under addition and scalar multiplication and is therefore a subspace of $\mathbb{Z}_p^n$. It is clear that the entire space $\mathbb{Z}_p^n$ is closed under addition and scalar multiplication and so it too is a subspace of $\mathbb{Z}_p^n$.

5. Let $\alpha \in \mathbb{Z}_p$ be nonzero and let $\mathbf{x} = \alpha 0$ and $\mathbf{y} = 0\alpha$. Then $\mathbf{x}\mathbf{y} = (\alpha 0)(0\alpha) = 00 = \mathbf{0}$, even though $\mathbf{x}$ and $\mathbf{y}$ are nonzero.

7. Yes. If $\mathbf{c}, \mathbf{d} \in \mathbf{E}$ then $w(\mathbf{c} + \mathbf{d}) = w(\mathbf{c}) + w(\mathbf{d}) - 2w(\mathbf{c} \cap \mathbf{d})$ is even as well. Hence, $\mathbf{E}$ is closed under addition and is therefore a subspace. No, for $w(111)$ and $w(100)$ are odd but $w(111 + 100) = w(011)$ is even. In the nonbinary case, the sum of two even weight strings may have odd weight. For example, $012 + 011 = 010$ over $\mathbb{Z}_3$.

9. No, for we have $1 + 1 = 0$ and so S is not closed under addition.

11. S is a subspace of $\mathbb{Z}_p^n$ for all p. When $p = 2$, $S = \mathbb{Z}_2^n$.

13. $\{0000, 1402, 2304, 3201, 4103\}$

15. $\alpha_1 = 1 + \beta, \alpha_2 = \beta, \alpha_3 = 0, \alpha_4 = \beta$, where $\beta = 0$ or 1.

17. $\alpha_1 = 1, \alpha_2 = 2, \alpha_3 = 3$.

19. If S is a subspace of $\mathbb{Z}_p^n$ of dimension k, then a basis for S has size k, say $\mathcal{B} = \{\mathbf{b}_1, \mathbf{b}_2, \ldots, \mathbf{b}_k\}$. Since any string $\mathbf{x}$ in S has the form

$$x = \alpha_1 \mathbf{b}_1 + \cdots + \alpha_k \mathbf{b}_k$$

for a unique sequence of coefficients $\alpha_1, \ldots, \alpha_k$, there are as many strings in S as there are sequences of coefficients. But there are p choices for each of the k coefficients in the sequence and so there are $p \cdot p \cdots p = p^k$ such sequences. Hence $|S| = p^k$.

Section 5.2

1. Yes, since $\mathrm{Rep}_r(n) = \langle \mathbf{1} \rangle$. The set $\{\mathbf{1}\}$ is a basis and $\mathrm{Rep}_r(n)$ is an $[n, 1]$-code with minimum weight n.

3. Yes, since C is closed under addition and scalar multiplication. A basis for C is $\{\mathbf{e}_2, \ldots, \mathbf{e}_n\}$, where $\mathbf{e}_k$ is the binary string with a 1 in the kth coordinate and 0s elsewhere. Hence $\dim(C) = n - 1$. The minimum weight of C is 1.

5. The formula $w(\mathbf{c} + \mathbf{d}) = w(\mathbf{c}) + w(\mathbf{d}) - 2w(\mathbf{c} \cap \mathbf{d})$ shows that if $w(\mathbf{c})$ and $w(\mathbf{d})$ have the same parity then $w(\mathbf{c} + \mathbf{d})$ is even; otherwise it is odd. For nonbinary strings, as an example, we have $120 + 110 = 200$ over $\mathbb{Z}_3$ and so this statement does not hold.

7. If not all of the codewords in C have even weight, then let $\mathbf{o} \in C$ have odd weight. Consider the subset E of C consisting of all codewords of even weight. The subset $O = \{\mathbf{e} + \mathbf{o} \mid \mathbf{e} \in E\}$ of C has the same size as E (for if $\mathbf{e}_1 + \mathbf{o} = \mathbf{e}_2 + \mathbf{o}$, then $\mathbf{e}_1 = \mathbf{e}_2$ and so the elements of O are distinct), Moreover, each codeword in O has odd weight (since "even + odd = odd"). We claim that O is the set of all codewords in C with odd weight. For if $\mathbf{d}$ has odd weight, then $\mathbf{d} + \mathbf{o}$ has even weight and so belongs to E; that is, $\mathbf{d} + \mathbf{o} = \mathbf{e}$ for some $\mathbf{e} \in E$, whence $\mathbf{d} = \mathbf{e} + \mathbf{o} \in O$. Thus, E is the set of all even weight codewords and O is the set of all odd weight codewords and $|E| = |O|$. Since $C = E \cup O$, we deduce that exactly half the codewords in C have even weight. A ternary linear code must have size a power of 3, which is odd. Hence, this statement cannot possibly be true for ternary codes.

9. $\{11111, 11101, 00110\}$ is a basis for C.

11. $C = \{0000000, 1001001, 0101011, 0010111, 1100010, 1011110, 0111100, 1110101\}$. (This is the set of all possible sums of the basis codewords, including the empty sum 0000000.) This is a $[7, 3, 3]$-code.

13. Use the generator matrix to write out all 16 codewords and check the weight of each codeword.

15. No, binary linear codes have size a power of 2.

17. Codewords in $C \oplus D$ have the form $\mathbf{c}(\mathbf{c} + \mathbf{d})$ so

$$c_1(c_1 + d_1) + c_2(c_2 + d_2) = (c_1 + c_2)(c_1 + c_2 + d_1 + d_2)$$

is in $C \oplus D$. Hence, $C \oplus D$ is closed under addition and is therefore a linear code.

Section 5.3

1. Since every string in $\mathbb{Z}_p^n$ appears exactly once in a standard array A for a linear (n, p^k)-code C over $\mathbb{Z}_p$, the number q of rows in A satisfies $qp^k = |\mathbb{Z}_p^n| = p^n$ and so

$$q = \frac{p^n}{p^k} = p^{n-k}$$

3. A standard array is

000	011	110	101
100	111	010	001

The string 111 is decoded as 011 and the string 100 is decoded as 000. Finally,

$$\mathcal{P}(\text{correct decoding}) = (1 - p)^3 + p(1 - p)^2 = (1 - p)^2$$

If 000 is sent but 001 is received, the received word will be incorrectly decoded as 101 (rather than reporting a tie).

5. A standard array for the ternary $[4, 2]$-code C_2 of Example 5.2.1 is

0000	0121	0212	2210	2001	2122	1120	1211	1002
1000	1121	1212	0210	0001	0122	2120	2211	2002
0100	0221	0012	2010	2101	2222	1220	1011	1102
0010	0101	0222	2220	2011	2102	1100	1221	1012
2000	2121	2212	1210	1001	1122	0120	0211	0002
0200	0021	0112	2110	2201	2022	1020	1111	1202
0020	0111	0202	2200	2021	2112	1110	1201	1022
1010	1101	1222	0220	0011	0102	2100	2221	2012
0110	0201	0022	2020	2111	2202	1200	1021	1112

7. The theorem holds with no change, since we are still computing the probability that the error string is a coset leader.

11. This probability is

$$(0.999)^{23} + 23(0.001)(0.999)^{22}$$
$$+ \binom{23}{2}(0.001)^2(0.999)^{21} + \binom{23}{3}(0.001)^3(0.999)^{20}$$
$$= 0.999999991$$

which is significantly larger than that of the Hamming codes in the previous exercise.

13. The number of strings of weight v or less is the volume of the sphere $S_p^n(\mathbf{0}, v)$, which is

$$V_p^n(v) = \sum_{i=0}^{v} \frac{n}{i}(p-1)^i$$

Since the number of cosets in any standard array for C is p^{n-k}, and since each of the $V_p^n(v)$ strings is the coset leader of a distinct coset, we must have $p^{n-k} \geq V_p^n(v)$. Since $M = p^k$ is the size of C, we have

$$M \leq \frac{p^n}{V_p^n(v)}$$

This is none other than the sphere-packing bound (for a linear code).

15. A generator matrix for such a code is

$$\begin{bmatrix} 1 & 0 & 0 & 0 & 1 & 0 & 0 & 0 & 0 & 0 & 0 \\ 0 & 1 & 0 & 0 & 0 & 1 & 0 & 0 & 0 & 0 & 0 \\ 0 & 0 & 1 & 0 & 0 & 0 & 1 & 0 & 0 & 0 & 0 \\ 0 & 0 & 0 & 1 & 0 & 0 & 0 & 1 & 0 & 0 & 0 \\ 1 & 0 & 0 & 0 & 0 & 0 & 0 & 0 & 1 & 0 & 0 \\ 0 & 1 & 0 & 0 & 0 & 0 & 0 & 0 & 0 & 1 & 0 \\ 0 & 0 & 1 & 0 & 0 & 0 & 0 & 0 & 0 & 0 & 1 \end{bmatrix}$$

17. For a burst of the first type, there are $p - 1$ choices for the first nonzero entry and p choices for each of the remaining $b - 1$ entries. Since there are $(n - b + 2) - 1 = n - b + 1$ choices for the location of the first nonzero entry, we deduce that there are $(n - b + 1)(p - 1)p^{b-1}$ possible bursts of this type. A burst of the second type is simply any string of the form $\mathbf{0y}$ where $\mathbf{0}$ is the zero string of length $n - b + 1$ and $\mathbf{y}$ is a nonzero string of length $b - 1$. But there are $p^{b-1} - 1$ such nonzero strings $\mathbf{y}$. Hence, $N + 1 = [(n - b + 1)(p - 1) + 1]p^{b-1}$. Finally, since $p^{n-k} \geq N + 1$, we have $p^k \leq p^n/(N + 1)$ and, taking logarithms,

$$k \leq n - \log_p(N + 1)$$

The result follows by substituting the computed value of $N + 1$.

Section 5.4

1. a) 1   b) 0   c) 3   d) 0

3. If $\mathbf{x}$ and $\mathbf{y}$ are in $A^{\perp}$ then, for any $a \in A$,

$$(\mathbf{x} + \mathbf{y}) \cdot \mathbf{a} = \mathbf{x} \cdot \mathbf{a} + \mathbf{y} \cdot \mathbf{a} = 0 + 0 = 0$$

and so $\mathbf{x} + \mathbf{y} \in A^{\perp}$. Thus, $A^{\perp}$ is closed under addition. If $\alpha \in \mathbb{Z}_p$ then

$$(\alpha\mathbf{x}) \cdot \mathbf{a} = \alpha(\mathbf{x} \cdot \mathbf{a}) = \alpha 0 = 0$$

and so $A^{\perp}$ is closed under scalar multiplication. Hence, $A^{\perp}$ is a linear code.

7. Since $\{\mathbf{a}_1, \mathbf{a}_2, \dots, \mathbf{a}_s\} \subseteq \langle \mathbf{a}_1, \mathbf{a}_2, \dots, \mathbf{a}_s \rangle$, the results of the previous exercise imply that $\langle \mathbf{a}_1, \mathbf{a}_2, \dots, \mathbf{a}_s \rangle^{\perp} \subseteq \{\mathbf{a}_1, \mathbf{a}_2, \dots, \mathbf{a}_s\}^{\perp}$. For the reverse inclusion, let $\mathbf{x} \in \{\mathbf{a}_1, \mathbf{a}_2, \dots, \mathbf{a}_s\}^{\perp}$. Then $\mathbf{x} \cdot \mathbf{a}_i = 0$ for all $i = 1, \dots, s$. Since any element $\mathbf{a}$ of $\langle \mathbf{a}_1, \mathbf{a}_2, \dots, \mathbf{a}_s \rangle$ is a linear combination of the $\mathbf{a}_i$'s, we have

$$\mathbf{x} \cdot \mathbf{a} = \mathbf{x} \cdot (\alpha_1 \mathbf{a}_1 + \alpha_2 \mathbf{a}_2 + \cdots + \alpha_s \mathbf{a}_s)$$
$$= \alpha_1(\mathbf{x} \cdot \mathbf{a}_1) + \alpha_2(\mathbf{x} \cdot \mathbf{a}_2) + \cdots + \alpha_s(\mathbf{x} \cdot \mathbf{a}_s) = 0$$

we see that $\mathbf{x}$ is orthogonal to every element of $\langle \mathbf{a}_1, \mathbf{a}_2, \dots, \mathbf{a}_s \rangle$, that is, $\mathbf{x} \in \langle \mathbf{a}_1, \mathbf{a}_2, \dots, \mathbf{a}_s \rangle^{\perp}$. Hence, $\langle \mathbf{a}_1, \mathbf{a}_2, \dots, \mathbf{a}_s \rangle^{\perp} = \{\mathbf{a}_1, \mathbf{a}_2, \dots, \mathbf{a}_s\}^{\perp}$.

9. $\mathrm{Rep}_p(n)^{\perp} = \{\mathbf{x} = x_1 \cdots x_n \in \mathbb{Z}_p^n \mid x_1 = -(x_2 + \cdots + x_n)\}$

11. Col1 = col2 and so there is a set of two columns that are linearly dependent, whence $d = 2$.

13. Every pair of columns is linearly independent but col1 + col2 = col4 and so $d = 3$.

15. $\begin{bmatrix} -1 & 1 & 0 & 0 & \cdots & 0 \\ -1 & 0 & 1 & 0 & \cdots & 0 \\ -1 & 0 & 0 & 1 & \cdots & 0 \\ \vdots & \vdots & \vdots & \vdots & \ddots & \vdots \\ -1 & 0 & 0 & 0 & \cdots & 1 \end{bmatrix}$

17. Suppose C is a binary self-dual code. If $\mathbf{c} \in C$ then $w(\mathbf{c}) = \mathbf{c} \cdot \mathbf{c} = 0$ in $\mathbb{Z}_2$, whence $w(\mathbf{c})$ must be an even integer. Since all codewords have even weight, we have $\mathbf{c} \cdot \mathbf{1} \equiv w(\mathbf{c}) \bmod 2$ and so $\mathbf{c} \cdot \mathbf{1} = 0$, which implies that $\mathbf{1} \in C^{\perp} = C$.

19. Since $C = C^{\perp}$, we have $\dim(C) = n - \dim(C^{\perp}) = n - \dim(C)$ and so $2 \cdot \dim(C) = n$, whence n is even. A generator matrix for C_n is the

$(n \times 2n)$-matrix

$$G_n = \begin{bmatrix} 1 & 1 & & & & \\ & & 1 & 1 & & \\ & & & & 1 & 1 \\ & & & & & & \ddots \end{bmatrix}$$

that has 0s everywhere except where indicated. It is clear that every row of G_n is orthogonal to itself and to all other rows and so $C_n \subseteq C_n^{\perp}$. But $\dim(C) = \frac{n}{2}$ and so $\dim(C^{\perp}) = n - \dim(C) = \frac{n}{2} = \dim(C)$ and so $C = C^{\perp}$.

21. The Gilbert-Varshamov inequality is $2^k < \frac{32}{3}$ and so we may take $k = 3$, whence $A_2(6, 3) \geq 2^3 = 8$, which is the exact value.

23. The Gilbert-Varshamov inequality is $2^k < 4$ and so we must take $k = 1$, whence $A_2(8, 5) \geq 2$. The exact value is 4.

Section 5.5

1. The number of rows is $|\mathbb{Z}_p^n| / |C| = p^n / p^k = p^{n-k}$. A generator matrix G has k rows and n columns so the number of rows in a syndrome-coset leader table is $p^{\#\text{cols } G - \#\text{rows } G}$. We cannot read the size of a syndrome-coset leader table directly from the size of a parity check matrix, however.

3. The parity check matrix for the Hamming $[7, 4, 3]$-code C_3 is

$$P_3 = \begin{bmatrix} 0 & 1 & 1 & 1 & 1 & 0 & 0 \\ 1 & 0 & 1 & 1 & 0 & 1 & 0 \\ 1 & 1 & 0 & 1 & 0 & 0 & 1 \end{bmatrix}$$

A syndrome-coset leader table for this code is

coset leader	syndrome
0000000	000
1000000	011
0100000	101
0010000	110
0001000	111
0000100	100
0000010	010
0000001	001

a) 1101101 is decoded as 1101001, b) 1111111 is decoded as itself, c) 0000001 is decoded as 0000000.

5. A syndrome table is a) 0110 is decoded as 0000, b) 2222 is decoded as 2122, c) 2012 is decoded as 1002.

7. If r is the number of coset leaders of weight 1 with odd weight syndromes, then $90 - r$ have syndromes of even weight. Now, a coset leader of weight 2 has the form $\mathbf{e}_i + \mathbf{e}_j$, and since $S(\mathbf{e}_i + \mathbf{e}_j) = S(\mathbf{e}_i) + S(\mathbf{e}_j)$, the weight of the former is odd if and only if the weights of $S(\mathbf{e}_i)$ and $S(\mathbf{e}_j)$ have different parity. Hence, there are $r(90 - r)$ coset leaders of weight 2 that have odd parity syndromes. But since the set of syndromes is precisely the set $\mathbb{Z}_2^{12}$ and since half of these strings have odd weight, we get $r + r(90-r) = 2^{11} = 2048$. But this has no solutions in positive integers, as can be seen by trial and error.

Section 5.6

1. Since C has size p, it has dimension 1. Let $\{\mathbf{c}\}$ be a basis for C, where $\mathbf{c} = c_1 \cdots c_n$. Hence, $C = \{(\alpha c_1) \cdots (\alpha c_n) \mid \alpha \in \mathbb{Z}_p\}$ and

$$G = [c_1 \quad c_2 \quad \cdots \quad c_n]$$

is a generator matrix for C. Now, if $c_i = 0$ for some i, then all codewords in C will have a 0 in the ith position, which implies that the minimum distance of C is at most $n - 1$. Since $d(C) = n$, we conclude that $c_i \neq 0$ for all i. Now, if $c_i \neq 1$, then we may multiply the ith column of G by c_i^{-1}, which will turn the ith entry into a 1. By doing this for all columns, we get the matrix $G' = [1 \quad 1 \quad \cdots \quad 1]$, which is the generator matrix for $\text{Rep}_p(n)$.

3. $G = \begin{bmatrix} 1 & 0 & 0 & 2 \\ 0 & 1 & 2 & 1 \end{bmatrix}$, $P = \begin{bmatrix} 0 & 1 & 1 & 0 \\ 1 & 2 & 0 & 1 \end{bmatrix}$.

5. $G = \begin{bmatrix} 1 & 0 & 0 & 1 \\ 0 & 1 & 0 & 1 \\ 0 & 0 & 1 & 0 \end{bmatrix}$, $P = \begin{bmatrix} 1 & 1 & 0 & 1 \end{bmatrix}$.

7. It is easy to see that a parity check matrix for E_n is $P = 1 \quad 1 \quad \cdots \quad 1$. The generator matrix that leads to this parity check matrix is

$$G = \begin{bmatrix} 1 & 0 & \cdots & 0 & 1 \\ 0 & 1 & \cdots & 0 & 1 \\ \vdots & \vdots & \ddots & \vdots & \vdots \\ 0 & 0 & \cdots & 1 & 1 \end{bmatrix}$$

which consists of an identity matrix of size $n - 1$ with an additional column at the far right consisting entirely of 1s.

11. Let $\mathbf{g}_1$ and $\mathbf{g}_2$ be rows of G and let $\overline{\mathbf{g}_1} = \mathbf{g}_1\epsilon$ and $\overline{\mathbf{g}_2} = \mathbf{g}_2\delta$ be the corresponding codewords in $\overline{C}$, where ϵ and δ equal 0 or 1. If $w(\mathbf{g}_1)$ and $w(\mathbf{g}_2)$ have the same parity, then $\epsilon = \delta$ and since in this case $\epsilon + \delta = 0$, we have $\overline{\mathbf{g}_1} + \overline{\mathbf{g}_2} = (\mathbf{g}_1 + \mathbf{g}_2)0$. But $w(g_1 + g_2)$ is even in this case, and so $\overline{\mathbf{g}_1} + \overline{\mathbf{g}_2} = (\mathbf{g}_1 + \mathbf{g}_2)0 = \overline{\mathbf{g}_1 + \mathbf{g}_2}$. Similarly, if $w(\mathbf{g}_1)$ and $w(\mathbf{g}_2)$ have different parity, then $\epsilon \neq \delta$ and so $\epsilon + \delta = 1$. Hence, $\overline{\mathbf{g}_1} + \overline{\mathbf{g}_2} = (\mathbf{g}_1 + \mathbf{g}_2)1$. But $w(\mathbf{g}_1 + \mathbf{g}_2)$ is odd in this case, and so $\overline{\mathbf{g}_1} + \overline{\mathbf{g}_2} = (\mathbf{g}_1 + \mathbf{g}_2)1 = \overline{\mathbf{g}_1 + \mathbf{g}_2}$. Thus, in either case, $\overline{\mathbf{g}_1} + \overline{\mathbf{g}_2} = \overline{\mathbf{g}_1 + \mathbf{g}_2}$. We may now extend this to any sum (i.e., linear combination) of the rows of G. Thus, if $\mathbf{c} \in C$ has the form

$$\mathbf{c} = \alpha_1 g_1 + \cdots + \alpha_k g_k$$

where $\alpha_i = 0$ or 1, then

$$\overline{\mathbf{c}} = \alpha_1 \overline{g}_1 + \cdots + \alpha_k \overline{g}_k$$

This shows that, by appending an even parity check to each row of G (thereby increasing the number of columns by 1), we get a matrix whose rows generate $\overline{C}$. It is easy to see that the rows are still linearly independent and so the new matrix is a generator matrix for $\overline{C}$.

13. By performing the following elementary row operations on P: a) add row 2 to row 1, b) add row 1 to row 3, c) add row 3 to row 2, we get the matrix

$$\begin{bmatrix} 0 & 1 & 1 & 1 & 1 & 0 & 0 \\ 1 & 0 & 1 & 1 & 0 & 1 & 0 \\ 1 & 1 & 0 & 1 & 0 & 0 & 1 \end{bmatrix} = [A \mid I_3]$$

which is in right standard form. Hence, a generator matrix for the code C is

$$G = [I_4 \mid -A^t] = \begin{bmatrix} 1 & 0 & 0 & 0 & 0 & 1 & 1 \\ 0 & 1 & 0 & 0 & 1 & 0 & 1 \\ 0 & 0 & 1 & 0 & 1 & 1 & 0 \\ 0 & 0 & 0 & 1 & 1 & 1 & 1 \end{bmatrix}$$

Section 5.7

1. The maximum value of k is 3. To encode **1**, we have

$$\begin{bmatrix} 1 & 1 & 1 \end{bmatrix} \begin{bmatrix} 1 & 0 & 0 & 0 \\ 0 & 1 & 0 & 1 \\ 0 & 0 & 1 & 0 \end{bmatrix} = \begin{bmatrix} 1 & 1 & 1 & 1 \end{bmatrix}$$

so **1** is encoded as 1111.

3. Note that $\{1000, 1101\}$ is a basis for C, whence the maximum value of k is 2. To encode $\mathbf{1}$ we have

$$\begin{bmatrix} 1 & 1 \end{bmatrix} \begin{bmatrix} 1 & 0 & 0 & 0 \\ 0 & 1 & 0 & 1 \end{bmatrix} = \begin{bmatrix} 1 & 1 & 0 & 1 \end{bmatrix}$$

and so $\mathbf{1}$ is encoded as 1101.

5. Since $\begin{bmatrix} 1 & 2 \end{bmatrix} G = \begin{bmatrix} 1 & 2 & 1 & 1 \end{bmatrix}$, we see that 1211 is source decoded as 12.

7. From the matrix P_3, we deduce that each codeword has the form $(\beta + \gamma)(\alpha + \gamma)(\alpha + \beta)(\alpha + \beta + \gamma)\alpha\beta\gamma$, where $\alpha, \beta, \gamma \in \mathbb{Z}_2$. From the matrix H, we deduce that each codeword has the form $(\beta' + \gamma')(\alpha' + \beta')(\alpha' + \gamma')\alpha'(\alpha' + \beta' + \gamma')\beta'\gamma'$, where $\alpha', \beta', \gamma' \in \mathbb{Z}_2$. These are the same, as can be seen by letting $\alpha' = \alpha + \beta + \gamma$, $\beta' = \beta$ and $\gamma' = \gamma$ and substituting.

9. a) The information set is $\{1,3\}$ as can be seen by underlining these positions in each codeword: $\{0000, 0110, 1001, 1010\}$. The underlined bits are 00, 01, 10, 11, which is all 4 binary strings of length 2. b) By simply writing down the strings of length 2 obtained by crossing out one coordinate, we get $\{00,10,01,00\}$, $\{00,10,00,01\}$ and $\{00,00,10,01\}$, none of which is all of $\mathbb{Z}_2^2$. Hence, C is not systematic. c) If a code is sytematic, we may easily encode source strings by embedding them in codewords without any changes.

Section 6.1

1. a) The syndrome is $S(1111000) = [100]$ and since $100_{binary} = 4_{decimal}$ the nearest neighbor codeword is $1111000 - \mathbf{e}_4 = 1110000$. b) The syndrome is $S(\mathbf{1}) = [000]$ and so the error string is $\mathbf{0}$, that is, $\mathbf{1}$ is a codeword.

3. $H_2(2) = \begin{bmatrix} 0 & 1 & 1 \\ 1 & 0 & 1 \end{bmatrix}$. The syndrome of 101 is $S(101) = [101]H_2(2)^t = [10]$ and since $10_{binary} = 2_{decimal}$, the word 101 is decoded as $101 - \mathbf{e}_2 = 111$.

5. $H_2(4) = \begin{bmatrix} 0 & 0 & 0 & 0 & 0 & 0 & 0 & 1 & 1 & 1 & 1 & 1 & 1 & 1 & 1 \\ 0 & 0 & 0 & 1 & 1 & 1 & 1 & 0 & 0 & 0 & 0 & 1 & 1 & 1 & 1 \\ 0 & 1 & 1 & 0 & 0 & 1 & 1 & 0 & 0 & 1 & 1 & 0 & 0 & 1 & 1 \\ 1 & 0 & 1 & 0 & 1 & 0 & 1 & 0 & 1 & 0 & 1 & 0 & 1 & 0 & 1 \end{bmatrix}$

The syndrome of 111110000011111 is $[111110000011111]H_2(4)^t = [1010]$ and since $1010_{binary} = 10_{decimal}$, the nearest neighbor codeword is $111110000011111 - \mathbf{e}_{10} = 111110000111111$.

7. Plugging into the left side of the sphere-packing condition gives (with the help of the binomial theorem)

$$r^{n-h}\left(1 + \frac{r^h - 1}{r - 1}(r - 1)\right) = r^{n-h}r^h = r^n$$

which is equal to the right side.

9. By performing the following elementary row operations to $H_2(3)$: a) add row 2 to row 1, b) add row 1 to row 3, c) add row 3 to row 2, we get the following parity check matrix in right standard form

$$P = \begin{bmatrix} 0 & 1 & 1 & 1 & 1 & 0 & 0 \\ 1 & 0 & 1 & 1 & 0 & 1 & 0 \\ 1 & 1 & 0 & 1 & 0 & 0 & 1 \end{bmatrix} = [A \mid I_3]$$

Hence, a generator matrix for $\mathcal{H}_2(3)$ is

$$G = [I_4 \mid -A^t] = \begin{bmatrix} 1 & 0 & 0 & 0 & 0 & 1 & 1 \\ 0 & 1 & 0 & 0 & 1 & 0 & 1 \\ 0 & 0 & 1 & 0 & 1 & 1 & 0 \\ 0 & 0 & 0 & 1 & 1 & 1 & 1 \end{bmatrix}$$

11. Following the hint, the matrix Q obtained from P by multiplying each column by the inverse of the most significant symbol has $n = \frac{p^h - 1}{p - 1}$ columns, each of which has most significant symbol 1. Since n is the total number of strings of length h with most significant symbol 1, the matrix Q has the same columns as the corresponding Hamming matrix. Hence, C is scalar multiple equivalent to that Hamming code.

13. First note that no word of weight 2 is a codeword, since it has distance 2 from the codeword 0. Consider the packing spheres $S_2^n(\mathbf{c}, 1)$ about each codeword. If $\mathbf{x}$ is a word of weight 2, then it must lie in one of these spheres, say $S_2^n(\mathbf{c}, 1)$. It is clear that $\mathbf{c} \neq 0$. Since $d(\mathbf{0}, \mathbf{x}) = 2$ and since $d(\mathbf{x}, \mathbf{c}) = 1$, we see that $d(\mathbf{c}, \mathbf{0}) = 3$, that is, $\mathbf{c}$ has weight 3. Thus, any word $\mathbf{x}$ of weight 2 lies in a packing sphere centered at a codeword of weight 3. Moreover, no such word $\mathbf{x}$ lies in more than one such sphere, since no word can have distance 1 from two distinct codewords. Also, there are precisely 3 words of weight 2 in each sphere with center $\mathbf{c}$ of weight 3. (These words are found by deleting one of the three 1s in $\mathbf{c}$.) Hence, the number of codewords of weight 3 is equal to the number of words of weight 2 in $\mathbb{Z}_2^n$ divided by 3, that is,

$$N = \text{\# codewords of weight 3} = \frac{\binom{n}{2}}{3} = \frac{n(n-1)}{6}$$

At first, it seems like the latter might not be an integer, but recall that n has the form $n = 2^h - 1$ and so

$$N = \frac{(2^h - 1)(2^h - 2)}{6} = \frac{(2^h - 1)(2^{h-1} - 1)}{3}$$

Now, $2^{h-1} - 1$ can always be written in the form $3u + r$, where $r = 0, 1$ or 2. But if $r = 2$, we get $2^{h-1} - 1 = 3u + 2$, or $2^{h-1} = 3u + 3$, which is not possible since 2^{h-1} is not divisible by 3. Hence, $r = 0$ or 1. But if $r = 0$ then $2^{h-1} - 1$ is divisible by 3 and if $r = 1$ then $2^h - 1 = 2(2^{h-1} - 1) + 1 = 2(3u + 1) + 1 = 6u + 3$ is divisible by 3.

15. We prove this by induction on h. For $h = 2$, the result follows simply by checking each pair of codewords. Assume it is true for h and let **c** and **d** be codewords in $S(h + 1)$. There are four cases to consider. If $\mathbf{c} = \mathbf{a}0\mathbf{a}$ and $\mathbf{d} = \mathbf{b}0\mathbf{b}$, then

$$d(\mathbf{c}, \mathbf{d}) = 2d(\mathbf{a}, \mathbf{b}) = 2 \cdot 2^{h-1} = 2^h$$

If $\mathbf{c} = \mathbf{a}1\mathbf{a}^c$ and $\mathbf{d} = \mathbf{b}1\mathbf{b}^c$, then

$$d(\mathbf{c}, \mathbf{d}) = d(\mathbf{a}, \mathbf{b}) + d(\mathbf{a}^c, \mathbf{b}^c) = 2 \cdot 2^{h-1} = 2^h$$

If $\mathbf{c} = \mathbf{a}0\mathbf{a}$ and $\mathbf{d} = \mathbf{b}1\mathbf{b}^c$, then, since $d(\mathbf{a}, \mathbf{b}^c) = 2^h - 1 - d(\mathbf{a}, \mathbf{b}) = 2^h - 1 - 2^{h-1} = 2^{h-1} - 1$, we get

$$d(\mathbf{c}, \mathbf{d}) = d(\mathbf{a}, \mathbf{b}) + 1 + d(\mathbf{a}, \mathbf{b}^c) = 2^{h-1} + 1 + 2^{h-1} - 1 = 2^h$$

The case $\mathbf{c} = \mathbf{a}1\mathbf{a}^c$ and $\mathbf{d} = \mathbf{b}0\mathbf{b}$ is similar and so in all four cases, we get $d(\mathbf{c}, \mathbf{d}) = 2^h$. This completes the proof by induction.

17. The table has $\frac{2^{24}}{2^{12}} = 4096$ rows.

19. By permuting columns if necessary, we can arrange it so that the first 8 columns $\mathbf{c}_1, \ldots, \mathbf{c}_8$ are linearly dependent. Elementary row operations can then be used to put the identity matrix I_7 in the upper left hand corner, with 0s below in rows 8 through 12. Moreover, the eighth column must be the sum of columns 1 through 7, for otherwise it is a sum of fewer columns, which would imply that fewer than eight columns are linearly dependent.

21. Since the codewords in $\mathcal{N}$ are formed by removing at most two 1s from words of weight at least 8, the codewords in $\mathcal{N}$ have weight at least 6. Plotkin's bound gives $A_2(16, 7) \leq 4 \cdot A_2(14, 7) \leq 4 \cdot 16 = 64$ and so $\mathcal{N}$ cannot have weight 7 or more.

Section 6.2

1. Assume that $\mathcal{R}(m)$ is linear. Let $\mathbf{c}, \mathbf{d} \in \mathcal{R}(m + 1)$. There are four cases to consider. *Case 1*: If $\mathbf{c} = \mathbf{uu}$ and $\mathbf{d} = \mathbf{vv}$, where $\mathbf{v} \in \mathcal{R}(m)$ then

$$\mathbf{c} + \mathbf{d} = \mathbf{uu} + \mathbf{vv} = (\mathbf{u} + \mathbf{v})(\mathbf{u} + \mathbf{v}) \in \mathcal{R}(m + 1)$$

   Case 2: If $\mathbf{c} = \mathbf{uu}$ and $\mathbf{d} = \mathbf{vv}^c$, where $\mathbf{u}, \mathbf{v} \in \mathcal{R}(m)$ then

$$\mathbf{c} + \mathbf{d} = \mathbf{uu} + \mathbf{vv}^c = (\mathbf{u} + \mathbf{v})(\mathbf{u} + \mathbf{v}^c) = (\mathbf{u} + \mathbf{v})(\mathbf{u} + \mathbf{v})^c \in \mathcal{R}(m + 1)$$

   Case 3: If $\mathbf{c} = \mathbf{uu}^c$ and $\mathbf{d} = \mathbf{vv}$, where $\mathbf{u}, \mathbf{v} \in \mathcal{R}(m)$ then

$$\mathbf{c} + \mathbf{d} = \mathbf{uu}^c + \mathbf{vv} = (\mathbf{u} + \mathbf{v})(\mathbf{u}^c + \mathbf{v}) = (\mathbf{u} + \mathbf{v})(\mathbf{u} + \mathbf{v})^c \in \mathcal{R}(m + 1)$$

   Case 4: If $\mathbf{c} = \mathbf{uu}^c$ and $\mathbf{d} = \mathbf{vv}^c$, where $\mathbf{u}, \mathbf{v} \in \mathcal{R}(m)$ then

$$\mathbf{c} + \mathbf{d} = \mathbf{uu}^c + \mathbf{vv}^c = (\mathbf{u} + \mathbf{v})(\mathbf{u}^c + \mathbf{v}^c) = (\mathbf{u} + \mathbf{v})(\mathbf{u} + \mathbf{v}) \in \mathcal{R}(m + 1)$$

   Thus $\mathcal{R}(m + 1)$ is closed under addition and so is linear.

3. This is certainly true for the matrix R_1. Assume the ith row of R_m consists of alternating blocks of 0s and 1s of length 2^{m-i} and the last row is $\mathbf{1}$. From (6.2.1) we see that the first row of R_{m+1} consists of alternating blocks of 0s and 1s of length $2^m = 2^{(m+1)-1}$ and the last row of R_{m+1} (which is row $m + 2$) is $\mathbf{1}$. The ith row of R_{m+1}, where $2 \leq i \leq m + 1$, consists of two copies of the $(i - 1)$-st row of R_m which, by assumption, consists of alternating blocks of 0s and 1s of length $2^{m-(i-1)} = 2^{(m+1)-i}$. Hence, R_{m+1} has the desired form.

5. $P = \begin{bmatrix} 1 & 1 & 1 & 1 \end{bmatrix}$

7. Interchange the fourth and fifth columns of R_3 and bring the resulting matrix to left standard form by row operations. Then interchange the fourth and fifth columns again to get

$$P = \begin{bmatrix} 1 & 1 & 1 & 1 & 0 & 0 & 0 & 0 \\ 1 & 1 & 0 & 0 & 1 & 1 & 0 & 0 \\ 1 & 0 & 1 & 0 & 1 & 0 & 1 & 0 \\ 0 & 1 & 1 & 0 & 1 & 0 & 0 & 1 \end{bmatrix}$$

9. $\alpha_1 = 1, \alpha_2 = 1, \alpha_3 = 0$, $\mathbf{x} - (\alpha_1 \mathbf{r}_1 + \alpha_2 \mathbf{r}_2 + \alpha_3 \mathbf{r}_3) = 11111101$ so $\alpha_4 = 1$, giving $\mathbf{c} = \mathbf{r}_1 + \mathbf{r}_2 + \mathbf{1} = 11000011$.

11. All of the majority logic equations give the correct value of the coefficients α_i.

13. We proceed by induction on r. For $r = 0$, we have $\mathcal{R}_0(m) = \text{Rep}_2(2^m)$, which has parameters $[2^m, 1, 2^m]$, as required. We have already seen that the first order codes $\mathcal{R}_1(m)$ have parameters $[2^m, m + 1, 2^{m-1}]$, also as required.

Assume that $\mathcal{R}_{r-1}(m)$ has parameters

$$[2^m, 1 + \binom{m}{1} + \cdots + \binom{m}{r-1}, 2^{m-r+1}]$$

for all $m \geq r - 1$ and consider the codes $\mathcal{R}_r(m)$. If $m = r$, then $\mathcal{R}_r(m) = \mathcal{R}_r(r) = \mathbb{Z}_2^{2^m}$, which has parameters $[2^m, 2^m, 1]$. Since

$$1 + \binom{m}{1} + \cdots + \binom{m}{m} = 2^m$$

these are the required parameters. We must now proceed by induction on m (for r fixed.) Suppose that $\mathcal{R}_r(m)$ has the required parameters

$$[2^m, 1 + \binom{m}{1} + \cdots + \binom{m}{r}, 2^{m-r}]$$

and consider the code $\mathcal{R}_r(m + 1) = \mathcal{R}_r(m) \oplus \mathcal{R}_{r-1}(m)$. Recalling the parameters of a $\mathbf{u}(\mathbf{u} + \mathbf{v})$-construction, we see that $\mathcal{R}_r(m + 1)$ has length

$$\text{len}(\mathcal{R}_r(m + 1)) = 2\text{len}(\mathcal{R}_r(m)) = 2 \cdot 2^m = 2^{m+1}$$

The size of $\mathcal{R}_r(m + 1)$ is

$$|\mathcal{R}_r(m + 1)| = |\mathcal{R}_r(m)| \, |\mathcal{R}_{r-1}(m)| = 2^{dim(\mathcal{R}_r(m)) + dim(\mathcal{R}_{r-1}(m))}$$

and so its dimension is

$$\dim(\mathcal{R}_r(m + 1)) = \dim(\mathcal{R}_r(m)) + \dim(\mathcal{R}_{r-1}(m))$$
$$= 1 + \binom{m}{1} + \cdots + \binom{m}{r} + 1 + \binom{m}{1} + \cdots + \binom{m}{r-1}$$

Since $\binom{m}{s} + \binom{m}{s-1} = \binom{m+1}{s}$, we can combine these terms to get

$$\dim(\mathcal{R}_r(m + 1)) = 1 + \binom{m+1}{1} + \cdots + \binom{m+1}{r}$$

Finally, the minimum distance is

$$d(\mathcal{R}_r(m + 1)) = \min\{2d(\mathcal{R}_r(m)), \dim(\mathcal{R}_{r-1}(m))\}$$
$$= \min\{2 \cdot 2^{m-r}, 2^{m-r+1}\} = 2^{m+1-r}$$

Thus, if $\mathcal{R}_r(m)$ has the required parameters, so does $\mathcal{R}_r(m + 1)$. This proves that $\mathcal{R}_r(m)$ has the required parameters for all $m \geq r$.

Section 6.3

1. The ISBN code is nonlinear since, for example, the sum of two ISBN code-words that both begin with a 5 has a 10 in the first position and so is not in $\mathcal{I}$. Since we may choose any digits for the first nine positions of an ISBN, there are 10^9 possible ISBNs. Hence, $\mathcal{I}$ has length 10 and size 10^9. Finally, the minimum distance of $\mathcal{I}$ is no smaller than the minimum distance of the linear code from whence it came and so $d(\mathcal{I}) \geq 2$. Since there are ISBNs of weight 2, we conclude that $d(\mathcal{I}) = 2$.

3. The code D is nonlinear since, for instance, $\mathbf{c} = 2000000045$ is in D but $5\mathbf{c} = (10)000000093$ is not in D. The minimum distance of D is not less than the minimum distance of the linear code from whence it came and so $d(D) \geq 3$. But there are codewords in D of weight 3 and so $d(D) = 3$.

5. a) 0111246792 (error $7e_4$) b) 4511156514 (error $8e_8$)
   c) 4535797953 (error e_{10})

7. Trasnpostiion erorrs are one fo teh msot common errosr made when tpying or wrtiing demcial digits.

9. The linear dependence of the first five columns of P shows that $c = (10)454(10)00000$ is in the code with parity check matrix P. Multiplying this by 3 gives a codeword with no 10s and thus a codeword in E of weight 5.

13. a) 1218681845 b) 7289985648 c) 1211923347 d) 1211637417 e) more than two errors

Section 6.4

1. Assuming the set $S_2 = \{1, 2\}$, there are only two Latin squares,

$$\begin{bmatrix} 1 & 2 \\ 2 & 1 \end{bmatrix} \text{ and } \begin{bmatrix} 2 & 1 \\ 1 & 2 \end{bmatrix}$$

and these are not orthogonal.

3. Write $q = 4u + r$ where $r = 0, 1, 2$ or 3. Suppose first that $r = 0, 1$ or 3. We want to show that if $2 \mid q$ then $4 \mid q$. If $2 \mid q$ then since $2 \mid 4u$ we must have $2 \mid r$. Hence, r cannot equal 1 or 3 and so must equal 0, that is, $q = 4u$, which is divisible by 4. For the converse, suppose that if $2 \mid q$ then $4 \mid q$. We must show that $r \neq 2$. But if $r = 2$ then $q = 4u + 2$ which is divisible by 2. It would follow that $4 \mid q$, in which case $4 \mid (q - 4u)$, that is, $4 \mid 2$, a contradiction. Hence, $r \neq 2$.

5. $\begin{bmatrix} 1 & 2 & 3 & 4 & 5 \\ 2 & 3 & 4 & 5 & 1 \\ 3 & 4 & 5 & 1 & 2 \\ 4 & 5 & 1 & 2 & 3 \\ 5 & 1 & 2 & 3 & 4 \end{bmatrix}$

7. Let $S_5 = \mathbb{Z}_5$. Let $\alpha = 1$, $\beta = 2$ in the proof of Theorem 6.4.4. Then $A = (i+j)$ and $B = (i + 2j)$. Hence,

$$A = \begin{bmatrix} 0 & 1 & 2 & 3 & 4 \\ 1 & 2 & 3 & 4 & 0 \\ 2 & 3 & 4 & 0 & 1 \\ 3 & 4 & 0 & 1 & 2 \\ 4 & 0 & 1 & 2 & 3 \end{bmatrix} \text{ and } B = \begin{bmatrix} 0 & 2 & 4 & 1 & 3 \\ 1 & 3 & 0 & 2 & 4 \\ 2 & 4 & 1 & 3 & 0 \\ 3 & 0 & 2 & 4 & 1 \\ 4 & 1 & 3 & 0 & 2 \end{bmatrix}$$

9. The other two transversals are $(1,1)$, $(2,3)$, $(3,2)$ and $(1,2)$, $(2,1)$, $(3,3)$.

Section 7.1

1. Any codeword $\mathbf{c} \in C$ is a linear combination of basis codewords

$$\mathbf{c} = \alpha_1 \mathbf{b}_1 + \cdots + \alpha_k \mathbf{b}_k$$

But if we denote the right cyclic shift of a string $\mathbf{x}$ by $\mathbf{x}'$, then

$$\mathbf{c}' = \alpha_1 \mathbf{b}_1' + \cdots + \alpha_k \mathbf{b}_k'$$

and so $\mathbf{c}'$ is also in C.

3. Referring to the hint, a generator matrix is $G = [-A^t \mid I]$, that is

$$G = \begin{bmatrix} 1 & 1 & 0 & 1 & 0 & 0 & 0 \\ 0 & 1 & 1 & 0 & 1 & 0 & 0 \\ 1 & 1 & 1 & 0 & 0 & 1 & 0 \\ 1 & 0 & 1 & 0 & 0 & 0 & 1 \end{bmatrix}$$

To see that C is cyclic, observe that the right cyclic shift of each row of G is orthogonal to each row of P and so is in C. It follows that the right cyclic shift of any codeword in C is in C.

5. E has generator polynomial $1 + x^2$. The polynomial $x + x^3$ also generates, but $1 + x + x^2 + x^3$ does not.

7. The polynomial $x^4 - 1$ factors into irreducible factors over $\mathbb{Z}_2$ as follows

$$x^4 - 1 = (1 + x)^4$$

Hence, the possible factors of $x^4 - 1$ are 1, $1 + x$, $1 + x^2$, $1 + x + x^2 + x^3$ and $x^4 - 1$. Thus, a complete list of binary cyclic codes of length 4 is $C_0 =$

$\langle\langle 1 \rangle\rangle = \mathcal{R}_4 = \mathbb{Z}_2^4$

$C_1 = \langle\langle 1 + x \rangle\rangle = \langle 1 + x, x + x^2, x^2 + x^3 \rangle = \langle 1100, 0110, 0011 \rangle$

$C_2 = \langle\langle (1 + x^2) \rangle\rangle = \langle 1 + x^2, x + x^3 \rangle = \langle 1010, 0101 \rangle$

$C_3 = \langle\langle 1 + x + x^2 + x^3 \rangle\rangle = \langle 1 + x + x^2 + x^3 \rangle = \langle 1111 \rangle$

$C_4 = \langle\langle x^4 - 1 \rangle\rangle = \{0000\}$

9. If $h(x) = h_0 + h_1 x + \cdots + h_k x^k$ then $h^r(x) = h_k + h_{k-1} x + \cdots + h_0 x^k$. Now, $x^n - 1 = h(x)g(x)$ and so

$$h^r(x)g^r(x) = x^k h(x^{-1}) x^{n-k} g(x^{-1})$$
$$= x^n[h(x^{-1})g(x^{-1})] = x^n(x^{-n} - 1) = 1 - x^n$$

and so $h^r(x)$ divides $x^n - 1$, which implies that $h_0^{-1} h^r(x)$ is the generator polynomial for a cyclic code D with generator matrix H, as given in the text. But this is also the parity check matrix for C and so $D = C^\perp$.

11. $\{0\}$, $\langle\langle 1 + x \rangle\rangle$, $\langle\langle 1 + x + x^2 + x^3 + x^4 \rangle\rangle$, $\mathbb{Z}_2^5$.

13. Let $C = \langle\langle -1 + x \rangle\rangle$. Then C is cyclic and has dimension $n - 1$. In addition, the parity check polynomial for C is $h(x) = 1 + x + x^2 + \cdots + x^{n-1}$ and so a parity check matrix for C is $H = [11 \cdots 1]$. It follows that every codeword in C has even weight and so $C = E_n$ (since the latter also has dimension $n - 1$).

15. Since a right cyclic shift of a word of even weight also has even weight, the code E is cyclic. Since $|E| = |C|/2$, the code E has dimension $\dim(C) - 1$, and so its generator polynomial $p(x)$ is a linear multiple of $g(x)$, and so $p(x) = xg(x)$ or $p(x) = (x + 1)g(x)$. But the former is not possible, since a generator polynomial has nonzero constant term. Hence, $p(x) = (x+1)g(x)$.

17. $C^\perp = \langle h^r(x) \rangle$ is equivalent to $\langle\langle h(x) \rangle\rangle$.

References

Cited Articles on Coding and Information Theory

Golay, M., Notes on Digital Coding, *Proc. IRE*, Vol. 37 (1949): 657.

Huffman, D. A., A Method for the Construction of Minimum Redundancy Codes, *Proc. IRE*, Vol. 40, No. 4 (1952): 1098–1101.

Kraft, L. G., A Device for Quantizing, Grouping, and Coding Amplitude Modulated Pulses, Q. S. Thesis, Electrical Engineering Department, MIT, 1949.

McMillan, B., Two Inequalities Implied by Unique Decipherability, *IRE Trans. Inform. Theory*, IT-2 (1956): 115–116.

Pless, V., On the Uniqueness of the Golay Codes, *J. Comb. Theory*, 5 (1968): 215-228.

Delsarte, P. and Goethals, J.M., Unrestricted Codes with the Golay Parameters Are Unique, *Discrete Mathematics*, 12 (1975): 211–224.

Sloane, N.J.A., Recent bounds for codes, sphere packings and related problems obtained by linear programming and other methods, *Contemporary Mathematics* 9, 153–185.

Snover, S.L., The Uniqueness of the Nordstrom-Robinson and the Binary Golay Codes, Ph.D. Thesis, Department of Mathematics, Michigan State University, 1973.

Books on Coding and Information Theory

Ash, Robert, *Information Theory*, Dover Publications, 1965.

Berlekamp, Elwyn, ed., *Key Papers in the Development of Coding Theory*, IEEE Press, 1974.

Hamming, Richard, *Coding and Information Theory*, Second Edition, Prentice-Hall, 1986.

Hill, Raymond, *A First Course in Coding Theory*, Clarendon Press, Oxford, 1986.

MacWilliams, F.J. and Sloane, N.J.A., *The Theory of Error-Correcting Codes*, North-Holland, 1977.

Peterson, W. and Weldon, E.J., *Error-Correcting Codes*, MIT Press, 1972.

Pless, Vera, *Introduction to the Theory of Error-Correcting Codes*, Second Edition, John Wiley and Sons, 1989.

Roman, Steven, *Coding and Information Theory*, Springer-Verlag, 1992.

Slepian, David, ed., *Key Papers in the Development of Information Theory*, IEEE Press, 1974.

Thompson, Thomas, From Error-Correcting Codes through Sphere-Packing to Simple Groups, *Carus Mathematical Monographs*, Mathematical Association of America, 1983.

van Lint, J.H., *Introduction to Coding Theory*, Second Edition, Springer-Verlag, 1992.

Index

Hilton/Holton/Pedersen: Mathematical Reflections: In a Room with Many Mirrors.

Iooss/Joseph: Elementary Stability and Bifurcation Theory. Second edition.

Isaac: The Pleasures of Probability. *Readings in Mathematics.*

James: Topological and Uniform Spaces.

Jänich: Linear Algebra.

Jänich: Topology.

Kemeny/Snell: Finite Markov Chains.

Kinsey: Topology of Surfaces.

Klambauer: Aspects of Calculus.

Lang: A First Course in Calculus. Fifth edition.

Lang: Calculus of Several Variables. Third edition.

Lang: Introduction to Linear Algebra. Second edition.

Lang: Linear Algebra. Third edition.

Lang: Undergraduate Algebra. Second edition.

Lang: Undergraduate Analysis.

Lax/Burstein/Lax: Calculus with Applications and Computing. Volume 1.

LeCuyer: College Mathematics with APL.

Lidl/Pilz: Applied Abstract Algebra. Second edition.

Logan: Applied Partial Differential Equations.

Macki-Strauss: Introduction to Optimal Control Theory.

Malitz: Introduction to Mathematical Logic.

Marsden/Weinstein: Calculus I, II, III. Second edition.

Martin: The Foundations of Geometry and the Non-Euclidean Plane.

Martin: Geometric Constructions.

Martin: Transformation Geometry: An Introduction to Symmetry.

Millman/Parker: Geometry: A Metric Approach with Models. Second edition.

Moschovakis: Notes on Set Theory.

Owen: A First Course in the Mathematical Foundations of Thermodynamics.

Palka: An Introduction to Complex Function Theory.

Pedrick: A First Course in Analysis.

Peressini/Sullivan/Uhl: The Mathematics of Nonlinear Programming.

Prenowitz/Jantosciak: Join Geometries.

Priestley: Calculus: A Liberal Art. Second edition.

Protter/Morrey: A First Course in Real Analysis. Second edition.

Protter/Morrey: Intermediate Calculus. Second edition.

Roman: An Introduction to Coding and Information Theory.

Ross: Elementary Analysis: The Theory of Calculus.

Samuel: Projective Geometry. *Readings in Mathematics.*

Scharlau/Opolka: From Fermat to Minkowski.

Schiff: The Laplace Transform: Theory and Applications.

Sethuraman: Rings, Fields, and Vector Spaces: An Approach to Geometric Constructability.

Sigler: Algebra.

Silverman/Tate: Rational Points on Elliptic Curves.

Simmonds: A Brief on Tensor Analysis. Second edition.

Singer: Geometry: Plane and Fancy.

Singer/Thorpe: Lecture Notes on Elementary Topology and Geometry.

Smith: Linear Algebra. Third edition.

Smith: Primer of Modern Analysis. Second edition.

Stanton/White: Constructive Combinatorics.

Stillwell: Elements of Algebra: Geometry, Numbers, Equations.

Stillwell: Mathematics and Its History.

Stillwell: Numbers and Geometry. *Readings in Mathematics.*

Strayer: Linear Programming and Its Applications.

Undergraduate Texts in Mathematics

Thorpe: Elementary Topics in Differential
Geometry.
Toth: Glimpses of Algebra and Geometry.
Readings in Mathematics.
Troutman: Variational Calculus and
Optimal Control. Second edition.

Valenza: Linear Algebra: An Introduction
to Abstract Mathematics.
Whyburn/Duda: Dynamic Topology.
Wilson: Much Ado About Calculus.